*COMRADE HALDANE
IS TOO BUSY
TO GO ON HOLIDAY*

SECRET 1563.

Intelligence Division,
Naval Staff,
Admiralty, S.W.1.

158B.

CUTTING FROM "PICTURE POST" OF 13.2.43. re HALDANE, J.B.S.

Advocate of Planned Science
Among the famous scientists present is Professor
J. B. S. Haldane, F.R.S.

PicturePost 13.2.43

all in England.

Parents - Jewish.

Doctor of Science of the German University, Prague.

Doctor of Medicine - German University, Prague.

Escaped January 1939. Family followed later in March of that year. Has since worked in the Department of

/ Zoology and ..

COMRADE HALDANE IS TOO BUSY TO GO ON HOLIDAY

THE GENIUS WHO SPIED FOR STALIN

Gavan Tredoux

Encounter Books
New York • London

© 2018 by Gavan Tredoux   http://jbshaldane.org

All rights reserved. No part of this publication may be reproduced, stored in a retrieval system, or transmitted, in any form or by any means, electronic, mechanical, photocopying, recording, or otherwise, without the prior written permission of Encounter Books, 900 Broadway, Suite 601, New York, New York 10003.

First American edition published in 2018 by Encounter Books, an activity of Encounter for Culture and Education, Inc., a nonprofit, tax-exempt corporation.
Encounter Books website address: www.encounterbooks.com

Manufactured in the United States and printed on acid-free paper. The paper used in this publication meets the minimum requirements of ANSI/NISO Z39.48–1992 (R 1997) (*Permanence of Paper*).

FIRST AMERICAN EDITION

LIBRARY OF CONGRESS CATALOGING-IN-PUBLICATION DATA

Names: Tredoux, Gavan, 1967– author.
Title: COMRADE HALDANE IS TOO BUSY TO GO ON HOLIDAY : the genius who spied for Stalin / Gavan Tredoux.
Description: New York ; London : Encounter Books, [2018] | Includes bibliographical references and index.
Identifiers: LCCN 2017039032 (print) | LCCN 2018006820 (ebook) | ISBN 9781594039843 (Ebook) | ISBN 9781594039836 (hardback : alk. paper)
Subjects: LCSH: Haldane, J. B. S. (John Burdon Sanderson), 1892–1964 | Lysenko, Trofim, 1898–1976. | Biologists—Great Britain—Biography. | Communists—Great Britain—Biography. | Espionage, Soviet—Great Britain. | Genetics—Soviet Union—History. | Science and state—Soviet Union—History.
Classification: LCC QH31.H27 (ebook) | LCC QH31.H27 T74 2018 (print) | DDC 570.92 [B]—dc23
LC record available at https://lccn.loc.gov/2017039032

Appendixes 1, 2, and 5 are reproduced from material held in the Haldane Papers, University College London, with permission from the estate of J. B. S. Haldane.

Appendix 3 is reproduced with permission from Immediate Media Co.

Appendix 4 and other quotations from the VENONA Intercepts are reproduced under Open Government Licence from material held in the Government Communications Headquarters Records, HW 15/43, National Archives, Kew, Richmond, England.

FRONTISPIECE: *A cutting from* Picture Post, *February 2, 1943, from J. B. S. Haldane's MI5 file. National Archives, KV 2-1832.*

At the front side of the Natural History Museum in Berlin there is a memorial plaque. It informs visitors about the fate of zoologist B. Arndt, who worked here and later died in a Nazi death camp. If similar plaques were installed on the All-Russian Scientific Research Institute of Plant Industry building in St. Petersburg, they would cover not just its façade, but also all the walls of the building.

—Eduard I. Kolchinsky, 2014

I am, so far as I know, the only person who has ever got duplicate determinations of urea by a volumetric method to agree to within one part in a thousand. And I am a better communist because of it.

. . .

I would sooner be a Jew in Berlin than a Kaffir in Johannesburg or a negro in French Equatorial Africa.

—J.B.S. Haldane, 1939

CONTENTS

Abbreviations   ix

INTRODUCTION   1

1. EARLY DAYS   9

2. WITH VAVILOV IN THE SOVIET UNION   33

3. THE THIRTIES   39

4. STALINOPHILIA   69

5. WAR ON ONE FRONT   89

6. IVOR MONTAGU AND THE X GROUP   101

7. THE FATE OF VAVILOV   121

8. EXPERIMENTS IN THE REVIVAL OF ORGANISMS   133

9. IT IS YOUR PARTY DUTY, COMRADE!   143

10. LYSENKO AND LAMARXISM 159

11. SOCIAL BIOLOGY 203

12. ANIMAL BEHAVIOR FROM LONDON TO INDIA 217

13. A CERTAIN AMOUNT OF MURDER 233

APPENDICES

APPENDIX 1. Why I am [a] Cooperator   245

APPENDIX 2. Haldane on the Nazi-Soviet Pact   299

APPENDIX 3. Self-Obituary   307

APPENDIX 4. VENONA Intercepts   313

APPENDIX 5. In Support of Lysenko   331

Notes   339

Early Gulag Memoirs and Descriptions   365

Bibliography   369

Index   383

ABBREVIATIONS

Cheka / GPU / OGPU / NKVD / MGB / KGB	The Soviet Security Police, which continually changed its name but not its nature.
CPGB	Communist Party of Great Britain.
GCHQ	Government Communications Headquarters, including signals intelligence.
GRU	Soviet Military Intelligence. Distinct from the NKVD, with its own espionage network.
MI5	UK Domestic Military Intelligence. Aka the Security Services.
VASKhNIL	The Lenin All-Union Academy of Agricultural Sciences.
VENONA	A highly secret program to decode Soviet embassy cables, a joint American and British project.

*COMRADE HALDANE
IS TOO BUSY
TO GO ON HOLIDAY*

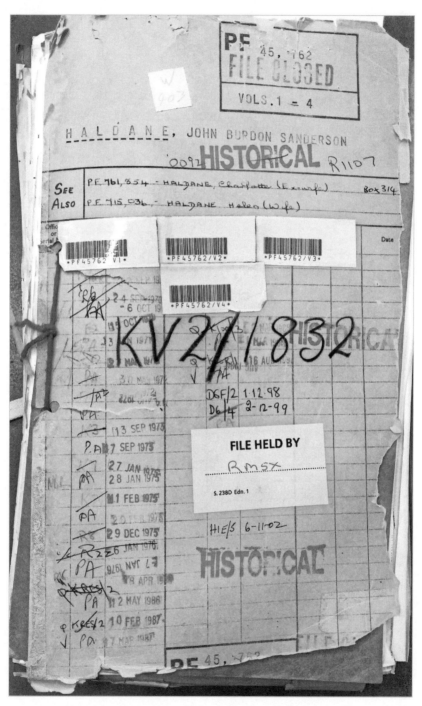

The cover of J.B.S. Haldane's MI5 file. National Archives, KV 2-1832.

INTRODUCTION

Shortly before noon on June 26, 1949, Professor John Burdon Sanderson ("JBS") Haldane elbowed his 6-foot, 245-pound frame into Harry Pollitt's office. The King Street headquarters of the Communist Party of Great Britain was crowded. JBS had known for some time that he was in serious trouble with the Party. His support for Trofim Denisovich Lysenko—a semi-literate peasant who had shinned his way up the Soviet patronage system to become Stalin's anointed authority on properly dialectical non-genetics—was insufficiently enthusiastic. The Party required less finessing and more commitment. It had come to this: a distinguished mathematical geneticist, physiologist, and Weldon Professor of Biometry at University College London (UCL) was consulting a boilermaker, Pollitt, about the correct line on the technical details of biological heredity.

MI5 were listening in as usual through the network of microphones that they had installed in the King Street building some years previously. Pollitt and the Party had been warned by the Soviets, who had long compromised MI5, that King Street was bugged. But several sweeps had failed to turn up anything. They created a "safe room." MI5 bugged that, too. They ripped up the floorboards but found nothing. So they carried on anyway, as if only half-aware of the fact.

2 COMRADE HALDANE IS TOO BUSY TO GO ON HOLIDAY

The transcript of the conversation that followed was duly filed in the dossier that MI5 had fitfully maintained on Haldane since his visit to the Soviet Union in 1928. There were some curious new developments.

> 11.53. Professor HALDANE came to see HARRY, who was expecting him. HARRY apologised for making him come upstairs. HARRY told HALDANE that the Soviet Academy of Science had asked him officially to approach HALDANE and invite him to go for a holiday to the Soviet Union.
>
> HALDANE said hastily with a lot of stammering that he would not be able to take a holiday this summer.
>
> HARRY said: "You won't?" in surprise and added that this was very important.
>
> HALDANE stammering more than ever, said he knew it was important but he had to keep his laboratory going.
>
> HARRY said he could choose the place he wanted to go to, they would offer him every facility and at the end of the holiday if he wished to have discussions with any of the comrades in whose work he was interested—
>
> HALDANE broke in saying he would have liked to go very much but he absolutely could not take a holiday.
>
> HARRY asked "what exactly that meant, not being able to take a holiday."
>
> HALDANE said that it meant that he could not be away for more than 3 or 4 days.
>
> HARRY said rather rudely that he saw the sort of a jam that HALDANE was in but "you are a Party man, you know". He suggested September.
>
> HALDANE said he was already booked up in September. He added that it was very awkward, but he could not help it in his rotten job. Next sentence difficult to hear because HALDANE stammered so much but it sounded as if he was explaining to HARRY that his secretary had left him.
>
> HARRY asked if she had left for political reasons.
>
> HALDANE said he thought not, she was a Party member.

HARRY asked if he would be able to take a [trip] when he had got a secretary.

HALDANE said it would be very difficult. He explained that he would have to train the girl, difficult to find the right type. He then told HARRY the history of the former secretary's departure, apparently a disagreement over when she should take her holiday. HALDANE said that money was a difficulty. Things were not as easy as they could be. He then said "by the way" and asked HARRY if [he knew] the name LANDARD (might be LANDAIN or just LAEDARL).

HARRY asked if he was a scientist.

HALDANE said no, an actor.

HARRY apparently knew nothing about him. Next few sentences very obscure, voices not at all clear.

HARRY said something about some money which was "just lying there" and HALDANE had only to sign for it and it would be paid into his account.

HALDANE said he knew about that but he did want to have too much money in the bank as long as he had a libel action to worry about.

HARRY told him to come along to him if wanted ready cash. He added that there were £46 lying there and that "Jack had over £200." This seemed to be a quotation of something that someone else said about HALDANE.

HALDANE thanked him but did not pursue the subject. He reverted to the subject of LANDARD, again asking if HARRY knew anything about him.

HARRY said no once more.

HALDANE told him that someone (name might be Joe ISMAEL) had rung him up said he had received a packet from LANDARD. Next sentence unintelligible. HALDANE said that he thought that "he" (presumably LANDARD) might be "an MI5 person trying to frame me."

HARRY apparently reassured him, for HALDANE said: "That's alright then." He had said he knew nothing about it.

HARRY then went back to subject of the Russian invitation and asked if October would be any use.

> HALDANE said no prospect of being able to go. He again spoke of the difficulty of finding a suitable secretary. Would prefer a party member.
>
> HARRY was clearly very annoyed, and told HALDANE this was just the sort of invitation he was always badgering "them" for.
>
> HALDANE apologised profusely.
>
> HARRY gave it up and asked HALDANE if he had had [an] interesting time in Czechoslovakia.
>
> HALDANE said he did not go. HARRY clearly annoyed about this too.
>
> HALDANE excused himself on grounds of ill health.
>
> HARRY warned him that he could not go on like this (not clear if this referred to overwork or HALDANE's treatment of invitations).
>
> HALDANE spoke again of LANDARD. This time he said that it was a chap called READ who had rung him up about LANDARD with this important packet.
>
> HARRY asked who READ was.
>
> HALDANE said he did not know.
>
> 12.28 HALDANE left.[1]

Up until now, Haldane had been an exemplary communist. He had supported Lysenko from the beginning. He had never turned down a free holiday to the Soviet Union before, or much-needed cash. It had not been necessary to offer him either, until now. Anyway, the physiology of frost resistance was not one of his research interests. There were, in the end, limits to his self-experimentation. What to do? His back hurt.

—•—

In his day, Professor J. B. S. Haldane was as well-known a scientist as one could hope to be. Magazines paid handsomely for his articles explaining science to the general public. Collected in books, these continued to sell for years. When he voiced his classically educated opinions, the newspapers listened, and the BBC transmitted them. Reckless physiological self-experimentation, learned from his father, created useful drama. "Prof" had the sort of presence as a general science popularizer and skeptic that Richard Dawkins and Stephen Jay Gould came to command half a century later. But Haldane had a far broader scientific reach and more panache. Technically, he was a mathematical population geneticist and evolutionary theorist, one of the founders of the Modern Synthesis that anchored Darwin to Mendel through statistical wizardry (impressive to those in the know, but an unpromising basis for broader fame). Along the way, he also took up communism.

It is hard to say exactly when Haldane became a communist. His influential biographer, Ronald Clark,[2] is partly to blame for this, but the vagueness started with Haldane's own complicated history of deception. Clark treated Haldane's politics as a personal eccentricity, not to be taken seriously; he wanted to believe the best about his subject. Since then, most people have preferred to play Haldane's politics down. He might be described a little vaguely as a Marxist, which has an academic, even philosophical ring. Or perhaps as a Bolshevik, which has a more romantic back-to-the-barricades flavor, vaguely archaic like the pipe clamped to his lips. Sometimes the adjective is merely "left-wing," which excludes few of the university professors that living readers can recall. Sometimes his communism is attributed only to youthful idealism, which is exactly back-to-front—he only fledged full-communist plumage in late middle age.

Above all, Haldane is almost never described as a Stalinist, which is the description that comes closest to the truth. This vagueness infuriated his former Communist Party comrade and friend Ivor Montagu, a lifelong unembarrassed Stalinist himself. After watching a BBC television documentary on Haldane in the late 1960s, Montagu complained to the *Labour Monthly* that "the picture that emerged safely was just one more stereotype of the eccentric professor, his contact with Communism the equivalent of accidentally dining off the lab-dissected frog instead of the packet of sandwiches." The BBC left out "any friend or associate from the dozen or so vital years of Haldane's work with the *Daily Worker* and the Communist Party." Montagu reassured his readers that in Haldane's case it was "no accident that he came to Marxism," since he "found in Engels a philosophy embodying his own approach to science and the relation between man and the rest of nature."[3]

In an unpublished fragment of autobiography, Haldane was careful to define these loose terms more precisely. "By the word communist I mean not merely one who sympathizes with the general aims of Communism, and occasionally supports it with his vote or money. I mean a member of the Communist Party, which is a section of the Communist International."[4]

Officially, Haldane did not join the Communist Party until mid-1942—a ruse that continues to pay off more than seventy years later, judging by how often this date is still repeated when Haldane's communism is referred to. By that stage, the Soviet Union was a new-found ally in the war, making his announcement seem unexceptional, a simple act of solidarity. Until he ducked from political view in the early 1950s, that made Haldane the most prominent scientific member of the Party, with all the prestige of a Fellow of the Royal Society, the Weldon Professor of Biometry at University College London, a former reader at Cambridge and fellow of New College, Oxford—not to mention his extended reach as a widely read popularizer of science, and his distinguished scientific pedigree. But he had really been an open sympathizer in public, and a concealed Party member in private, for many years prior to announcing his membership.

Haldane's motive for finally coming out in 1942 was probably defensive—his wife, Charlotte, had just defected on her return from a trip to the Soviet Union.[5] We cannot say for sure precisely when JBS and Charlotte were first recruited as underground members of the Party, but it was probably no later than 1936 or 1937, and may have been far earlier than that.

—•—

MI5 spent nearly thirty years paying intermittent attention to J. B. S. Haldane's doings, starting with his trip to the USSR in 1928. They opened his mail when it involved other persons of interest. They made copies or excerpts of anything that looked promising. Every time he left the country, his luggage would be discreetly searched, and his companions would be noted. Reports would be filed from harbors and airports, sometimes even describing his appearance and demeanor on the day. Nothing ever turned up. Sporadic reports from foreign intelligence organizations would be inserted diligently, in proper chronological order. Sources would report from his public meetings with summaries of his speeches. All duly filed.

Telephone conversations between leading communists would be tapped and transcripts mentioning Haldane placed in his file, though they never went to the length of tapping Haldane's own phone. Reports would be cross-referenced from other files kept to monitor his connections, such as Hans Kahle and Ivor Montagu. But most of MI5's work involved simply reading the newspapers, inserting clippings into his file, again in chronological order, underlining key phrases, circling paragraphs, and adding helpful photographs. Every once in a while, typed summaries would be made of the contents thus far.

At no stage did they tail Haldane or conduct interrogations of his associates to find out more. Nor did they make any effort to use the information they had to discredit him. For some brief moments they hoped, vainly, that he might turn friendly and cooperate with them. In the meantime, official inquiries about his affiliations would be carefully answered, noting that he was a known communist of long standing, in the past under concealment. These answers would always fairly reflect the limited knowledge they had. "Fair play" held sway.

The lethal fountain pens and Miss Moneypenny were, apparently, reserved for MI6 and SMERSH. As counter-espionage work went, this was the most routine, perfunctory kind. MI5 was never able to deduce much from it. There is a charming naïveté to it all.

Yet the mass of material they gathered, especially their collection of telephone intercepts, is invaluable once the connections that are latent in them are understood. To get anywhere, some erroneous assumptions need to be discarded first, such as the idea that the Communist Party of Great Britain was a political party pursuing ideals. It was, as MI5 recognized only very late in the game, set up to act as nothing more or less than a remote channel for Moscow. They fully caught on to Haldane only in the late 1960s, when a few decoded Soviet Embassy messages emerged from the remnants of the VENONA program. By then he was safely dead. After that, the information lay dormant in their files. Occasionally someone would thumb through the material, vainly looking for living connections that might have been missed. But the harsh truth is that the VENONA intercepts were really unnecessary for making the deductions that MI5 needed to make. Open societies are no good at this kind of thing.

—•—

This book is not a biography of Haldane. Nor is it principally concerned with his scientific work, except insofar as that interacted with his communism. A great deal of new information is presented here, derived from MI5 files on Haldane and his associates, declassified VENONA signals intelligence, and the Haldane Papers at University College London. Many other minor sources have been used to give a clearer picture of his context. Haldane's own writings are extensively referenced to establish the (at times incredible) opinions he voiced about the Soviet Union, Stalin, and related topics. The reader may easily suspect that these are the products of misleading summarization; therefore many quotations have been provided. The appendices contain complete VENONA decryptions relevant to Haldane and the X Group that he was a member of. The reader will also find there Haldane's incomplete autobiography, up to 1938, which has never been published before, and other useful primary material.

1. EARLY DAYS

Haldane's home life was not especially politicized. His father, John Scott Haldane (1860–1936), was a mild-mannered professor of physiology at Oxford, though he had been educated at Edinburgh, with moderately socialist views and a mania for self-experimentation —extending to the use of his own children as supplementary guinea pigs. His attractive mother, Louisa Kathleen (1863–1961), was a quiet but convinced conservative, perhaps even doctrinaire. JBS was born at Oxford on Guy Fawkes Day in 1892, and allusively nicknamed "Squawks" by his parents. A sister, Naomi—later a successful novelist and far-left political activist—followed in 1897.

The Haldane family tree led to many eminent men and women on both sides, creaking with the weight of them. His father, John Scott, had three notable siblings. The Liberal Lord Chancellor Richard Burdon Haldane (1st Viscount Haldane) (1856–1928) was a lawyer, politician, and occasional Hegelian philosopher. Sir William Haldane (1864–1951) was a lawyer who rose to the senior post of crown agent for Scotland. The writer and suffragette Elizabeth Sanderson Haldane (1862–1937) was entirely self-educated but published extensively, including studies of Hegel and Descartes. John Scott's father was the distinguished Scottish lawyer Robert Haldane (1805–1877) of Auchengray, who had married Mary Elizabeth Burdon-Sanderson, from Northumberland, a daughter of the well-known nonconformist

Richard Burdon (later Burdon-Sanderson) (1791–1865). John Scott's maternal uncle, and hence JBS's great-uncle, was John Scott Burdon-Sanderson, Waynflete Professor of Physiology at Oxford, later Regius Professor of Medicine. One can go on like this for some time—JBS himself claimed to be able to trace his male ancestry without interruption back to the thirteenth century.

JBS's mother, Louisa Kathleen, was a daughter of the leisurely geographer (and amiable hypochondriac) Coutts Trotter (1831–1905), a Fellow of the Royal Geographical Society, who wrote reviews in the periodical press. William Robertson Smith persuaded him to contribute articles on Polynesia, which he had traveled in, to the *Encyclopædia Britannica*. Her mother, Harriet Augusta Keatinge, was the daughter of a judge of the high court of Dublin, the Right Honourable Richard Keatinge (1793–1876). Keatinge had married Harriet Augusta Joseph (1792–1869), who, as JBS's maternal great-grandmother, provided an obscure connection that might have interested JBS had he been aware of it. Harriet Augusta Joseph was a cousin of the great Victorian mathematician James Joseph Sylvester (1814–1897), who was actually born James Joseph and later added the surname Sylvester. Her father, Samuel Joseph, was Sylvester's uncle. Haldane was therefore homeopathically Jewish himself. It is likely that JBS was not aware of this, as he would have delighted in even a small dose of a very potent ability—Joseph Sylvester and Arthur Cayley are commonly considered the two most eminent British mathematicians of the nineteenth century.

As a child, JBS was clumsy and accident prone, developing a hernia in the crib, fracturing the base of his skull as a boy, breaking his arm when a young teenager, cutting his foot badly when bathing, and more besides. These accidents persisted throughout his life. His mother called him "Jack," no doubt to distinguish him from his father John, also known as the "Senior Partner." He started his schooling in Oxford at the Preparatory School run by the "Skipper," Mr. Lynam. He quickly established himself as their top pupil, demonstrating a tenacious memory and excellent analytical skills. His vocabulary and reading skills were far in advance of his peers. He was also outsize, always the largest and strongest physical specimen in his class, but never good at team sports.

JBS went up to Eton as a King's scholar in 1905, after coming first in the scholarship exams, which in those years stretched over a full four days. He preferred to solve the higher-marked questions first. Success meant that the fees were modest, but he was exposed to bullying until his physical strength could protect him; he was twelve years old. One week he was caned by the senior boys on every night, including a bastinado. Of course, "the diet was monotonous and the cooking shocking."[1] He fagged—that is, acted as a boy of all work—for Julian Huxley.[2] He was eventually elected to Pop and made the captain of "School" (as opposed to the "Oppidans," or paying pupils), but he was personally unhappy throughout, which disappointed his mother. "There was a very great deal of homosexuality, occasionally reaching the point of sodomy," and worse, "where there was much disparity of age the younger boy was not always a free agent."[3] Bitter memories predominated, colored by social alienation in later years. "The Eton Society, or 'Pop' included the most distinguished and popular athletes. The shapely youths who were alleged to assuage the desires of this august body, often in return for presents, were known as 'Pop bitches'. Some of them have since risen to positions of high distinction and trust."[4] One of JBS's contemporaries was Harold Macmillan, the future prime minister. In after years, JBS suggested to an acquaintance, Woodrow Wyatt, that Macmillan had been expelled for buggery.[5] Julian Huxley thought Haldane "always eccentric, even as the boy of thirteen whom I remember in College at Eton."[6]

Nevertheless, JBS conceded that by the time he left Eton in 1911 at age eighteen, he had received a good education; the Oppidans' fees paid for competent instruction. "I could read Latin, Greek, French and German. I had won a mathematical scholarship at Oxford. I knew enough chemistry to take part in research work, enough biology to do unaided research, and I had a fair knowledge of history and contemporary politics."[7] In 1908 he read a paper to the Scientific

Society on "Respiration."[8] (The previous year it had been "Parasitic Worms," which was described as a "delightful paper."[9]) In 1909 he won the Chemistry Prize[10] and in 1910 he was second on the English Essay and History Prizes.[11] That year, Haldane and Clarke also read a paper to the Scientific Society on "The Structure and Functions of the Blood."[12] He developed "a mild liking for the Anglican ritual and a complete immunity to religion," but when at games was "utterly bored."[13] Records from 1908 show two cricket batting innings with a highest score of 2, and one *not out*, for a run average of 2.[14] Before he left, one of his duties as Captain of School had been to receive the king there; but his mother complained that her son generally found the duties expected of him an imposition. It had been her idea to send him there in the first place, despite the misgivings of his father. She was pleased, though, that he joined the Volunteers.

With his mathematical scholarship in hand, Haldane went up to Oxford in 1911. At New College, he read a year of mathematics and took a First Class in Mathematical Honours Moderations in 1912. A term of biology under Edward Stephen Goodrich (1868–1946), a former artist noted for his skillful blackboard drawings, was followed by a switch to Greats, that is, classical literature and philosophy. In this he took another First Class in the Honours Moderations of 1914. The switch to Greats may have been due to a feeling of his father, who distrusted "mere mathematicians," or it may have been showing off. In any case, if Haldane had taken mathematics seriously, as his primary interest, he would during that era have been drawn to Cambridge—G. H. Hardy would only be lured to New College eight years later, when Haldane would get to know him in a different capacity. Still, he found the abstractness of Greats and the discipline acquired in the composition of essays useful. "The successful Greats Man, with his high capacity for abstraction, makes an excellent civil servant, prepared to report as unemotionally on the massacre of millions of African natives as on the constitution of the Channel Islands."[15]

Haldane thought of himself at this time as "a liberal with leanings further left." He particularly remembered that he was "considerably influenced by my contemporary Herron (killed in 1915) who was a syndicalist."[16] Alec Rowan Herron was the son of a ship broker and entered New College in 1911 with a scholarship in modern history, after attending Gresham's School in Norfolk. Herron was active in the Oxford Union and was a friend of the young Harold Laski. He served during the war as a 2nd lieutenant in the King's Royal Rifle Corps and was killed during an attack near Givenchy on March 10, 1915. Following Herron would have made Haldane a mild socialist, of a kind not unknown among the dons, who included A. J. Carlyle and G. D. H. Cole. JBS relished some early agitation at least.

> My only serious political gesture was, I think, in May 1913. Oxford was then served by horse trams, which could readily be overtaken by a runner, but went definitely quicker than most people can walk. Neither the drivers nor the conductors earned so much as £1 per week. Wishing to remedy this state of affairs, they struck. Their places were taken by blacklegs. On the first three evenings of the strike trams were stopped and the horses taken out. The police made baton charges, and finally order was restored. I was unable to participate in these riots, I think because I was in training for a race. On the fourth evening the streets were quiet. I walked up and down Cornmarket Street chanting the Athanasian creed and the hymeneal psalm 'Eructavit cor meum'[17] in a loud but unmelodious voice. A large crowd collected. The police ineffectively pushed pious old ladies into the gutter. The trams failed to penetrate the crowd and their horses were detached and wandered off in an aimless manner.
>
> The strike was successful, and as the trams could no longer yield a profit, they were replaced by motor omnibuses, which were capable both of higher speed and higher wages. I was subsequently martyred by the proctor to the extent of two guineas. This was, I suppose, the first case for over three centuries when a man was punished in Oxford for publicly professing the principles of the Church of England.[18]

Since the tram could not proceed due to his physical blockade of it, he was punished for obstruction and not for recitation. Tricks of this type set in early. Other reports mention less delicate physical battles between Oxford undergraduates and the police, with stiff fines handed out by the magistrates.[19] The tram drivers blamed the students for drawing the strike out. Though the strike quickly collapsed and was by any reasonable definition a failure, contrary to Haldane, stories about his involvement in it continued to do the rounds in New College for another decade at least. Christopher Hollis, who was there in the early 1920s, recalled one anecdote that had Haldane causing some surprise by emerging from under a manhole cover over a drain in Oxford Street.[20] (It was not certain what the advantage of this tactic was supposed to be for the strikers, but Hollis thought it was considered unbecoming conduct by the authorities. Haldane was quite comfortable in tunnels.) Other Oxford socialists seized on the strike in a more genteel way. Magdalen's G. D. H. Cole rushed out a pamphlet (*The Tram Strike*) arguing that wages should be based on need rather than on profit. The university's Fabian Society was usually excited by industrial unrest, but seemed to hold aloof in this case. One of its members was Harold Laski, a friend of Herron's and soon a friend of Haldane's, too, in a circle that included Victor Gollancz.[21] Both Laski and Gollancz would play leading roles in radical British politics in later life, but it is hard to tell how much influence they had on Haldane's thinking. It was certainly not lasting. Much later, in a letter to Laski's biographer Kingsley Martin, Haldane remembered the Harold Laski of his Oxford days as a compulsive liar.[22] Since they had fallen out over politics, this may have been spite. But in those days, they were allied at the Oxford Union, where Haldane argued that "this house approves of the principles of eugenics."[23]

Later that year, on Guy Fawkes Day, Aldous Huxley attended "Jack's" twenty-first birthday party at the family home in Cherwell, Oxford.[24] They played "Nebuchadnezzars," a variant of charades peculiar to the Haldane family.

I have been having a good dose of one part of the north lately, in the shape of the Haldanes, who carried off their double event birthday party on Friday and Saturday. Friday was their dance . . . in honour of Jack's twenty-firster . . . and on Saturday was the common or garden birthday party. The dance was very amusing . . . the Haldanes always contrive to know and invite very good people to their functions: however, I must get to dance better, or otherwise everyone suffers. But the Saturday party is the really wonderful affair: I came at 4.30 and left at 11.30: the first half of the time was occupied in eating tea and playing other essentially childish games, for the benefit of the hordes of infants: the H's have a most admirable device for breaking the ice. They turn the whole party . . . about forty, with the children . . . into a large and empty room, in the middle of which stands a bran pie, where bran is replaced by confetti. Everyone having dived in the pie and removed something, one proceeds to take the confetti and throw it at everyone else. After half an hour of this . . . ones hair, pockets, stomach and inmost underclothing being completely filled with confetti . . . shyness, as such, almost completely ceases to exist. One then comes out into an ante-chamber, where seven highly trained, amateur officials remove as much of the confetti from one's hair and outer garments as is humanly possible. . . . Fireworks ensue, then (children dismissed) supper and afterwards, the most magnificent Nebuchadnezzars, and finally a good form of blind-man's buff, where everyone stands round the room in a circle and the blind man walks up and prods someone, telling him at the same time to make a noise . . . such as the sound of rain falling on mu[d]—and the speaker has to be recognised by the sound of his voice.[25]

When he left in 1914, Haldane did not get a science degree from Oxford, nor did he ever complete one in later life (though he received many honorary degrees once his name was made)—an idiosyncrasy that he was proud to proclaim. His father had already given him a long apprenticeship in science at home, involving him there in his laboratory, his notorious self-experiments, and his many industrial excursions. At age six, JBS helped his father capture samples of sooty air on the London underground by leaning out of the moving train to stopper vacuum bottles—his father recommended more ventilation. Even earlier, at the tender age of four, JBS had been taken down a mine, and remembered being terrified. Recurring Easters in Cornwall made the tin mines, which his father tested for firedamp and air circulation, feel like home. Visiting North Staffordshire, father and son were lowered down a nonoperational shaft in a bucket on a chain. John Scott then hoisted his son to the roof of a tunnel gallery, there to declaim Mark Antony's speech from *Julius Caesar* (he knew it) while breathing methane—all to see how long young Jack would last before he passed out. He made it from "Friends, Romans, countrymen" to "the noble Brutus." Lowered back down to ground level, he recovered quickly in better air.

As a teenager, he even went deep-sea diving off the coast of Scotland, in an ill-fitting leaky suit meant for adults, as one of his father's human "rabbits." He consented to being sealed in a "coffin" with only his head protruding in order to measure volumes of air breathed in response to more carbon dioxide (panting) and less oxygen (normal). This experiment suggested that we breathe more to expel carbon dioxide than we do to inhale oxygen. Self-vivisection seems to have been involved, too. When it became clear that his son loved to make calculations, his father turned to him for arithmetical and then mathematical help. Work on problems of ventilation on the new naval submarines meant that JBS was asked along to keep track of the soda-lime. John Scott: "Do you know the formula for soda-lime?" Jack: "$CaHNaO_2$." All these practical activities were summed up by Haldane as "bottle washing" for his father, and meant that he could get by without needing more extensive scientific training. But necessity may pass for virtue. There was universal agreement among his colleagues in later life—John Maynard Smith, N. W. Pirie, C. D. Darlington, Julian Huxley, Peter Medawar, to name a few—that

Haldane was clumsy in a laboratory and not a practical experimenter where precision was needed. The only exception to this was with gas measurement equipment, which he was an expert at from long use in his father's lab at home.

What JBS did know a lot about by the time he left Oxford, without requiring formal tuition, was genetics. Although he had heard A. D. Darbishire speak on Mendel's laws in 1901, he was only eight years old at the time and could not have retained more than the impression of excitement. Teenage experiments with a large stock of guinea pigs, when he was home from Eton, had set him thinking and reading up on the new literature about genetics; this interest was kept up when he went up to Oxford. There he noticed that one of the papers he had read (by Darbishire himself, on mice) contained data that didn't make sense.

The frequencies of some traits—albinism, pink eyes, and pigmented coat—were wrong; some seemed to vary together. This was known to happen in plants and was thought to be caused by "reduplication," but had not so far been noticed in animals. ("Reduplication" was soon replaced with the concept of "linkage," where genes coexist close to each other on chromosomes.) The result was Haldane's first genetics paper, which he read to one of E. S. Goodrich's seminars at Oxford in 1912. Next, he got his own data to bolster this finding by breeding mice in collaboration with a friend at New College, Alexander Dalzell Sprunt, with some assistance at home by his young sister, Naomi.

Later, when Haldane was serving at the Western Front, he hastily submitted preliminary results for publication.[26] His coauthor, A. D. Sprunt,[27] "a man of considerable promise," had died of his wounds at Neuve Chapelle on March 17, 1915, after leading a charge on March 10. The following day, Haldane wrote a letter to the leading geneticist William Bateson (1861–1926) of the John Innes Horticultural Institute, explaining the work they had done on mice and asking, "If I am killed could you kindly give my sister help if she needs it."[28]

War had been declared shortly after Haldane graduated from

Oxford. Plans for a six-week walking tour of the continent had to be abandoned. Instead, he volunteered to join the Black Watch, the family regiment of the Haldane clan. A commission as 2nd lieutenant was immediately forthcoming, since he had been in the Officers' Training Corps while at Oxford. Four months of training in Scotland at Nigg, the headquarters of the 3rd Reserve Battalion, followed. There he learned bombing techniques.

After transferring to the 1st Brigade of the Black Watch, he landed in France in January 1915 as their bombing officer. The brigade was located in the region of Neuve Chapelle and Festubert. Given the freedom to shift for himself, he made smoking compulsory in their bomb repair shop—to strengthen nerves, he said. He noted national differences between Scots Guards and Indian troops under bombardment by trench mortars. "The Guards swore with great fluency, dodged round the traverses, and were rarely hit. The Indians stood and waited to be killed, which they were. They apparently thought that the bombs were devils, and could not be dodged." But he was thoroughly enjoying it. "April was one of the happiest months of my life."[29] Minor wounds he received to his arm at this time went unmentioned in later life.[30]

After the first gas attack in May, JBS was withdrawn to assist his father, who had been asked to devise countermeasures. His assistance involved inhaling chlorine gas to test respirators, arguably less dangerous than being at the front. A working respirator was put together using carbonate of soda, but manufacturing it in larger quantities at home was complicated by the inadvertent substitution of "caustic" for "carbonate" by the War Office, leading to many burned fingers.[31]

JBS was quickly appointed a gas officer to the Black Watch. He was just in time for the Battle of Aubers Ridge of May 9, arriving in the afternoon. (This was not the Battle of Festubert, which began on May 15, by which time he was already wounded.) As he ran toward the front line—"the moderate dose of chlorine which I had inhaled prevented me from expanding my lungs"—the Black Watch ahead of him went over the top. The Germans were replying with a massive bombardment.

Imagine the loudest bang you have ever heard, say a clap of thunder from a house struck in your immediate neighbourhood. Now imagine this prolonged indefinitely, a solid bang without intermission. And behind this, like the drone of a bagpipes behind the individual notes, a sound as of devil-driven tramcars taking a sharp corner.

Lesser bombardments had frightened me. This entirely novel sound intoxicated me. I ran forward through the monstrous black bursts of smoke and fountains of earth and bricks where the German shells were exploding . . .

I woke up, and began to scrape the earth off me. I noticed blood on my face and hands, and pains in various places. I realised that I had been hit. This struck me as funny, an automatic psychological defence reaction of considerable value. I ran on to a house and took stock. I was wounded in the right arm and left side, but my face was only scraped.[32]

Since all the ambulances were full, he had to walk back for help, once the shelling was over. After a few miles he met the Prince of Wales, who was touring the battle scene in his personal motorcar, and gratefully accepted a lift. Shrapnel was left permanently in his arm. "Lord Haldane's nephew wounded," reported the *Times*.[33] Shell-shocked, he was sent back to England to recuperate. Organizing a grenade school helped him to regain his nerves. Covert (but unwelcome) intervention by his family delayed his return to France. Assigned a desk position for a while, he kicked his heels.

In September 1916, Haldane went with the 2nd Black Watch to Mesopotamia, where he commanded a sniper unit on the banks of the Tigris. The company of his fellow officers there was congenial. "Our conversation was often fairly intelligent. We would discuss incidents in mediaeval Scots history or topics in elementary physics, with a vehemence which was encouraged by the complete absence of reference books."[34] It was recorded that "Captain Haldane did excellent work sniping and kept the enemy well in hand."[35] He liked the work. "I get a definitely enhanced sense of life when my life is in moderate danger."[36] This stint was cut short when a warehouse fire ignited some of their own bombs and he was wounded in the leg and, when it would not heal, invalided to India. After recovering, he ran the Central Bombing and Stokes Howitzer School in Mhow, south of Indore, demonstrating *inter alia* the use of rifle-fired Mills bombs. A bout of jaundice—which he attributed to compulsory meat rations in a hot climate—cut this short. Rest in the Himalayas allowed him to return to London and a cameo role there in Military Intelligence. This was less romantic than it sounded, "mainly routine snooping into other peoples' affairs in the hope that something may turn up."[37] (As in his own case, a decade later.)

Demobilization came in January 1919, accompanied by a 25 percent disability pension. But there would be no stirring lines from Haldane about poppies or foreign fields. "I liked the war, or rather those brief periods of it when I was actually in the front line."[38] The violence appealed to him. "When I got the opportunity of killing other people during the war I enjoyed it very much, though it is now more fashionable to say that one hated every moment of it. If I were ashamed of that particular skeleton (which is really a quite respectable relic of primitive man) I should hide my real motives from myself, invent excellent moral reasons for violence, and go forth in holy anger and pious grief to smite the wicked, or at least encourage others to do so. As it is I view that kind of moral indignation in myself and others with profound suspicion, and try to work off my steam in other ways."[39]

JBS proceeded, at the age of twenty-six, to take up a fellowship at New College in Oxford. Supplementing the early practical training his father had given him with some quick reading, he taught physiology at New College and got by without a scientific degree for the rest of his career. He seems to have remained politically inactive at this time, except for the occasional street-fighting incident.

> And in June 1919 I acted as chucker-out at a meeting addressed by George Lansbury and Austin Harrison to protest against the original version of the Treaty of Versailles, which was even more

unworkable, and an even more flagrant breach of Wilson's promises to Germany, than the final form of the treaty. The interrupters, who were the sort of people who now hail Hitler, threw tomatoes. I had my tactical scheme prepared. I approached one of the smaller ones from behind, placed a finger in each nostril, and dragged him backwards, hooked and struggling like a salmon, and too agitated to hit me in a vital spot. The rest followed, but before they rescued him they were half way to the door.

I am no boxer or wrestler. My tactics are to grapple with a man and use my weight to bump him against a wall or floor. I remember being down once or twice, and there was some rather half-hearted fighting with chairs before we cleared the Corn Exchange. On my way home the interrupters counter-attacked. I took refuge in a jeweller's doorway between two plate-glass windows, determined to break them if attacked. My opponents did not assault this position, and on the approach of a policeman, who scented danger to property, they dispersed.[40]

Sometime around the early 1920s, he was introduced to Lady Ottoline Morrell's offbeat circle at Garsington Manor, probably by the family friend and Old Etonian acquaintance Julian Huxley. Huxley later became a fellow traveler for a while in the 1930s, but at that time was a zoological Fellow in New College. Huxley had been inducted into the country retreat of the Bloomsbury set at Garsington during the war and remembered Haldane as "an odd character" in those days, who "enjoyed displaying" his memory by "reciting Shelley and Milton and any other poet you chose, by the yard."[41] He once had to escort Haldane all the way downstairs to the front door to escape a long recitation of Homer in Greek. He also recalled Haldane's penchant for sampling the scientific literature broadly.

> [H]e used to make me feel embarrassed—I was a zoologist, he a physiologist—by asking me if I'd read so-and-so on something-or-other in such-and-such a learned journal, always in my own field. I never had, but I always assumed that he had. But fate

caught up with him, in the shape of the Lancelot Hogbens, who were staying with us. Haldane asked me one of those humiliating questions—had I read X on something-or-other: but before I could answer, Mrs. Hogben said 'Have you?'—and he confessed, 'Well, actually I haven't—but I *did* note the title. . . . '[42]

Soon Haldane mixed with the likes of John Maynard Keynes, Lytton Strachey, Clive Bell, and Virginia Woolf—he already knew Julian's literary brother, Aldous. The eccentric Ottoline—"Very tall, fantastically dressed and enjoying the wearing of exotic clothes acquired in her travels . . . with something equine in her long intelligent face"[43]—had captured Bertrand Russell for a while in earlier years.

It could be dangerous to linger at Garsington too long. Ottoline stopped speaking to D.H. Lawrence when she appeared without much disguise as Hermione Roddice in *Women in Love*—Lawrence was bemused by her fury. Haldane, too, was hurt when he reappeared in Aldous Huxley's satire *Antic Hay* (1923) as that strident materialist the physiologist Shearwater, with a "large spherical head" and no time for the newspapers, since he is "chiefly preoccupied with the kidneys."

> 'The kidneys!' In an ecstasy of delight, Coleman thumped the floor with the ferrule of his stick. 'The kidneys! Tell me all about the kidneys. This is of the first importance. This is really life'.[44]

JBS also recognized his baleful reflection in Mr. Scogan from *Crome Yellow* (1922), a character who may have been based on tabletalk for Haldane's own book *Daedalus* (of which more later). In his "fluty voice," Scogan dreams of "the goddess of Applied Science" abolishing sex.

> An impersonal generation will take the place of Nature's hideous system. In vast state incubators, rows upon rows of gravid bottles will supply the world with the population it requires. The family system will disappear; society, sapped at its very base, will have

to find new foundations; and Eros, beautifully and irresponsibly free, will flit like a gay butterfly from flower to flower through a sunlit world.[45]

Aldous bit through the muscle. Two decades later Haldane returned fire in his autobiography—ineffectually, since it was never published—with a humorless taunt that Aldous was muddled and embittered by the eye infection that led to his near-blindness. "It took not only superb natural gifts, but a staphylococcus, to make the man who most perfectly voices the spiritual muddle of the English middle class intellectuals."[46]

In his last year as a fellow at New College, JBS walked with his sister, Naomi, on a tour through the hilly Auvergne region of central France. They lodged inexpensively in small cafes with attached rooms along the way. Their relationship swung between extremes in later years, with long periods of estrangement. On this tour, it seems to have been at its closest. One day, after drinking too much wine, they dozed off in an abandoned quarry. According to Naomi, soon to be a novelist, they "turned dizzily towards one another. And suddenly Jack was shocked to his respectable Haldane soul. I wasn't. But that was all. The next year he had gone to Cambridge."[47]

Haldane moved to Trinity College, Cambridge, in 1923, as a reader in biochemistry. Soon he met up with the Honourable Ivor Montagu (1904–1984), forming a connection that had important consequences much later in his career. Montagu was the youngest son of Lord Swaythling, from a family of immensely wealthy Jewish bankers, originally named Samuel, that had been ennobled in 1907. He was an undistinguished zoology student at Cambridge from 1921 to 1924, going up to King's College as a Fabian but quickly progressing to the stronger stuff disseminated by Marxist circles. Ivor fondly recalled that he first met Haldane

> in Market Square, Cambridge, on General Election night. An illuminated screen was fitted up which announced constituency results on lantern slides as soon as they were declared. Conserva-

tive win after Conservative win came up, each differing only by the size of its majority, and as each appeared on the screen J.B.S. Haldane booed. The rowdies did not like the boos and, headed by one or two soldiers in uniform, they surrounded and tried to rush Haldane. He stood his ground, and such was his stability and mass that, as they approached, each successive wave fell back as from a cliff.[48]

They struck up a friendship, but although Montagu was recruited to a Communist Party cell at Cambridge around this period, or just after, he later insisted that Haldane was himself not yet a Marxist.

Montagu stated in his memoirs that he lost touch with Haldane for some years after leaving the university in 1925, which may well be true. After going down with a pass degree, Montagu set off on one of several purportedly zoological trips to the Soviet Union. This attracted the attention of MI5, who marked down his fondness for saluting correspondents as *tovarich* (comrade) and then routinely monitored him, intercepting mail and telephone calls and following his public career with some interest. Their case-note description of Montagu reads "broad-shouldered, over six foot, has dark curly hair and is of distinctly Jewish appearance. His eyes are dark brown and his complexion is pale. He is generally rather dirty and untidy." Another entry in his file, from 1926, noticed his first visit to the Soviet Union, and that "Montague [*sic*] has for some time been known to associate with the inner ring of the Communist Party."[49] He was already in touch with Willi Münzenberg, who operated a well-funded propaganda empire of front companies for the Soviets, including film distributors.[50] Montagu then began distributing pro-Soviet films through his fledgling film company, Brunel and Montagu, and the London Film Society, which he helped to found in order to evade British censorship. In the meantime, he had taken up international table tennis with so much enthusiasm that he eventually raised suspicions in intelligence circles that he was using his foreign ping-pong contacts as a cover for espionage. "They write interminably to IVOR MONTAGU about Table Tennis and trying out of

Table Tennis balls . . . but even in England, which is not known for sanity in this respect, we find it hard to believe that a gentleman can spend weeks upon weeks upon weeks testing Table Tennis balls."[51] He appeared and reappeared in a slew of Soviet front organizations, such as the Friends of Soviet Russia, and the Society for Cultural Relations between the Peoples of the British Commonwealth and the USSR, which organized chaperoned tours of the Soviet Union for likely friends. Presently the *Daily Express* was gleefully reporting that "Banker-peer's son acts as communist cashier."[52]

Since he was independently wealthy, Montagu was able to experiment widely beyond table tennis, trying journalism, script writing, film editing, and public political activism. Along the way, he encountered Leon Trotsky, Alfred Hitchcock, and the Soviet director Sergei Eisenstein. He may have been sent to spy on the recently exiled Trotsky, whom he visited in Turkey in 1930. The former leader of the Red Army did not trust him, and wrote that Montagu was "paralyzed by his adherence to the party" in an intercepted letter to a British follower.[53] A film project with Eisenstein in Hollywood failed after six months, but Montagu successfully worked with Alfred Hitchcock, through the Gaumont Film Company, on the early spy thrillers *The Man Who Knew Too Much* (1934), *The 39 Steps* (1935), *Secret Agent* (1936), and *Sabotage* (1936).

At a cocktail party in 1933, the left-wing Labour MP "Red Ellen" Wilkinson introduced Montagu to the energetic Soviet agent Otto Katz, a Münzenberg associate also known as "Andre Simone" and by many other names—though Montagu may already have been known to him.[54] This piqued the interest of MI5, since they had been following Katz for some time. In the same year, Montagu assured a public meeting that the British Metropolitan-Vickers engineers on trial in Moscow for "wrecking"—in one of Stalin's early show trials—were without doubt guilty, and that the British government knew it, but were using the trial as an excuse to break off relations with the Soviets.[55] In 1937, he wrote a sarcastic letter to the *New Statesman* about the Great Terror show trials of that year, opening hopefully "Sir, It is amazing what a mess people get themselves into when they try to explain the 'Moscow confessions' by any cause other than the guilt of the accused"—only hitting the mark by accident when doubting that foreign legal representation would have been helpful to the victims.[56]

It took MI5 another thirty years to discover that, during the Second World War, Montagu had graduated from spy films to the real thing, successfully unifying theory with practice as an active member of the "X Group," an espionage apparatus run by the Soviet military intelligence agency, GRU. But that evidence required a lucky break that their intermittent snooping could never turn up.

Haldane stayed on at Cambridge till 1932, but by 1924 he had already established a liaison with the female journalist who would ultimately push him much further left, toward the Communist Party proper. This was the year in which he would turn thirty-two. They met after he published a skillful piece of speculative science futurism called "Daedalus, or Science and the Future," in the American *New Century* magazine (August 1923). This is one of his best known and most admired pieces, a quiverful of well-informed eugenic speculations that still penetrate today. In Haldane's telling, Daedalus is a god-killing geneticist, having slain Minos, son of Zeus, and conducted breeding experiments in the labyrinth. The "breeding experiments" are an allusion to the insemination of Minos' wife by a bull, with the aid of Daedalus' hollow wooden cow, in which she lay for copulation; she gave birth to the minotaur. When Minos cut off Daedalus' "funding" for these "experiments," Daedalus slew him.

Charlotte Burghes, *née* Franken, a journalist for the *Daily Express*, latched onto these reproductive themes, and early in the following year secured an interview with Haldane at his rooms in Cambridge. JBS was immediately impressed by her accuracy, integrity, and more besides. An affair followed, which Haldane openly flaunted around the college. After forcing a messy divorce from her husband, John McLeod Burghes, which involved a private investigator and a room at the Adelphi Hotel, Haldane was required to pay Burghes £1,000. He also had to fight for his position at Cambridge after being dismissed for "gross immorality" by a committee of six senior colleagues, the *Sex Viri*, in 1926. An aggravating circumstance was that Charlotte already had a four-year-old son, Ronnie, at the time of the affair. On appeal, after some assistance from Maynard Keynes, Haldane was reinstated on procedural grounds. He relished this victory and the recurring opportunities for *Sex Viri* puns.

Bertrand Russell's second wife, Dora, left a description of the Haldanes as they were then, down at the Russell retreat, Carn Voel, in Cornwall. "J.B.S. Haldane came with his wife Charlotte, just after his battle with Cambridge University about the accusations of 'moral turpitude' on account of her divorce. Ronnie, her son, was with them, and I could not help feeling some anxiety for him, with such a dominant personality for a stepfather. Charlotte was lovely, with her waist-length mane of black hair."[57]

After quickly marrying in 1926 at Haldane's insistence, the pair set up quarters just outside the town of Cambridge in Roebuck House, an eight-bedroom former inn on Ferry Lane, in Old Chesterton on the Cam. They attracted their own fringe set from Cambridge and beyond to their famous "open houses." Along with some of Haldane's own students, like Martin Case, who lived at Roebuck House for a time, the company included authors, actors, and aesthetes such as John Davenport, Kathleen Raine, Michael Redgrave, Malcolm Lowry, William Empson, Hugh Sykes Davies, Wynyard Browne, Douglas Cooper, T.H. White, and many others. Anthony Blunt, the sexual adventurer and future keeper of the queen's art collection, who would be exposed many years later as a devastatingly effective Soviet spy, was in attendance. JBS called these gatherings "Chatty's addled salon" and, perhaps anticipating another Aldous Huxley, preferred not to attend. Charlotte was the principal attraction. She soon advanced on Malcolm Lowry, "the most romantic undergraduate of that period in Cambridge," but Lowry seems to have been too frightened to go further than the furtive confidences that Charlotte responded to.[58]

The marriage became frustrated after Charlotte discovered that she could not have children by Haldane, who may have been hampered by a war injury requiring a truss. Gossip had Charlotte complaining "what can one do with a man who carries his balls around in a bag."[59] Years later, she told her daughter-in-law, Betty Burghes, that JBS was impotent.[60] Life with Haldane's temper was not easy, as she later complained to the BBC.

> If he read something in the *Times* that annoyed him he would begin to bite his lips. Shortly afterwards he would find a pretext—any pretext—to pick a quarrel with me. I would try to postpone the nagging by saying 'Yes dear', and 'excuse me for a moment, will you? I have to go to the bathroom' and I would lock myself in there, knowing him to be standing outside biting his lips, clenching and unclenching his fists until I reluctantly came out, when the nagging would continue until his temper was sufficiently relieved; for the time being.[61]

She found other outlets, writing novels inspired by the science that she found impossible to follow at Haldane's level but could work with emotionally, and pursuing several affairs. She was also able to boost Haldane's own reputation in print as an increasingly prominent intellectual and science popularizer, through her knowledge of the journalism business, distributing his articles through the Science News Service she started.

Charlotte was a radical feminist of an idiosyncratic kind, perhaps to the left of Haldane at that stage, and she seems to have accelerated his drift further left.

> Our interest in politics had always been a strong one. We both were and had been, long before we had met, socialists and . . . 'left-wing intellectuals'. Temperamentally also, we were strongly inclined to radicalism; both of us were psychologically counter-suggestible types, but also capable of enthusiastic interest in the social experiments now beginning to take place in the worlds of politics and economics. So it was that we began to feel more than

slight curiosity about the Soviet Union, and in the theories of Marx and Engels which had inspired Lenin and Trotsky, and on which this State was founded.[62]

On coming to Cambridge in 1923, Haldane had joined the biochemical laboratory of Frederick Gowland Hopkins (1861–1947), who would receive a Nobel Prize in 1929 for his discovery of vitamins, or "accessory food factors," as he called them. Being the reader in biochemistry, Haldane was his second in command. "Hoppy" was a chain-smoking socialist, though apparently that meant no more than indiscriminately signing any petition that came his way.[63] Many of those who passed through his lab were budding radicals. Barnet "Doggy Woggy" Woolf (1902–1983),[64] born in the East End of London to impoverished Lithuanian Jewish immigrants, was one of the earliest and more adamant of these.

Woolf had joined the Communist Party in 1920 as a founding member while still a scholarship student at Cambridge. Ivor Montagu used to meet him in the rooms of the economist Maurice Dobb, an influential and openly communist don at Trinity College, a rarity in the 1920s.[65] They were joined there by Allen Hutt, Philip Spratt, A. L. Morton, and John Desmond Bernal, among many others, who would all become prominent in the cause. Woolf became involved with Charlotte Haldane, who said that she made a protégé of the "able biologist and impassioned communist." "It was he who first indoctrinated me with Marxist theory" because his "contempt for the ancient University, and, particularly, for most of its dons and undergraduates, found sympathy with me."[66] "Doggy," who had joined the Hopkins lab after graduating in 1926, became close to JBS too, and must have been an important influence on his growing attraction to communism. JBS later referred to him in his book of children's stories, *My Friend Mr. Leakey*: "When I want to borrow money I always go to a friend called Dr. Barnet Woolf, who thinks it is wicked to pay interest for it, I owe him two pence halfpenny at present."[67] Woolf remained in the Hopkins lab until the early 1930s, when he left to experiment with drama, poetry, and songwriting, but resumed a career from 1939 onward as a medical statistician.

James Murray Luck (1899–1993) was another contemporary under "Hoppy." He remembered Haldane as a "walking encyclopedia" and avid reader, familiar with the details of everybody's research, who believed in "being one's own rabbit." "As such, he swallowed in three days a 3.5-liter aqueous solution of 85 grams of calcium chloride to induce a marked acidosis. Robin [Hill] and I were responsible for analyzing the great man's urine. He developed an acidosis that was noteworthy." When the acidosis was at its peak, they went swimming in the river Cam. "Soon, a punt, bent on descending the river, made its approach. Seated therein were Hoppy, who had been knighted but recently, Lady Hopkins, and two distinguished-looking guests. Haldane at once swam under and around the punt, describing in his booming voice his experiment on acidosis: 'I am now excreting the most acid urine that has ever been excreted.'" According to Luck, learning physiology from Haldane was unorthodox. "At the first session, I remember he started off with the query, 'How big do you think my liver is?' He weighed 100 kg. We answered with widely different percentages of his body weight. 'How much blood do you suppose I have?' Answers: A few pints up to a few gallons. 'How may one determine the blood volume?' And so on."[68]

Other Hopkins lab alumni included younger biochemists like N. W. "Bill" Pirie and Joseph Needham, both of whom knew Haldane well at this time. Pirie, who took a Potemkin tour to the Soviet Union in the 1930s and wrote for the Marxist journal *Modern Quarterly*, would later write the Royal Society obituary of Haldane.[69] The gullible Needham (who succeeded Haldane as reader in the Hopkins lab in 1933, after JBS left the lab for University College London) was successfully manipulated by the Soviets in a bizarre piece of cold-war propaganda concerning germ warfare during the Korean War.[70] He was a permanent fixture of Soviet-sponsored "Peace" organizations after that. Immersed in this atmosphere, Haldane had repeated opportunities to absorb far-left alienation, a sign of the broader radicalization of intellectual life in Cambridge that was already under way during the 1920s but that accelerated rapidly with the arrival of the Great Depression and the 1930s.

The geneticist Sydney Cross Harland, who will often reappear in this story, left a curious reminiscence of Haldane from the spring of 1927, when he first met him. "I used to go pub crawling with him and his wife in the West End of London. On one occasion a man sitting near made a sneering remark about Mrs. Haldane, who was talking in a loud and rather shrill voice. Without saying anything, Haldane walked over to the man, took him by the scruff of his neck and the seat of his trousers, and propelled him through the door into the street. Then he sat down again. 'Do you think I did right?' he asked." Harland remembered that when Haldane was "agitated or excited," then "the first half of each sentence was spoken while breathing out, and the second half while breathing in." Even though he could be "brusque and even rude if he was bored or irritated," and "liked to say things to shock people," his company was stimulating. "When I upset one of his theories, he used almost to brag about it, rather as if he had done it himself. In conversation he could throw new light on almost any topic that arose." Harland disputed the idea that Haldane had no sense of humor, but he never heard him tell a joke or laugh out loud—"although at times he would make an attempt to smile graciously, it was rather an evanescent and flickering sort of performance."[71]

2. WITH VAVILOV
IN THE SOVIET UNION

On March 22, 1928, Haldane received an invitation from Madame Polotseva of the Society for Cultural Relations with the USSR to visit the Soviet Union, in order to attend the Third Congress of Russian Physiologists, Biochemists and Pharmacologists in Moscow, from May 28 to June 22. With this communication, Haldane's surveillance by MI5 was initiated, since the Society for Cultural Relations was known to be a vehicle for the Soviet propaganda agency VOKS (All-Union Society for Cultural Relations with Foreign Countries). Henceforth his mail, telephone conversations, public speeches, newspaper articles, and border crossings would receive intermittent attention, eventually amounting to a large dossier—routine snooping of the kind that he had himself performed for military intelligence at the end of the last war. Haldane accepted the invitation at once. It was to be his only trip to Russia; Charlotte would go one more time, with a different result, in 1941.

Like all such tours, their visit was carefully planned and chaperoned by VOKS. Beside the conference, Haldane was invited to lecture at the Moscow and Leningrad institutes for genetics run by Nikolai Vavilov (1887–1943).[1] The trip stretched through the summer, during which Vavilov and VOKS ensured that the Haldanes got first-class treatment, with plenty of champagne, caviar, and conducted sight-seeing: a rare private viewing of Lenin's mummy in Red Square; the Kremlin museums; the diamond and sable crown of Ivan the Terrible; the gilded carriages of the tsars; their summer palaces in Leningrad; ballets at the Bolshoi; and operas in Leningrad.

Vavilov, who in 1928 was "a handsome man in his early forties with little dark twinkling Tatar eyes,"[2] with a baritone voice, had trained as a geneticist before the First World War. He had even spent some time, between 1913 and 1914, with William Bateson in England at the John Innes Horticultural Institution. He had also made several expeditions abroad to gather seed samples before the revolution of 1917. After the Bolshevik seizure of power, he had adapted quickly to the new conditions and had continued to assemble a world-leading seed collection at his institute, with frequent forays all over the world in search of plant varieties. In 1928, Vavilov already had more than 20,000 types in this growing collection, the result of expeditions that eventually covered more than forty countries on five continents. He had received one of the first Lenin Prizes in 1926 and was the director of the Institutes of Plant Breeding and Genetics. Haldane, now a part-time adviser on genetics at the John Innes, was also an admirer of Vavilov's novel theories about the origin of cultivated plants like wheat. Based on the patterns of genetic variation in his collection, Vavilov conjectured that these had been selected first by accident from wild varieties in mountainous areas before they were introduced into the great valley-based civilizations, such as those on the Euphrates and the Nile. Vavilov was thus one of the first genetic archaeologists or gene geographers.

Because science in the USSR was then carved into institutional fiefdoms run by individuals with political patrons, Vavilov had a near monopoly in his field, controlling all of plant science.[3] This covered some four hundred research institutes and experimental stations and tens of thousands of workers. Although Haldane always referred to him as a Soviet scientist, Vavilov had been a geneticist well before the Revolution, and was only Soviet and Marxist now by pure necessity, as were all the scientists who continued to work in the USSR. With few exceptions, the others had left of their own accord or had been forcibly exiled or even executed by Lenin and his successors.

Like all the shepherded tourists of the Soviet era, the Haldanes were shown model institutions constructed and run only for that purpose. Charlotte was taken to the "Red October" chocolate factory, with seemingly superb day-care facilities for working mothers. She could not help noticing that the infants had a "fierce possessiveness" about their enamel chamber pots; socialism still had much to abrade. Later, she would regret that she was so easily taken in. "I strongly suspect that many 'Intourists' who have returned full of enthusiasm for such show-places had as little previous experience by which to judge them as myself."[4] Their will to believe supplied the push that VOKS gratefully pulled on. Charlotte thought that "People talked frankly with foreigners, and the O.G.P.U. was not at all obtrusive in everyday life."[5] But three years earlier, Haldane's friend John Maynard Keynes had on his own visit already noted "a policy which finds a characteristic expression in spending millions to suborn spies in every family and group at home."[6]

In addition to attending the Third Congress of Russian Physiologists, Biochemists and Pharmacologists, the Haldanes were introduced in person to several scientists we will return to later. Vavilov's junior colleague Georgii Dmitrievich Karpechenko (1899–1941) was a plant cytologist at the Institute of Applied Botany near Leningrad, and had also worked at the John Innes. Solomon Grigorievich Levit (1884?–1938)[7] was a Jewish physician from Lithuania who had joined the Communist Party in 1919 and pioneered human medical genetics in the USSR as a member of the communist Society of Materialist Physicians. In 1928 he was attached to the Moscow University Clinic and in the next decade would start pioneering twin studies into human hereditary conditions. Professor Aleksei Nikolaevich Bach (1857–1946) had been a revolutionary under the old regime, living abroad in Switzerland before returning to Russia after 1917. He would later run the Institute for Biochemistry in Moscow in concert with Aleksandr Oparin and serve on the Central Executive

Committee of the USSR. Lina Solomonovna Stern (1878–1968) was another biochemist and physiologist, a Jewish Latvian who had been educated in Switzerland but returned to the USSR in 1925. Stern had been invited to Moscow by Aleksei Bach, who had known her when he was exiled in Switzerland, and she accepted out of sympathy for the stated ideals of the new regime. In 1928 she was the head of the Department of Biochemistry at the Mechnikov Institute of Infectious Diseases in Moscow. The Haldanes were welcomed into the homes of Stern and Levit and established a close rapport with them.

In later years, two of these scientists were ultimately shot after brutal torture; one was sentenced to death after being tortured, but reprieved only to die of starvation and abuse in prison; while another was arrested, tortured, nearly shot, imprisoned, and then exiled. Only one escaped wholly unscathed.

Charlotte later wrote that after this tour she had more misgivings than JBS about the USSR. "To him, its outstanding characteristic was the Soviet attitude to science and scientists . . . the scientists and the factory workers were the most favoured classes."[8] This theme is repeated so often in JBS's own references to the USSR (he never wrote at length about the tour) that we must reluctantly accept that he really believed it. For example, on returning home, he addressed the Fabian Society:

> . . . assuming the present Russian *régime* to last for another fifteen years, you will, for the first time in the history of the world, have a scientifically educated governing class at the head of a great State. What the result of that will be I do not pretend to know. It will, undoubtedly, be interesting. It may be a little too interesting for this country! . . . Among the small fraction of the Russian population who read seriously, science and politics take the place which are taken in England and the United States by religion and sport. . . . The public papers are full of science. . . . Workers' classes there are on the most magnificent scale, and they have real science and real experiments of a type which are not allowed

by law to be demonstrated to medical students in this country. They attempt in every possible way to link up science with politics.⁹

Another comment deserves close attention. "They have altered the ruling class. They did not try to educate the old one." Here "altered" is a euphemism for mass execution, torture, internal enslavement in labor camps, forced external exile, demotion to menial positions, constant hounding and overall loss of the usual privileges of citizenship, a campaign of persecution that extended through biological kinship to reach even distant family. Members of the ruling class became known as "former people."¹⁰ A euphemism was required. His audience understood what he meant.

The linkage of science and politics had already taken a form and direction Haldane did not understand. Charlotte noticed the cost. "Nearly all ex-middle-class and middle-aged or old people showed visible signs of the strain. They were underfed, shabby, cowed, unless they belonged to the intellectual élite, such as technicians, teachers, scientists, or politicians."¹¹ JBS had not read Lenin's plainspoken letter to Gorky. "The intellectual forces of the workers and peasants are growing and getting stronger in the struggle to overthrow the bourgeoisie and their accomplices, the intellectuals, the lackeys of capital, who think they're the brains of the nation. In fact, they're not its brains, they're its shit."¹²

There were signs that the supposed new "élite" were in as much danger as the old. While the Haldanes were in Moscow for the Congress, the Shakhty Trial was in progress in the Hall of Columns and dominated the news. This was a precursor of the great show trials to come. Engineers, or "specialists," were accused of "wrecking" mining in the Donetz Basin, in which the North Caucasus town of Shakhty was set. In reality, the Bolsheviks had severely damaged the coal mining industry by neglecting capital investment and imposing a regime of increasing production quotas, while degrading the management of the mines. Fifty-three engineers were selected to take the blame; they were tortured to confess to a fantastic conspiracy involving the former "capitalist mine owners," who either had been "altered" by the Bolsheviks a decade earlier or had fled abroad, but had, out of spite, somehow persuaded the engineers to deliberately flood the mines. Five of these "specialists" were sentenced to death, including the leading expert in his field, Peter Palchinsky. Forty-four were sent to the rapidly emerging system of forced-labor prison camps that came to be known as the Gulag, and the remainder were acquitted.

We know that JBS followed the Shakhty trial. He would have discussed it with his friend Professor A. N. Bach, who was one of the "public prosecutors" at the trial.[13] To illustrate Soviet devotion to science, Haldane told the Fabians that "a shop-window display of books bearing on the trial of the Donetz colliery engineers was 'starring' a volume on the relation of geology to economics." Haldane made no other comment on the case.[14] By contrast, the *New York Times* considered Walter Duranty's reports, that "most" of the accused deserved their fate, news fit to print.[15] "Those of course were the days when we looked to the Soviet Union for light and hope," according to Haldane's sister, Naomi. "In the thirties we refused to believe any criticism of the Soviet Union because that put us on the same side as those for whom money power is the standard: not love, not comradeship."[16]

3. THE THIRTIES

After his trip to Russia, it would still be some time before Haldane publically displayed the shift in his politics that was under way. He started to refer to the Soviet Union more often, usually in very favorable terms, but took care to leave some room for doubt. An article he published in *The Nation* in 1931 shows this technique fully fledged. It involved a great deal of "one might be as bad as the other" fudging.

> Today the old civilization of Europe which we share is adapting itself with some difficulty to the new conditions created by modern industrialism. But it is also threatened by two new types of civilization on its east and west, namely, communism and Americanism, which claim to be improvements on it. Both of these interest me intensely, and I think that we could copy some features of each with advantage. I should like London to have as good operas as New York[1] and as good biological teaching for the average person as Moscow. But I do not desire that London should adopt either of their standards of personal liberty. I follow with immense interest the fierce and sometimes bloody struggles of the American and Russian governments against wets and whites respectively, in which they display a vigor and intolerance to be found only in young and growing civilizations. I am particularly interested in the Five year plan of economic expansion in Russia.

If it succeeds it will prove that socialism is a practicable system, and I shall probably live to see some form of socialism adopted in England. If it fails, Russia may revert to capitalism, and socialism all over the world experience a great setback.

I cannot accept the American and Communist ideals because both are too exclusively economic. They agree in taking economic efficiency to be the principal human virtue, even though in one case the benefits go mainly to private individuals, in the other to the state. They are both moving toward the mechanization of life and the standardization of man. Now I am not much interested in machinery, and very much so in life.[2]

But on July 29, 1931, MI5 intercepted a letter from Maurice Dobb to Rajani Palme Dutt (1896–1974),[3] then considered to be the leading theoretician in the Communist Party of Great Britain. Dutt was a founding member of the Party back in 1920 and a Comintern representative. The letter proposed a new journal of Marxist ideology, "to break the hold of bourgeois ideology, to explore and uproot the assumptions underlying current thought and practice." Dobb insisted that the "small working" editorial board "must be composed of persons completely accepting the Marxist line." He did not need to explain to Dutt that the Marxist line to be completely accepted was more specifically the line of the Communist Party, as opposed to heretics like the Trotskyites. Dobb also proposed forming a board "to act in an advisory capacity to the editors." One of the names he suggested was J. B. S. Haldane, along with V. Gordon Childe, Roy Pascal, Piero Sraffa, and Lancelot Hogben, who were all fellow travelers, and the Dutt brothers themselves, a fail-safe addition.[4] To be considered worthy of this company was high praise for a Marxist, and suggests that, even if Haldane had not fully realized it yet, he was making the grade. From then on, MI5 considered him to be within the orbit of the Communist Party.

Haldane drew political attention again the following year. On February 8, 1932, a traveling exhibit of posters on Soviet education was hosted at Cambridge in the YMCA hall, at which he was the opening speaker. It is likely that his connection to the exhibit was based on political sympathy. When the YMCA belatedly realized that their hall was being used to display anti-religious propaganda, they closed the exhibit down, resulting in some newspaper coverage.[5]

The gifted American mathematician Norbert Wiener spent a year's sabbatical at Cambridge in the early 1930s and got to know Haldane well, after noticing his science-fiction story "The Gold-Makers" in *The Strand Magazine*.[6] Wiener declared that he had "never met a man with better conversation or more varied knowledge." The Wieners visited Roebuck House. "I used to go swimming with him in a stretch of the River Cam, which passed by his lawn. Haldane used to take his pipe in swimming. Following his example, I smoked a cigar and, as has always been my habit, wore my glasses. We must have appeared to boaters on the river like a couple of great water animals, a long and a short walrus, let us say, bobbing up and down in the stream."[7] Wiener was fiercely left-wing himself, and would later, after Hiroshima, attract attention from the FBI for publically denouncing nuclear-weapon research.

Haldane's election to the Royal Society in 1932 underlined his professional frustration at Cambridge. Brash, unpopular, and controversial, after a decade he had still not been offered a fellowship there, despite several attempts on his behalf. His habit of stentorian discussion of intimate personal details at High Table, accompanied by a gallon jar of urine from his latest experiment, set down among the college silver, did not help. Nor, one suspects, did his notoriety at the Philosophical Society for cracking open walnuts by placing them on a table and hammering them with his forehead. Attacking organized religion at the Union as a purveyor of fear and shame to the emotional and defective, by means of the dregs of the universities, may have been unsettling to some.[8] The Australian anatomist James Thomas Wilson, who had known John Scott Haldane well, considered JBS at this time "clever—sometimes too clever not always profound, & frequently wrongheaded & bitter."[9] Nor had he made much headway in his part-time position, which he had held since 1927, as an adviser on genetics at the John Innes Horticultural Institute. But the Royal Society praised his work of the past decade on "mathematical evolution" and his swelling output of forty-three published papers, electing him in 1932. His sponsors "from personal knowledge" included F. G. Hopkins, R. A. Fisher, A. V. Hill, R. C. Punnett, and Charles Sherrington.

Although he published hundreds of scientific papers in biochemistry, genetics, cosmology, statistics, ethology, and other fields, and served as a voluble lightning rod for many of his colleagues, Haldane is still known mainly for his work on theoretical population genetics.[10] This research ran through the 1920s and early 1930s, resulting in nine major papers and a classic book of published lectures, *The Causes of Evolution* (1932). The book combined less formal and wider-ranging, sometimes philosophical, commentary with an appendix summarizing the mathematical theory. The problem he addressed had been raised by the rediscovery in 1900 of Mendelism, with its discrete, particulate genes as units of heredity. It was now clear how variation was preserved and then refreshed by recombination. The paradox of blending inheritance, the inexorable homeopathic watering-down of characteristics that had perplexed Darwin, was dissolved. But the question was whether Darwinian natural selection was strong enough to be an important driver of evolutionary change. "A satisfactory theory of natural selection must be quantitative. In order to establish the view that natural selection is capable of accounting for the known facts of evolution we must show not only that it can cause a species to change, but that it can cause it to change at a rate which will account for present and past transmutations."[11]

Haldane showed that the natural selection of genes was effective by constructing difference, or "recurrence," relations between discrete generations of idealized populations, where the variation was acted on by natural selection of those genes. He then solved those equations. His results showed that selection of genetic variation inexorably drove evolution by altering the frequencies of the genes in populations, gradually shifting their distributions. He progressively relaxed the assumptions he had made at first, and then proved that the results still held true for the new recurrence relations that emerged, solving those too. The non-linearity of the recurrences meant that he had to invent methods that gave numerical approximations instead of exact solutions. "Some of my calculations led to surprising results. Thus it appears that, as a result of the survival of the fittest, a population may become less fit, just as the effect of gravity on a spinning top is to make it stand up, instead of falling down."[12] At the same time, R. A. Fisher and Sewall Wright were independently working on broadly similar ideas, which meant that there were now three different ways to attack the problem, starting from different assumptions. This successful combination of natural selection with Mendelian genetics was called the "Modern Synthesis" by Julian Huxley.[13]

In 1933, Haldane left his readership at Cambridge to accept a chair as Professor of Genetics, and later Weldon Professor of Biometry, at University College London. Perhaps this setback fed his emerging alienation from the institutions, traditions, and history that had produced him. While Charlotte house-hunted in London, disappointed and frustrated by her Roebuck coterie and supposing herself "a rebel against Cambridge intellectual and social snobbery," JBS spent three months teaching at Berkeley in California at the end of 1932. They then moved to London in 1933, taking 16 Park Village East, "an almost hidden secret corner of Regent's Park" facing the canal.[14] Charlotte at least was back on home ground. They would be seen more often at Kleinfeldt's Fitzroy Tavern between Bloomsbury and Soho, in the questionable company of painters such as Augustus John and Nina Hamnett, their old friend Malcom Lowry (when he was sober enough to go out drinking), their jazz-piano-playing former student Martin Case, his brother Ralph, and occasionally Bertrand and Dora Russell.[15]

At the John Innes Institute, the cytologist Cyril Darlington interacted with Haldane frequently. Darlington found his conversation stimulating and later credited Haldane with helping to spark, "in long youthful discussions," the ideas that finally led to the exploration of social biology and genetically informed history in his

Evolution of Man and Society—"although I parted company with him over communism."[16] In Darlington's view, Haldane's reliance on the genetic data gathered by others allowed him to make rapid advances by cross-fertilizing ideas between related fields, but was a "source of weakness" in the long term, since he could not establish his own comprehensive empirical research program; he was restricted to the role of a skilled interpreter, providing "quick effects" and "unusual connexions." Darlington thought him "induction blind," blinkered by his unusual talent for deductions from mathematical models, an exactness that often led to excessive caution. Most of all, JBS was always an outsider. "Haldane had his own view of friendship. Men, he had told me a few years earlier, were of no interest to him save in an intellectual sense. This was a view which revealed something more than he could have understood. It showed him as a man emotionally undeveloped: a man lacking that irrational gift of friendship so widely needed and so warmly felt in managing human affairs."[17]

Throughout the 1930s, Haldane's open involvement in politics grew exponentially. The Relief Committee for the Victims of German Fascism and the Berlin Reichstag Fire trial drew him onto public platforms in 1933. At this time, British universities actively recruited persecuted academics fleeing the Nazis. "Herr Hitler supplied me with two first-rate junior colleagues, Drs. Grüneberg and Philip, who were of Jewish origin. I had some difficulty in finding them salaries, but when this was done through the generosity of the Rockefeller Foundation, we got a real school of Genetics started in London."[18] In the self-obituary that he recorded for the BBC in 1964, he recalled another refugee, who had applied to him in person. "Among the people who came in was a man called Chain.[19] We talked for an hour or two about the work he had been doing, and I said: 'I don't think I can help you much, but there is a man called Florey[20] at Oxford who is certainly interested in this kind of stuff, and I would advise you to have an interview with him.' Chain did, and, as perhaps you know, Chain and Florey shared the Nobel Prize for the isolation and preparation of Penicillin."[21]

The Nazis were fodder for Willi Münzenberg, who discreetly ran the Relief Committee for the Victims of German Fascism and the subsequent *Brown Book* campaigns launched to protest the Reichstag Fire trial, at the behest of Moscow.[22] Haldane's appearance on their platforms was coordinated by Münzenberg's associate, the OGPU agent Otto Katz, whose letters to Harry Pollitt containing the arrangements were intercepted by MI5.[23] Haldane may not have been aware then of this backroom direction. Writing a few weeks before her death in 1969, Charlotte Haldane believed he was not all-in by then. "Around 1934–5 J.B.S., in my presence, told Professor Hyman Levy and Dr. Barnet Woolf (both then CP members) that he was a dialectical materialist, but not a communist. They thereupon instructed him that such a position was logically and practically impossible."[24]

Haldane's public positions on foreign and domestic policy, now appearing in print so regularly that it no longer seemed odd for the Professor of Genetics at UCL to volunteer them so eagerly, showed that he kept accurately synchronized with the party line regardless. Until 1936, that was to be officially *for* "Peace" and *against* rearmament, with the rationale that only the industrialists would benefit, and that war was "imperialist." In a piece on the causes of war, which the BBC declined to broadcast but he published in the *Daily Herald* anyway,[25] he elaborated.

> I notice that Dean Inge[26] has no use for the idea that armament manufacturers ever frighten us to make us buy their wares. . . . If you want to catch war-raisers find out who is making money out of wars and rumours of wars. . . . As long, then, as we have massive unemployment, there is a very good reason for war. Every unemployed man or woman is a cause of war. . . . if we really want peace, we must examine all the causes of war, economic and technical, as well as psychological and political.[27]

And so the government was advised from University College that they could save money and gain security and friends by building air-raid shelters instead of weapons. "If we had protection of this kind, we could afford to reduce our expenditure on ships and airplanes, and people in other countries would be less afraid of us."[28] Whereas in the 1920s Haldane had settled that "the man with a gift for thought on scientific lines is of more use to his fellows in the laboratory than out of it" and that "the scientific mind is still best employed outside politics," he was now tackling Professor A. V. Hill in *Nature* about politically active scientists. "Prof. Hill condemns the irrational character of certain modern political movements. May it not be that the remedy for this lies simply in the application of scientific thought to political and moral problems?" Scientists "have not merely the right, but even sometimes the duty, to interest themselves in more controversial matters."[29] His own contribution, for now, was to start a long-running public clamor for gas masks and especially deep air-raid shelters, a recurring Haldane theme.

In 1936, his father, John Scott Haldane, to whom JBS had been close, died at Oxford of pneumonia, with "a look of intense interest on his face as though he were taking part in some crucial experiment in physiology which had to be monitored."[30] JBS paid tribute to him shortly before his own death. "I am sometimes asked to whom I owe most in my scientific career, and I have no reasonable doubt about that. I owe most to my father, the late J.S. Haldane. He was, like me, only more so, a dabbler."[31] Carrying the ashes back to Scotland, JBS made a scene with his sister, Naomi, and Aunt Bay, insisting on traveling third-class on his own. "We were told what was what about the capitalist class travelling first." Naomi's own political sympathies were strongly left-wing, but she still had a sense of the absurd. She noted that JBS now tried to eschew luxuries, "except for the spiritual luxury of open quarrelling with our mother," which she could never do herself. He began writing a life of his father, but whenever Naomi asked him about it, "he flared into anger." She interpreted his radicalization as a form of penance for their "class origin."[32]

Both of the Haldanes were drawn deeply into the Spanish Civil War of 1936–1939, in which the Communist Party of Great Britain (CPGB) was prominent. Charlotte and JBS had visited Spain in the early 1930s and, as early as February 1935, JBS had spoken on behalf of the Spanish Aid Committee in Worcester.³³ This was during the several years of political turmoil that preceded the war and came to a head only in February 1936, when the narrow election victory of a coalition of socialists, anarchists, and communists, intent on radical reforms, plunged Spain into chaos. The coalition ("Republicans") had not won the majority of the popular vote but managed to gain a majority of seats. Civil war between the Republicans and Nationalists followed when the military revolted. Haldane immediately sided with the Republicans, and raged in public against the non-intervention policy pursued by Britain and France, who were afraid that the war would escalate internationally. The war caused a bitter argument and break between JBS and his mother, Kathleen, which was never fully repaired.³⁴

Charlotte later remembered that JBS had offered to openly join the Communist Party around this time, but that Harry Pollitt advised him to keep his affiliation unofficial, since he would be more useful if he was known publically as only a sympathizer.³⁵ Her son, Ronnie Burghes, then only seventeen, joined the Communist Youth League and volunteered to fight in the British Battalion of the International Brigade, which was raised to aid the Republicans. Charlotte soon met with Harry Pollitt to give her permission for her underage son to fight, and to try to ensure that her son would have a gas mask. She recommended JBS to Pollitt as an expert on gas warfare. JBS could certainly call on his previous experience of gas warfare and counter-measures in the First World War, and in 1925 had even published a heterodox book on the subject, which he cryptically called *Callinicus* after the seventh-century Syrian from Heliopolis who invented a forerunner of napalm called "Greek fire" (hardly gas warfare). JBS assured his readers that the humanity of gas warfare was highly underrated compared to the alternatives.

It seems likely that Haldane took Pollitt's advice and was a secret member of the party by 1936 at the latest, since he was in Spain by December 1936, as an adviser to the Republicans on gas countermeasures. That month, an article by Charlotte for the *Daily Worker*, the organ of the CPGB, declared that "my two men are in Spain" and that "The Brigade is the greatest romance of modern politics."[36] She was all in, though she implies that she only joined the Party secretly in 1937. In January 1937, an intercepted letter to Clemens Palme Dutt again suggested JBS as an editorial board member of a new Marxist journal, this time *The Modern Quarterly*, which he would be associated with through the early 1950s.[37] And then JBS publically declared himself a Marxist in 1937, publishing "A Dialectical Account of Evolution" in *Science and Society*—at the ripe age of forty-four, going on forty-five.

JBS never played an important role in the Spanish conflict. Gas was a difficult weapon to deploy effectively and was not used by either side. But the war put him in touch with some enigmatic characters who would reappear in London after it was over, and it colored his ideology. On his first trip, from mid-December 1936 to mid-January 1937, he could do little more than observe air raids and their effects in Madrid, offer opinions on the construction of bomb shelters, tour the trenches, lecture about gas warfare, help out at the blood transfusion clinic of the communist Canadian doctor Norman Bethune, where he stayed for two weeks, and sample the modest biological research still being pursued in Republican Spain.

International Brigade members were struck, as was his wife, Charlotte, by the ill-fitting black leather outfit that Haldane roamed around in, augmented by a tin hat with a broken strap that he had kept from his days twenty years previously in the Black Watch. "As it was the only tin hat in the whole of Republican Spain, it attracted a good deal of attention from passers-by, and twice sentries saluted us respectfully, obviously impressed," according to a foreign correspondent he befriended, Virginia Cowles. Haldane seemed like an eccentric to Cowles, and not completely serious when he told people that he was "Just a spectator from England. Enjoyed the last war so much I thought I'd come to Spain for a holiday." She tells how Haldane led her on a madcap tour of the frontline trenches of Madrid, only to abandon her in order to clamber out over a hill in open view—she was horrified—to get to a spot with a better prospect. She only caught up with him later, after being led to safety by a Spanish soldier.[38]

The diminutive Jewish South African photographer Vera Ines Morley Elkan encountered Haldane at a hostel in Madrid, in the unisex washroom. "He was huge beside me." As the hostel was cold and miserable, he offered her a place with him at the marginally more comfortable apartment of a decamped nationalist lawyer. The apartment had been taken over by the Canadian Blood Transfusion Unit. She took photographs of the operations there, which included gruesome direct body-to-body transfusions without anesthetic, involving "quite a bit of screaming and quite a bit of wriggling."[39] Haldane used members of the unit to help him test gas masks, and made propaganda broadcasts at the radio station. The unit was run by Dr. Norman Bethune, who, Elkan remembered, had a large supply of Canadian whiskey on hand. Haldane helped Bethune perform some transfusions, reusing the experience for later propaganda.

> I believe that every healthy communist ought to be ready to give his blood, and not only to comrades. A pint of good red communist blood is better propaganda for the party of Marx and Lenin than a gallon of Tory beer for the National Government. For this reason, it is worth finding out to what group one belongs. . . . A Spanish comrade was brought in with his left arm shattered. He was as pale as a corpse. He could not move or speak. We looked for a vein in his arm, but his veins were empty. Bethune cut through the skin inside his right elbow, found a vein, and placed a hollow needle in it. He did not move. For some twenty minutes I held a reservoir of blood, connected to the needle by a rubber tube, at the right height to give a steady flow. As the new blood entered his vessels his colour gradually returned, and with it consciousness. When we sewed up the hole in his arm he winced. He was still too weak to speak, but as we left him he bent his right arm and gave us the Red Front salute.[40]

The Red Front salute! On January 4, 1937, all of the foreigners at Bethune's blood transfusion clinic, including Haldane and Bethune himself, were arrested by the Republican Secret Police. Most of them were released, except Bethune's Swedish lover Kajsa Rothman, who was detained for a while longer, and one Hurturg (possibly "Harturg" or "Hartung"), an Austrian who was then shot by the Republicans.[41] Curiously, Haldane referred to this incident in only an elliptical and flippant way, saying that "a number of foreigners" were "rounded up" and that "I think, one was shot."[42] Elkan must have been arrested too, but makes no mention of the incident. After a few weeks Haldane left for London. According to Elkan, "he left behind a pair of pajamas, which was a Godsend . . . they went about three times around me, but I lived in them because they were beautifully warm at night."[43]

Back in London, Haldane spoke at a series of public meetings in aid of the Republicans, which were duly monitored by MI5. Addressing a crowd of a thousand at Euston Road on February 25, he was defensive. "Haldane opened his speech by referring to the scorn with which his scientific colleagues regarded his association with a man such as Pollitt."[44]

On the basis of its non-intervention policy, the government prevented the *Daily Worker* correspondent "Frank Pitcairn" (Claud Cockburn) from going to Spain to cover the war. Haldane volunteered to go in Cockburn's place, arriving in Madrid on March 25. On this second visit, Haldane saw the hard-drinking Norman Bethune again, inspected trenches, and once more helped to broadcast propaganda on the government radio station, EAQ. The American geneticist and future Nobel laureate Hermann Joseph Muller was also at Bethune's clinic at this time. Muller was ostensibly researching transfusion of cadaver blood, but was really ducking the storm developing in the Soviet Union, where he had worked, since 1933, with Solomon Levit, Israel Agol, and others that Haldane knew well.[45] The Soviets had turned increasingly hostile to mainstream genetics, and Muller had volunteered to go to Spain to avoid the maw of the NKVD. On the way to Spain, Muller wrote to Julian Huxley that, judging from the "tone and content" of his letters, Haldane was "at present having his political opinions impressed upon him with a rubber stamp" and that therefore he would not disillusion him about the Soviet Union, since that might push him even further into the pro-Soviet camp.[46] Muller would return after eight weeks to Moscow to pack his things quickly and leave for good. He chose this roundabout route, instead of leaving directly from Spain, to avoid casting suspicion on Vavilov, who had organized the Spanish trip to get Muller out of sight.

Haldane assured *Daily Worker* readers that the International Brigade soldiers were holding up. "They have suffered heavy losses and have been in the trenches for many weeks. But their spirit was wonderful."[47] But Haldane did more than passive reporting. It must have been around this time that the future Soviet Red Orchestra spy and defector Alexander Allan Foote met him. "I remember Professor JBS Haldane, when for a short period he served with the Brigade as a private soldier, standing in a trench brandishing a tiny, snub-nosed revolver and shouting defiance at the advancing Franco infantry. Luckily for science, we managed to repel the rebel attack and the Professor was spared for his further contributions to world knowledge."[48] Fred Copeman—an imposing and intelligent brawler, who had been raised in a workhouse but now commanded the British Battalion of the International Brigade in Spain—was more blunt when he was interviewed forty years later:

> . . . and old Haldane was there, and more bloody nuisance than he was worth. He was a big fat fellow who wore a little leather jacket with only one button on. I don't think he ever changed his bloody shirt in four months you know, and the button used to be on his big fat old belly, you know, and it kinda stuck out, and he had funny dirty old trousers; and yet, he was a brilliant scientist. He insisted on being in the front line, and he had a little tiny revolver, I doubt that would hit that bloody window if you tried, but he would hop up on the step holding this bloody thing, and I would go up and every time I would say 'What bloody good

do you think you are? First of all you're taking two blokes room, two blokes can sit where your fat arse is, so get down out of it and get back to Brigade Headquarters'. I'm being told politically JBS Haldane must not get killed, he's too valuable, keep him out the line. He was all the time in the bloody line! After about 3 months I said, I had a long talk with him, I said 'look you've done enough bloody talking old fat man', I said 'you've got to go home, you've got to go home! They want you there, you're important there now. You've been in the line, you can talk and a thousand will listen to you'. So he finally caved in and went back home."[49]

The Republican Government wanted maximum value from their celebrity visitors. But Copeman's memory was faulty: Haldane was there for less than a month on this second visit, and his final tour in December would be just as short. His own memories of this visit were rosier. "I remember one evening when the conversation in the battalion head-quarters dug-out passed from adultery to telegony. The commanding officer (a former mutineer in the navy) upheld the view that the foetal and maternal circulations anastomose in the human placenta, and a fascist attack unfortunately terminated my effort to convince him that he was wrong."[50] The former mutineer was Copeman, who had fonder memories of Haldane's lectures to the troops, which were always popular, drawing up to a hundred men each night. "He would lecture on anything and of course the lads would always try and get him onto a lecture to do with sex, men always do, you know." *Lady Chatterley's Lover* was a favorite theme. "I got a lot out of that old boy."[51]

Haldane also met up with the physiologist Juan Negrín, minister of finance in the Republican cabinet. They may have met for the first time in 1933, at a physiological conference in Madrid, or on Haldane's visit in December 1936. Negrín was from the Canary Islands and had established a successful university career, first as a professor of physiology and then as an administrator. The minister of finance was an extravagant gourmand and womanizer who lived large. Reports circulated that he ate out up to three times a night, a luxury enabled by bulimia. Officially Negrín was a socialist, but the Soviet defector Walter Krivitsky[52] stated that the Soviets viewed him as more pliable than Prime Minister Francisco Largo Caballero, who had resisted "liquidation" of radicals and anarchists like the POUM (the Workers' Party of Marxist Unification). As minister of finance, Negrín enabled the secret transfer to Stalin of Spain's treasury of gold, one of the largest in the world and worth some 500 million dollars. Stalin gratefully accepted it for "safe-keeping" and as collateral for arms shipments. The gold was never returned, and exchange rates were manipulated to grossly inflate the prices of the arms supplied, many of which, though not all, were obsolete.[53] The Soviets had early on resolved to do no more than prolong the conflict as long as possible, in order to tie up any Western powers unwise enough to get involved. Making a lot of money along the way was an unanticipated bonus.

Before Haldane left Spain, he met Martha Gellhorn, a foreign correspondent, at Bethune's blood transfusion clinic in Morata on April 5, and presumably through her or Bethune, her lover Ernest Hemingway. On April 9, Haldane and Gellhorn watched a Nationalist assault through binoculars from a house on the edge of Casa de Campo—"like college kids on an outing," as Gellhorn put it.[54] A few days later, Hemingway, Herbert Mathews, and Virginia Cowles, who were all foreign correspondents, watched the battle from a building christened the "Old Homestead" by Hemingway. Cowles was amused.

> We heard footsteps coming up the stairs and looked around to find Professor J.B.S. Haldane. He greeted us with his usual cordiality and looked round for a place to sit. The house was gutted with pulverized furniture, old clothes and broken pictures. From the débris he dragged a dilapidated red plush chair, placed it in the middle of the room, and sat down in full view of the battlefield. He put his elbows on his knees and adjusted his field-glasses. Hemingway warned him it was dangerous to remain exposed, but Haldane waved him aside. A few minutes later Hemingway spoke again: 'Your glasses shine in the sun; they will think we

are military observers'. 'My dear fellow, I can assure you there isn't any danger here in the house'. Ten minutes later there was a loud whistle as a shell plunged into a flat next door. Two more screamed overhead and we all went down on the floor—all except Haldane, who scrambled down the stairs and disappeared. We were shelled for fifteen or twenty minutes, and when at last we got back to the [Hotel] Florida we found him sitting in the lobby, drinking beer. 'Hallo,' he called amiably, 'let's have a drink'. We did; and more than one.[55]

Perhaps it was through Hemingway that Haldane met Hans Kahle (1899–1947), though it may have been the other way around. They established a close relationship that would stretch over the next decade. Kahle, who was born in Berlin in 1899 to Dr. Karl Kahle and Maria Caroline Duebener, was commander of the 11th International Brigade, including the "Ernst Thälmann" battalion; he usually referred to himself as "Lieutenant Colonel."[56] Tall for his day at 6 feet, ½ inch, with brown hair and brown eyes, he had served as a lieutenant in the German Army during the First World War from 1916 to 1918, but he was captured and held until 1920. After the war, he was, so it was said, "in business" in Mexico from 1921 to 1927, returning to Germany by 1927 or 1928. He was an open member of the Communist Party of Germany from then onward, and a member of the communist paramilitary group Roter Frontkämpferbund, which was led by Ernst Thälmann.[57] By July 1935, he was in Moscow under the aegis of the "International Red Help" group, having traveled there via the UK. Apparently this was his second visit to the USSR. In Moscow, he edited a German newspaper, *D.Z.Z.* By this time he was married to Gertrude Ernestina Kahle (1913–?). MI5 believed they were married in Berlin in December 1933.[58] He entered Spain in 1936 via Switzerland and France.

In *For Whom the Bell Tolls*, Hemingway based the sympathetic character Hans on Kahle.[59] A photograph survives of the two from this period.[60] Another International Brigade member, Gustav Regler, later claimed that Hemingway was so enamoured of Kahle during the Spanish Civil War that he proposed to write a book about him.

> He was a Communist, commanded the Eleventh Brigade, was precise in his orders, understood the unstable Spaniards, mingled the methods of Potsdam with those of Alcazar, obeyed his Party, because obedience flowed downwards to the troops, but leavened it at staff meetings and conferences with an almost French irony. He resisted onsets of melancholy with a formal bearing which dominated all staff activities and meals. He liked looting, but handed everything he found in the castles over to the legal Government, departing from this principle only with a big china vase which he took with him from one field to another in a packing-case. He wore silk shirts and during lulls in the fighting went to Madrid, where he slept in the Empire bed of a film star who had fled, swam in her pool and slowly drank her cellar dry. During critical periods he scorned all feminine consolations, but as soon as things eased up he was to be seen again at Gaylord's or at the theatre, which was kept open despite the bombardment.[61]

Kahle's activities included working on the German Freedom radio station in Spain. MI5 were informed by multiple sources that Kahle was in the service of, and perhaps even the head of, the Soviet OGPU in Madrid at this time, and that he was responsible for numerous liquidations there; elimination of ideological enemies quickly became the major focus of Soviet involvement in Spain.[62] But this information was incorrect in one important respect: the OGPU in Madrid was headed by Alexander Orlov.[63] Kahle, like all International Brigade commanders, was a Soviet military intelligence (GRU) man. But it is true that many members of the Brigades were tortured and executed for political reasons, and Kahle would have been involved in these "repressions." Another International Brigade commander, André Marty, is said to have executed more than five hundred of his own men.[64]

The British Brigade soldier Walter Greenhalgh remembered being visited near the front line by Haldane, who was "particularly interested in the effects of bombing and so on." They put him up on a camp bed in their office.

> . . . in the middle of the night—Haldane was a huge man, almost filled the door you see—he's kicking at me you see. 'Yes, yes?' 'What are those things in there?' he says. I said 'what things?' He said 'those things against the wall'. 'Oh', I said, 'those are Pete's shells'. I said, 'when the enemy sends over a dud shell, Pete goes out there and he collects it, and he brings it back here, and one day he's going to open it up because he firmly believes that there is a message of solidarity, that these shells have been sabotaged'. 'What? You're stark raving mad! That's enough explosive in there to blow the town to pieces! Madness, absolute madness!' And off he went, we never saw him again. He disappeared . . . Anyway the next day along came a lot of engineer men, they loaded all these shells into a truck and then they took them all away and they sort of blew them up outside in a field.[65]

When Haldane finally arrived back in London from Boulogne on April 16, MI5 archly noted that "the condition of the clothing he was wearing suggested that he had recently been exposed to the weather."[66] Back in Spain, Norman Bethune was now considered to be a fascist spy by a Republican cabal, and was therefore persuaded to leave on a fund-raising tour in North America at the end of April. Then Bethune was simply denied re-entry into Spain, though this doesn't seem to have shaken his faith. He would die in 1939 of a septic cut on Chairman Mao's Long March, in northern China. Years later, H.J. Muller wrote to a colleague that if the Canadian Blood Transfusion Unit had not broken up, then "all of us . . . would still be there (under the ground)."[67]

In the meantime, Charlotte had taken up the Dependents Aid Committee Fund on behalf of the *Daily Worker*, to support the volunteers they were recruiting. Since the British government had forbidden its citizens from fighting in Spain, this was, strictly speaking, illegal, though the law seems to have been weakly enforced. After March 1937, she was based in Paris as an underground Party member, facilitating the transfer of the volunteers to Spain, warning them about the dangers of venereal disease, reporting to the Party on their "political and personal reliability," and calling herself "Rita." She noticed that the volunteers were forced to surrender their passports to the Party—these valuable documents were later used by the NKVD for espionage purposes. Her circle in Paris included a young American she called "Jack"—Arnold Reid,[68] a former editor of the magazine *New Masses*, who appears to have been another of her lovers. Later Reid would be killed in Spain, apparently "sold down the river by his own party" due to ideological differences, a verdict attributed by Charlotte to the CPGB apparatchik Bill Rust.[69] After three months in Paris, at the end of May 1937, she returned to her Dependents Aid Committee fund-raising in London.

MI5 were curious at this time about a house that the Haldanes had leased in the country—Mascalls in Broad Chalke, Salisbury, Wiltshire, described by MI5 as "A Cromwellian house of about 10 rooms standing in about two acres of grounds."[70] They spent occasional weekends there, and it was looked after by a "handyman" when they were away. At times Charlotte went there alone. It is possible that the house was used for clandestine purposes, though Norbert Wiener recalled visiting the Haldanes for a few weeks before the Spanish Civil War, at an "old stone house" set in "a country of delightful walks and views." Wiener thought that the Haldanes had chosen it because the downs of Sussex, which Charlotte took a liking to, were too pricey.[71] MI5 also noticed several weekend trips to Paris by JBS—for example, he returned to England from Boulogne on May 15 and October 8, 1937. These trips were too brief to have included Spain itself, and were likely to have been visits to the Communist International setup that Charlotte had been based in, but MI5 did not make any explicit deductions.

JBS briefly returned to Spain for a final visit with Harry Pollitt, arriving sometime in December 1937 and leaving in early January

1938. There had been changes. His friend Juan Negrín had replaced Largo Caballero as prime minister in May 1937. According to Walter Krivitsky, who described Negrín as a "willing collaborator" and "just the type to suit Stalin's needs," this had been an NKVD objective for some time.[72] Under Negrín's administration, the NKVD operated much more freely than before, and the Soviets called in their favors. The POUM had been "liquidated" with Negrín's consent, and its leader, Andreu Nin, had been "disappeared." Alexander Orlov's men had tortured Nin in one of their secret prisons and shot him—the warrant for this had been signed by Stalin himself and was found in the Soviet archives decades later.[73] His body was never found. Other factions were also "liquidated," leaving the Communist Party in a commanding position.

Negrín invited JBS and Pollitt to visit the front at the Battle of Teruel and to view the cabinet meetings held "on the ground floor of an ordinary house."[74] Haldane had romantic notions about Negrín, writing of the Spaniards that "Being an heroic people, they will only give their allegiance to an heroic man."[75] In Barcelona, Haldane observed the escalating air raids and was shown the air-raid shelters in the hills dug 55 feet down, "a labyrinth of passages about 7 feet high by 4 feet broad."[76] He noticed that air raids produced differing reactions. He had found Madrid nonchalant, but not the Catalans. "After an air-raid on a village in December 1937, people ran out into the country with such vigour when they saw another squadron approaching that I could not help joining them, though I did not run as far as some, and stopped as soon as I saw a ditch."[77] He visited the exhibition in Barcelona on the Battle of Madrid. "As a foreigner in a city under bombardment, and infested with spies, I did not consider it healthy to take too many notes."[78] He described spending a night (January 2, 1938) roughing it with evacuees in Tarancon, "on the road from Madrid to Valencia."

> I had managed, after a long walk through snow, to get a place on an empty food lorry from Madrid as far as Tarancon, along with another man and a woman. Night fell, and the ground was

covered with snow. After an hour or so's search, we found a farm on the outskirts with a big dug-out, which was used as sleeping quarters for women and children. I also got a small loaf by telling the authorities that I was a very important person. I slept with about twenty other people in a stable where there were also three mules, which helped to warm the place, and which had a pleasant smell.[79]

Haldane re-entered England on January 6, 1938, carrying numerous small parcels he said were "Presents from members of the International Brigades to their families."[80] Whereas his earlier reporting from Spain had been upbeat morale-boosting stuff for the Republican cause, now he wrote unconvincingly that "I hope . . . that the people of Britain will never see what I have seen in Spain."[81] Stephen Spender, who was drawn into the Communist Party for a time in 1937, said that Haldane actually relished the details of this, his brand-new war. "During the Spanish Civil War I was one evening at a Christmas party given by his sister, Naomi Mitchison, when Haldane appeared, having just returned from Spain. Haldane seemed quite unhappy until the children's charades were stopped and he could regale the guests with stories of his violent Spanish adventures." Spender noticed that Haldane "seems to enjoy displays of violence," and that a few years later, when air-raid shelters were being bombed during testing, "Professor Haldane insisted on sitting in one of the shelters whilst high explosives were dropped nearby."[82] Recall Haldane's own reaction to being bombarded on the Western Front during the First World War, where the "entirely novel sound intoxicated me."[83]

In late January 1938, Charlotte made her own trip to Spain, without JBS. She chaperoned Paul Robeson, the Negro folk singer and enthusiastic fellow traveler of the Soviet Union, and his wife, Ellie. Robeson made a ten-day tour and was pleased to croon "Ol' Man River" to the International Brigade in his baritone. Charlotte's job probably included keeping the Party informed of all doings. After the tour was over, she stayed on for nearly two more weeks, and visited the British Battalion at Teruel with Bill Rust. The fact that she was then granted an interview in Barcelona by the prominent Stalinist Republican personality Dolores Ibárruri underscores the importance that the Republicans attached to their foreign friends. She would meet up with Ibárruri again in Russia during the next war.[84]

JBS continued to appear in public through 1938 to support the Spanish cause and to lambaste the foreign policy of the government. (Naomi said that "his great gesture to demonstrate a point was to slap his thigh."[85]) MI5 reported that, at the Merseyside Aid to Spain Committee, in front of 1,000 attendees in St George's Hall, Liverpool, he "described the Prime Minister as the greatest advocate of Communism in this country, intimating that if he continued acting in this weak-kneed manner in which he was doing when intimidated by Fascist bullies, the people of this country would rise in open revolt to assert their rights."[86] He opened a screening of his old friend Ivor Montagu's propaganda film *Spanish Earth* to the more select audience of the Farnham Left Book Club.[87] But the developing defeat of the Republicans proportionately deflated the overseas campaign on their behalf. Now Haldane's focus turned increasingly to air-raid precautions in Britain, which quickly developed into a form of monomania.

Charlotte's marriage to JBS seems to have been in trouble by this time. She rarely mentions him in her autobiography, *Truth Will Out*, between mid-1937 and 1939. On returning to England from Spain, she was pushed out of the Dependents Aid Committee Fund by Harry Pollitt. Later in 1938, she set off on a long mission for the Communist International, billed as a *Daily Worker* correspondent, to Chiang Kai-shek's communists in China. The purpose of this extensive trip was never clearly explained by her, since the reportage alone could not have justified the expense. When she returned in early 1939, she found that her "personal affairs" had "undergone a change" since she had left. Fred Copeman, the sardonic battalion commander from Spain, who had replaced her at the head of the Dependents Aid Committee Fund, lived at the Haldane residence in London at around this time.

Good woman Charlotte was. Oh, she was a good one. Mind you, you do in life see things. Charlotte must have been 45 or 50 but she acted as if she was 16 you know. She would doll around that bloody house in Regent's Park with castanets. She was getting a bit fat by now you know. Everything wobbled. And she had about half a dozen lads from the Brigade living there, and I used to sit back—on the one hand there was the old boy the Professor who had the chair at London University in Biology, and whose biology was very practical when it came to the good looking girls you know. He wasn't short of that. And on the other hand here was Charlotte trying to look as nice as these lovely little students. The lads were quite delighted because—but I used to sit back and think Oh Charlotte![88]

Aside from Charlotte's numerous love affairs, JBS had met Helen Spurway, a student of his at UCL who had made rapid progress. Afraid of drawing damaging publicity, the Communist Party, via Bill Rust, exerted its discipline over both of the Haldanes to block a divorce. All boundaries between personal and political life were long erased. JBS had submerged intellectually into Marxism and the Party.

In an exchange provoked by his essay "A Dialectical Account of Evolution," which he had first published in 1937, Haldane confessed that he had been working on his Marxist interpretation of biology for a while. "The process took me some six years, so it was hardly love at first sight." This account was obviously his ticket to the Marxist game, proof of his credentials as a serious worker of their technique, and he wanted to be confident that he had something convincing. He worked out the details in a broader setting the following year in *The Marxist Philosophy and the Sciences* (1938), which attempts to reveal that mathematics, astronomy, sociology, and biology are all thoroughly dialectical, when read correctly. He begged off tackling economics. His chief inspiration here was Engels, since Marx wrote little about science. Engels had published *Anti-Dühring (Herr Eugen Dühring's Revolution in Science)* in 1878, and added more material in his unpublished fragment *Dialectics of Nature*—Haldane wrote a preface to a translation of this in 1939.[89]

The gist of the analysis can be extracted from Haldane's attempt to explain one of the central Marxist dogmas, "the unity of opposites," early on in *The Marxist Philosophy and the Sciences*. Haldane gives two examples, both of which fail.

1. "For example, if I say, 'John Smith is a man,' I am asserting the identity in a certain context of a particular, John Smith, and the universal, man. This identity has led philosophers into very great difficulties. . . ." Haldane wants to argue that John Smith is therefore a "universal" and a "particular" at the same time, with the particular being the "opposite" of the universal. But a particular is *not* the opposite of a universal; "John Smith is a man" means only that John Smith is an *element* of the set of men, not that he *is* the set of men. There is no "unity of opposites" in this, and the "difficulty" is just a confusion.

2. "Again, I say that the wood of which this table is made is hard, or it would not support things, and soft, or it could not be cut. Two opposite qualities are united." Now, if something can be exhaustively divided into two categories A and B, then we may well say that B is the opposite of A because not-A implies B, and it must be one or the other. But the opposite of "hard with respect to supporting things" is "soft with respect to supporting things," not "soft with respect to cut-ability." Another confusion.[90]

Haldane nevertheless went to great lengths to squeeze, press, fold, iron, soak, pummel, and fillet evolutionary biology into the round hole of dialectical materialism, with its limited toolbox of concepts such as "thesis," "antithesis," "synthesis," "constant flux," and "contradiction," and its mechanistic idea that these are the steps by which all history progresses, as if on an eternal revolving ratchet. His "solution" is summarized in the table below.

None of the terms that feature in dialectical materialism have any precise meaning, so that this sort of categorization is limited only by the patience of the audience and the ingenuity and determination of the practitioner; almost any domain will do. Contrast this to Haldane's revealing statement, in another context, that the theoretical work that R. A. Fisher and he conducted on evolution "made us investigate the exact meaning of a number of words in common currency. Thus Darwin wrote about the survival of the fittest, but never defined fitness exactly. Fisher and I have had to do rather complicated calculations about natural selection. So we had to define fitness fairly rigorously."[91]

Just how broad these dialectical categories must be is driven home by the realization that absolutely everything, real, conceptual, or imaginary, must be placed in one or the other category, since dialectical materialism "explains" everything. Haldane embraces this flexibility: "the negation of a negation may mean several rather different things. Why not? If a formula is to be applicable to natural events in the spheres of physics, chemistry, biology, psychology and economics, and also to our thought about them, then it must be elastic."[92] But elastic classification, or in this case reclassification, does not explain anything unless it adds to, rather than subtracts from, what we already

HALDANE'S DIALECTICAL CATEGORIES FOR EVOLUTIONARY BIOLOGY

Thesis	Antithesis	Synthesis
Heredity	Mutation	Variation
Variation	Selection	Evolution
Selection of the fittest	Consequent loss of fitness	Survival of noncompetitive species

know. Abstraction is useful when it allows a body of theory to be developed that is applicable to anything that satisfies its definitions, so that the theory does not have to be reinvented for every instance or special case of the abstraction. Correct classification makes that theory available and applicable to the case at hand. The best examples come from mathematics, which in modern times has proceeded by revealing unsuspected commonality through continued abstraction and unification; but precise definitions are unavoidable there, and the right abstractions have required very hard work to get right.

What biological consequences follow from the one-size-fits-all elastic dialectical classification? Haldane argued that he personally had only been able to "see" some consequences of selection-mutation equilibria, like the frequency of hemophilia, after describing the concept of an equilibrium verbally in terms of dialectical materialism. He conceded that "I do not claim that these results could not have been obtained without a study of Engels. I merely state that they were not reached without such a study, and that so long as I find dialectical materialism a valuable tool in research, I propose to state the fact." In short, *personal revelation*.

John Maynard Smith—who was a student and close friend of Haldane and a communist at Cambridge in the 1930s and for many years afterward—found little of value in Haldane's Marxist treatment of biology. The problem the Marxists faced was that "classical Mendelian genetics was damned undialectical. . . . What it says, if you think about it, is that genes determine development in an embryo, but development has no influence on genes."[93] Thesis-antithesis-synthesis is most naturally interpreted here as the idea that development affects heredity, as in Lamarck's acquired characters, but it just so happens that it doesn't work that way. But we know that this is so, and Haldane knew that it is so, despite and not because of dialectical materialism. Haldane was able to avoid that sort of interpretation only because he was working backward from the known facts and already knew which pitfalls to steer clear of. But his interpretation has some obvious objections of its own.

Calling variation the "synthesis" of the "antithesis" mutation and the "thesis" heredity is empty, given that variation is exactly what mutation produces; there is no synthesis. Stabilizing selection, by far the most common kind of selection, does not produce evolution; it produces stasis, the opposite of evolution, since it immediately eliminates the variation introduced by selected-against mutation, leaving things as they were before. And in the rare cases when evolution really is the result, because the mutation is selected for, that is immediately provided by the "antithesis"; merely relabeling it "synthesis" is not helpful. As A. P. Lerner put it, this approach "clearly does not improve the interpretation of biological facts for students who are not suffering from an overpowering emotional urge to embrace the dialectic."[94]

Engels felt confident that dialectical materialism could conquer mathematics, too, and Haldane's treatment of this excursion-without-compass-or-map anticipates his later response to the innovations of Soviet biology under Lysenko. Engels had some trouble with calculus, which he attempted to teach himself late in life, from scratch, apparently working from a textbook published in the 1790s. This text still employed the system of infinitesimals, which gave the right results if used in the right ways, but lacked rigorous justification. Augustin-Louis Cauchy would supply a rigorous reworking of calculus in the 1820s by discarding infinitesimals and formalizing the notion of limits. In the twentieth century, infinitesimals even re-emerged in a completely different rigorous formalization as a result of advances in mathematical logic and model theory. But Engels did not know about Cauchy, and persuaded himself that infinitesimals and consequently differentiation were "contradictions" in mathematics, which should just be embraced as part of the dialectical process.

Meanwhile, Marx had also been trying to work mathematics out from scratch, armed only with the dialectical method, and had concluded after much labor that $y = ax$ could be differentiated, which he conceived of as a kind of symbolic computation, by first writing it as $y/x = a$, and then imagining that at $x := 0$ (where ":="

means "assigned") this gave $o/o := a$. Since $dy/dx = a$, as the textbooks taught, he concluded that mathematicians had simply defined $dy/dx := o/o$ as the next symbolic computational step—something like the progression of history. Engels wrote to Marx enthusing that this demonstrated how the "contradictions" in mathematics simply had to be accepted, since the ratio of "vanished quantities," o/o, made no sense. "I compliment you on your work. The matter is so perfectly clear that we cannot be amazed enough how the mathematicians insist with such stubbornness upon mystifying it. But that comes from the one-sided way of thinking of these gentlemen."[95] The poly-sided Engels found many other "contradictions" in mathematics involving things he did not understand—for example, complex numbers, and the fact that square roots are powers of ½.

Haldane's account of all this, from his preface to the English translation of *The Dialectics of Nature* (1940), needs considerable unpacking and unwrapping.

> Engels' remarks on the differential calculus, though inapplicable to that branch of mathematics as now taught, were correct in his own day, and for some time after. He points out that it actually developed by contradiction, and is none the worse for that. To-day 'rigorous' proofs are given of many of the theorems to which he refers, and some mathematicians claim to have eliminated the contradictions. Actually they have only pushed the contradictions into the background, where they remain in the field of mathematical logic. Not only has every effort to deduce all mathematics from a set of axioms, and rules for applying them, failed, but Gödel has proved that they must fail. So the fact that the calculus can be taught without involving the particular contradictions mentioned by Engels in no way impugns the validity of his dialectical argument.[96]

It is not true that Engels, who was writing in the late 1870s, was "correct" in his own day. As we have seen, he was at least forty years out of date. Nor is it true that the formalization of calculus was "developed by contradiction" unless that means nothing more than that a rigorous treatment was postponed until one could be found. The fact that Cauchy and others worked hard to place calculus on a rigorous footing shows rather that they did *not* accept a "contradiction." The idea that mathematicians have only "pushed" the "contradictions" a level up into mathematical logic is an early, and very egregious, example of the sort of misuse that Gödel's theorems have suffered.[97] This sense of "contradiction" is not *something we don't understand yet*, or *something Engels does not understand*, but logical inconsistency within a formal system, the statement 'p and not-p'. The fact that a formal system powerful enough to construct integer arithmetic with both addition and multiplication cannot be used to prove its own consistency—so that the system *cannot* derive 'p and not-p'—is *not* an invitation to embrace contradictions in mathematics. For example, if someone were to discover an inconsistency in Zermelo-Frankel set theory today, considerable effort would go into devising a new system. Otherwise that theory would literally prove everything, since a false statement implies anything, and the system would be no use—at least, to a non-dialectician. And calculus is not merely "taught" using rigorous methods today—as if this is a matter of pedagogical style—but was completely rebuilt to use those methods by 1830, even if the new system took a little longer to reach England.

The thing to notice here is the way in which Haldane engages in special pleading on behalf of Engels. Since Haldane was a skilled mathematician himself (putting the Gödel flummery to one side as a bit of overambitious dabbling), he could certainly recognize the concerted mathematical illiteracy displayed by Engels, who was out of his depth here at the shallow end. Jean van Heijenoort, who was once Trotsky's bodyguard in Mexico but became an eminent logician after he renounced Marxism and gave up politics, wrote an amusing essay that surveys the impressive dimensions of this illiteracy—it has even been said that van Heijenoort finally gave up Marxism when he encountered Engels on mathematics.[98] Instead, Haldane screened Engels behind an evergreen hedge that few of his readers would have been able to poke through, leaving them with the idea that Engels was not bad for his era, even partly correct, maybe well ahead of his time in some ways, with an argument valid in parts. Later, he would say much the same about Trofim D. Lysenko.

Another interesting episode throws some light on Haldane's knowledge of Soviet practices at this time. In an intercepted letter to the Marxist historian Dona Torr, dated November 3, 1938, Haldane thanked her for lending him a copy of a Soviet edition of the *Dialectics of Nature*. He was taking it with him to Spain, but he had spotted some excisions from an earlier edition he had seen. "I am inclined to think that somebody in Moscow had quietly suppressed some passages in the earlier edition. If not I must have imagined them. Granted that they were passages where Engels backed the wrong horse in a scientific controversy, but I don't think we have any right to suppress them without mentioning the fact. As a matter of fact his mistakes were most interesting, and as he did not publish the stuff he cannot be held responsible for the mistakes."[99]

4. STALINOPHILIA

On December 1, 1937, the *Daily Worker* announced "J.B.S. Haldane to Make Science Plain for You . . . a special and exclusive series of articles dealing with science from the point of view of the progressive thinker. . . . He treats his subject from the point of view of the active fighter for Socialism." In case these public announcements were not enough, a few months later the MI5 informant M/2, who had been talking to Marjorie Pollitt, reported that "it is clear that J.B.S. Haldane is in the closest touch with the more intellectual leaders of the Communist Party. He apparently sees Harry Pollitt and Emile Burns quite frequently."[1] Haldane would go on to write hundreds of articles for the *Daily Worker*, up until the 1950s. Most of these were written in the style of popular science that he had based on H.G. Wells and perfected by the early 1930s—only now with Marxism, the Soviet Union, Lenin, and Stalin drawn in at every possible opportunity for inordinate praise. He republished most of these articles in book form, adding an occasional footnote or addendum, so that they were more generally available.

Haldane ranged far and wide to find interesting topics, from biology to physics, cosmology, mathematics, medicine, physiology, statistics, geography, chemistry, and beyond. Connecting all these topics to politics sometimes required acrobatics. For example, when explaining the rudiments of calculus, Haldane simply asserts that the

Soviets know more about it: "To judge from the technical books which sell by tens of thousands in the Soviet Union, a bigger fraction of the people understand it there than here. In a society where workers are encouraged to understand their work it is natural that it should be widely studied."[2] Or, when considering statistical samples, that the Soviets got more political sampling done: "in the Soviet Union today all citizens get at least some political education from attending public meetings to decide on local as well as national affairs."[3] An article on controlled experiments presented an opportunity for flights of fancy about Soviet "experimentation" and Stalin's *History of the Communist Party*.

> Marxism became more scientific as it developed in Leninism and Stalinism. The Bolsheviks had a correct political and economic theory. But that was not enough. They experimented on a great scale. For example they tried state farms and collective farms. The latter worked better in most cases. . . . They were able to compare the results of different experiments, for many types of productive relation are possible under Socialism. Above all, as any reader of the *History of the C.P.S.U.* finds out, they learned from their mistakes in a truly scientific manner. Leninism is not only a historical science like geology. It is an experimental science because Leninists make history as well as studying it.[4]

Psychology? "Though Marxism will certainly benefit from advances in individual psychology, it is already a genuine and scientific analysis of human behaviour in the mass."[5]

Geology? "Marxists think that history is moving towards universal Socialism and the breakdown of barriers not only between classes but between nations. And one solid block of two hundred millions in the Soviet Union is working towards these goals."[6]

Probability? "You can never be sure that you will not die before you get home, or that when you do you will not find that your wife has gone mad or your child been run over. Society should be organised so as to make such events as these as unlikely as possible. It is a striking fact that in the Soviet Union the theory of probability plays a big part in planning."[7]

Weed control? "In a socialist country all applied science is public service, and the gap between theory and practice is narrowed. That is one reason why more and more scientists are becoming socialists."[8]

Newts? "In the Soviet Union, aquaria were very well developed. . . . In 1934 it was easier to buy small tropical fish in Moscow than it was in London."[9] Even in the case of flying ants, we are informed along the way that "under socialism the average citizen will have more, not less, private property than to-day, besides his or her share in the public property."[10]

But what about mining? "It was not merely uneconomic to take steel for thousands of miles from Ukraine to Siberia, when it could be made on the spot, or to take raw cotton from Tashkent to Moscow, and send cotton goods back again. Stalin saw that the formerly subject peoples had a right to industrial development."[11]

And gas? "Near Moscow the most remarkable of all the developments of gas liquefaction is in progress. Some of the coal seams are near ground level and only a foot or two thick. The Soviet people do not want miners to work in such narrow seams [so they turn the coal into gas.] . . . It is wonderful to think what this will mean from the human angle. Instead of coalmines working under dangerous and dirty conditions, and gas-works which belch out smoke and can be smelt a mile away, there will be spotlessly clean factories with machinery controlled by a few skilled men and women. Under capitalism the worker is subordinated to the machine. Under socialism the machine becomes the servant of man as it should be."[12]

Surely not cooking? "Rational cooking on a factory scale is more likely to start in the Soviet Union than elsewhere."[13]

Apparently, in the Soviet Union even small children, when they were not savoring "rational cooking on a factory scale," were scientists too. "In Britain many school children learn science, but they have little chance of making any discoveries for themselves. In the Soviet Union some children make discoveries. . . . [M]any birds fly south in autumn . . . to discover what routes they took the children in hundreds of country schools started trapping birds. The traps were carefully designed so as not to hurt them. Each bird had a numbered ring put on its leg and was then let loose. Results began to come in when the same bird was caught two or three times in the same place."[14] But this is just a foretaste of the scientific enlightenment that is to come. "With the combination of scientific education and leisure to which we may look forward as Leninism spreads over the world, we can look forward to a day when about one person in twenty will be a naturalist, and many mysteries of nature will be a mystery no more."[15]

There hardly seem to be any limits. In the context of an article about the mathematician G. H. Hardy, a self-declared enemy of all applications of mathematics, universal versatility emerges instead.

> Lenin aimed at something quite different, a society of 'men who can do everything,' where professors can and do mend their own cars, and mechanics investigate the scientific principles involved in their jobs. The Soviet Union has not yet reached this goal, but it is well on the way to it. Foreign engineers who brought over complicated machinery to the Soviet Union were often horrified because the workers insisted on taking it to bits, even if it took a month to put it back together again. This meant a delay in starting production, but it meant that the workers were able to mend the machines if they went wrong, and often to improve on their design fairly quickly. Soviet intellectuals are mostly drawn from the ranks of the manual workers, and proud to show that they are still capable of tackling skilled jobs. Clearly a society of this kind can switch from peace to war, and will switch over from war to peace again, much more quickly than our own; quite apart from the fact that there are no landlords and other vested interests to impede these changes. . . . This is why Marxism is rightly called scientific socialism. It is based not merely on an analysis of the breakdown of capitalism, and what is needed to replace it, but on a study of how social changes actually do occur. Before 1917 it had not had an adequate experimental test. But Lenin, Stalin and

their collaborators, have given it the test which every scientific theory requires, and it has proved to work in practice. No wonder that Marxism is spreading among scientists to an extent that alarms many people.[16]

In case human instincts might seem like a hurdle here, Haldane is confident that man's very nature can be remade, quickly. Indeed, this had already happened. "Human character can be changed in one generation. The younger generation in the Soviet Union will work together for the public good. They regard the struggle for one's own interests which is inevitable under capitalism, as being not so much wicked as ridiculous selfishness."[17] And so when considering human needs, the Soviet Union pointed the way to prosperity, though the example may be harder to emulate elsewhere.

> The Soviet Union was so backward technically at the time of the revolution that it was only entering the age of universal plenty in 1941. Britain could enter it within a year of establishing socialism.... For the younger people in the Soviet Union, brought up under socialism, mostly take it for granted that it is pleasant and honourable to work for the community. They are therefore ripe for communism. We will not be so till socialism has taught us the same moral lesson.[18]

Fortunately, "the example of the Soviet Union will lead other countries to adopt socialism.... If we are not afraid of applying the ideas current in the Soviet Union to our own affairs, we too have a splendid future ahead of us."[19] Part of the secret of this "universal plenty," apparently, was planning.

> There are two S's in U.S.S.R. One stands for Socialist and implies planning from above; the other for Soviet, means initiative and criticism from below. The two-way traffic of ideas is as vital to science as it is to industry. Without planning, Soviet science would not have shown the greatest growth in a generation which has ever been recorded, and, to take only one example, the

Red Air force would have been many years behind the Luftwaffe in design. Without democracy the Soviet scientists would have made few original discoveries, and would not have had the initiative needed to tackle war problems.[20]

Haldane was quite definite about who was responsible for the universal plenty. "No living man has a clearer grasp of the nature of society than Stalin, who has played a leading part in two great changes, the overthrow of capitalism, and the building up of socialism."[21] But these developments did not come at the expense of freedom, which was rapidly improving under communism, approaching American levels. "I am inclined to think that the average man has a greater freedom of movement in the United States than anywhere else, and that this freedom is increasing most rapidly in the Soviet Union, where it is already fairly high. This, if correct, is due to the great development of transport and the real high wage in the U.S.A., and to the system of holidays with pay and workers' holiday resorts in the U.S.S.R., together with the fact that, as there is no unemployment there, workers tend to move very freely from one job to another."[22]

Freedom of movement in the land of universal plenty was accompanied by freedom of expression, up to a point.

> In the Soviet Union . . . [l]egally there is fairly complete freedom of speech. And actually there is a good deal. I have heard a man say that he could not see much difference between Stalin and Nicholas. A member of an important Soviet merely replied that there was quite a big difference. But on the whole custom is more stringent than law; so that there is somewhat less verbal criticism of the government than in England.[23]

One supposes that extra-judicial sentencing by a troika might be considered "custom." But Haldane nevertheless conceded *some* limits to freedom of speech. "In the Soviet Union any attempt to start an opposition journal would probably meet with practical rather than legal difficulties."[24] Writing elsewhere, he conceded somewhat obliquely that there had indeed been *some* ideological imposition, but no longer. "I do not think that non-Marxist biologists need fear an attempt to impose Marxist dogmas on science, such as probably occurred in some quarters in the Soviet Union between 1922 and 1932."[25] Shortly after the end of the Second World War, when republishing some of his earlier essays, Haldane added even more qualifications.

> [T]he overthrow of the class state has meant a period of 'dictatorship of the proletariat' with considerable restrictions on freedom, in the Soviet Union, and would probably do so elsewhere.... I wrote this essay during the winter of 1939–40.... The Nazis had bought, blackmailed, or persuaded members of men in every country to become their supporters. The only state which was taking adequate measures against them in 1938 was the Soviet Union. These measures, like other measures of national defence, involved certain restrictions on freedom, which appeared to many of its foreign friends to be excessive. They did so to me in 1939 when I wrote this essay. I do not now think that they were so. I have no doubt that these restrictions will disappear as the Soviet Union feels itself safer. This is made highly probable by the fact that in the Soviet Union alone among belligerent nations there have been very substantial increases in freedom during the war ... freedom in the transmission of opinions is strongly on the upgrade in the Soviet Union.[26]

If Haldane thought in 1938 that "restrictions" on freedom were excessive in the Soviet Union, he showed no sign in print. But he remained sensitive to this line of criticism, if only as a debating technique: "it is claimed that in practice all power is in the hands of the Communist party and its sympathisers. In practice, however, parliaments are also controlled from outside.... The plain fact is that over most of the world such parliaments as survive are at least as subservient to Big Business as is the Supreme Soviet in Moscow to the Communist Party."[27]

Religious freedom was thriving, too: "there is about as much religious freedom in the Soviet Union as in Great Britain . . . I do not know of any legislative step in that direction taken in the last ten years in Britain which is comparable with the enfranchisement of priests in the Soviet Union in 1937. . . . Of course, in so far as they are followers of Engels, Communists are definitely enjoined not to use the police [to attack religion, but] . . . this does not mean that they should not argue against religion or suppress organizations which use religious forms for political ends."[28] The war demonstrated that the people themselves were not unhappy. "The heroic resistance of the Soviet Peoples has shown to the world what I knew before, that by and large, they are pretty satisfied with their way of life, and certainly not longing for deliverance from it."[29]

The unstoppable praise for Stalin and the USSR that flowed from Haldane's pen raises the question of how much he really knew about it. A single visit in 1928 of one month is not much to go on in 1938 or 1947. He did not read Russian, so he was completely dependent on translations for information. He continually complained that it was hard to get hold of translated scientific papers. As he never quotes any source, even for the purpose of criticism, that is not pro-Soviet, one suspects that he just repeated whatever the Soviets provided him with. There is some justification for this suspicion. In the Haldane Papers at UCL an "information" leaflet issued by a Soviet propaganda agency survives, with a summary of new and marvelous "achievements" in the USSR, paragraph by paragraph. Haldane annotated "use" next to those paragraphs he found interesting.

When Haldane wrote his rosy descriptions of Soviet science, concerns about the fate of Vavilov had been circulating for a while. The state-sanctioned ideology had increasingly turned against orthodox genetics, and a peasant scientist by the name of Trofim D. Lysenko had steadily gained influence, winning Stalin's support. Leading geneticists had been arrested and, after confessing under torture to lurid conspiracies and deviant behavior, had been either shot or exiled to deadly terms of forced labor. Coherent reasons for this turn against genetics have to be teased out of, and reconstructed from, the confused propaganda and general opprobrium that the Party ideologues rained down on the subjects and personalities involved. Lysenko was just one aspect of this process.

Orthodox genetics was unacceptable to the Party because, in essence, it was just too *slow*, implying that organisms had an essential nature that was resistant to change, requiring repeated natural or artificial selection over a long period. This understanding frustrated an underlying communist theme, the idea of *perfectibility*—of man, of nature, and of society. It also contradicted the idea that all of the social "superstructure" was determined by the "modes of production" of its economic base. Reflexively, Marxists preferred to explain everything in terms of social class. Perfectibility through Communist Party policy set itself the task of remaking society, and even creating new kinds of humans, as they put it. Until the 1930s, this tension had remained latent. However, the second wave of eugenicists from 1910 to 1930 were mostly, though not all, left-wing radicals of Haldane's type. The eugenicists believed in a radically different program for perfectibility that involved improving the human stock slowly through selective breeding, gradually but inexorably altering the gene pool in desired directions.

Russia and the Soviet Union itself had an active eugenics movement up until the early 1930s.[30] It was pioneered by Nikolai Konstantinovich Koltsov (1872–1940), a zoologist and geneticist who had been strongly influenced by Francis Galton and the English school of eugenics. Koltsov had been educated at Moscow University, graduating from there in 1894, after which he held a professorship from 1895 to 1911. He had spent some years abroad in Germany, France, and Italy, and was politically a socialist. The influential Institute for Experimental Biology had been founded at his initiative in January 1917. As a foretaste of things to come, Koltsov's career was almost cut short in 1920 when he was arrested as part of one of Lenin's prototype show trials. The "anti-Soviet Tactical Center" affair implicated a collection of politicians, intellectuals, and scientists in an imaginary espionage plot. Along with Koltsov, the accused included the historian Sergei Melgunov and Leo Tolstoy's daughter Aleksandra. While he was on trial, Koltsov kept detailed observations of his weight to test the relationship between anxiety and weight loss. Only an appeal from Maxim Gorky, an admirer of Koltsov, saved him from execution.

The Russian Eugenics Society was founded by Koltsov in the early 1920s to promote the scientific study of human genetics. At that time, as in the West, knowledge of human genetics and remedial measures to reduce genetic disease or otherwise direct the gene pool were not clearly distinguished. Koltsov's aims were entirely scholarly, and early twin studies as well as genealogical investigations were conducted. By 1925, more than one hundred twin pairs in Moscow were under observation. Koltsov's remedial eugenics was essentially Galtonian, a religious urge to favor the wellborn, more by selection of positive traits than selection against defects, which would instead be crowded out.

Koltsov's musings on the possibilities for genetic alteration would have appealed to Haldane. Both recognized that changes need not be improvements. Koltsov illustrated this by speculating that, if Martians really did descend on Earth to colonize it, they might decide to "domesticate" man, in the same way that man domesticated wolves. Men could rapidly be selected to their needs and tastes. Any tendency to independence could be weeded out. Instead timidity or slavishness might be selected for, "because humankind has always had and has now and will for a long time have inborn slaves." Distinct breeds of physical laborers, musicians, and craftsmen suggest themselves. Who could predict the aesthetic tastes of the Martians for form and shape? They might breed men for merely decorative purposes, or breed them to test the laws of genetics.[31]

In parallel, a Society for the Study of Racial Pathology was founded in 1928 to study differences in patterns of disease exhibited by isolated subpopulations, particularly in areas like the Caucasus—in the 1950s Haldane would suggest a similar line of inquiry for endogamous Indian castes. Other members of the Society included Solomon Levit, who had trained in medicine and founded a Laboratory of Human Heredity that year. Born in Lithuania to a poor Jewish family, Levit was a Communist Party man, but had been converted from Lamarckism by a course in fly genetics. Nevertheless, he remained a member of the Society of Materialist Physicians. His institute conducted twin studies and collected medical genealogies to study the etiology of racial pathologies. Levit was also closely associated with a fellow Lithuanian Jew and Party member, Israel Agol, who pursued genetics at the Moscow Zoo Biological Institute.

Koltsov's protégés by now included two eminent geneticists, Sergei Chetverikov (1880–1959) and Nikolay Timofeev-Ressovsky (1900–1981), though neither shared his enthusiasm for eugenic engineering of the gene pool, nor were they Party men. But political changes were under way, and fine distinctions were not in order. Chetverikov was arrested by the OGPU in 1929 and sentenced to internal exile without trial. In 1930, Stalin turned more decisively against eugenics and the general idea of innate human nature. Along with its leading advocates, eugenics quickly disappeared from the Soviet world and was subject to ritual abuse. Human intelligence testing, also taken to imply limits on perfectibility, was banned by 1936.[32] Linking social phenomena to biology was henceforth forbidden (a taboo that became a staple of left-wing thought worldwide, especially in the social sciences). "Biologizing" became a term of abuse: it might be materialism, but it was the wrong kind of materialism, a "bourgeois-reactionary-fascist" and "Menshevizing idealist" kind. Marxism, correctly interpreted, understood "modes of economic production" and "class" only. Koltsov came under open attack in the official organs by the Marxist philosopher Isak Prezent, who was later to play a much bigger role in the war on genetics. In April 1931, the Society of Materialist Biologists (OBM) issued a report demanding that Koltsov be "unmasked" as a "reactionary."[33] All this was duly solemnized by the *Great Soviet Encyclopedia* of 1932. The five-year plans were going to reconstruct society and it was defeatist to suppose that there were obstacles they could not overcome. Koltsov was denounced as the instigator of "fascist eugenics." The Russian translation of R. A. Fisher's classic *Genetical Theory of Natural Selection*[34] included only the first seven chapters; the remaining four, dealing with "social biology," were quietly dropped.

In the interim, Solomon Levit reorganized his lab as a Genetics department and moved it to the Medical Biological Institute. In 1931 both he and Agol were awarded Rockefeller grants to study abroad. They spent the year with Hermann Muller in Austin, at the University of Texas, with a summer interlude at Cold Spring Harbor on Long Island. On returning to Moscow in early 1932, they discovered that Levit had been under public attack *in absentia* as a "Menshevizing idealist." He was promptly removed as the director of the Medical Biological Institute, but managed to get himself reinstated by enlisting Maxim Gorky's personal aid (interventions of that kind almost always proved latently fatal). Muller, who had been politically influenced by Levit and Agol toward communism, followed them to the USSR in 1933. He hoped to persuade Stalin to adopt a program in positive eugenics through measures like sperm banks for the gifted. But he had not read the *Great Soviet Encyclopedia* carefully enough. Nor, it seems, had Solomon Levit, who continued to pursue human genetics by amassing twin pairs, going from 600 in 1933 to 700 the next year, increasing to some 1,700 by 1937.

Parallel developments were under way in plant genetics. Agriculture was particularly important to the "Great Break" that Stalin had initiated in 1929, and plant genetics was, in principle, part of the same enterprise as human genetics. To create their new society, the Soviets proposed to "revolutionize" agriculture (they ended up by utterly destroying it). The idea that plant genetics could limit their ambitions was a festering sore, though it took some time to burst open. Trofim D. Lysenko exploited this developing tension by promising a "socialist" alternative steeped in claims about homegrown, folksy, peasant science. All that resonated with Stalin and the Party, who reflexively despised traditional specialists.

Lysenko, the son of a peasant, was born in 1898 in the Ukraine. He received a rudimentary education at a vocational school for gardeners in Uman and some practical training as an agronomist at the Kiev Agricultural Institute, graduating from there in 1925. Early on he learned the value of publicity, making the pages of *Pravda* for work he had done on legumes in Azerbaijan, with some ambitious claims of success. He would build his career on a fantastic elaboration of his next subject, the effect of low temperatures on the maturation of plants. It had been known since 1858 (and even before that) that slightly earlier maturation of plants could be produced by exposing seeds to low temperatures to provoke germination before the growing season, under the right conditions. This "vernalization" process was reinvented by Lysenko, who made increasingly ambitious, and always shifting, claims for its benefits.

Lysenko's initial claim was that the traditional process of sowing winter wheat in the fall—leaving it to germinate over winter under the snow, and only harvesting it in the summer after substantial attrition—could be sped up. If the seed was exposed to water in the winter, to induce germination, and then kept at a low temperature and planted only in the spring, it could be harvested in summer. Doing this would greatly increase the yield by avoiding attrition, or so he claimed in 1928 to have found experimentally. Like all of his later claims, there was never any hard evidence for this increased yield, since at no stage of his long career did Lysenko conduct anything resembling a controlled scientific experiment, a procedure he found unnecessary and confusing. He simply declared success, and once again made the pages of *Pravda*. This was the right tactic, from a political point of view, since quick fixes for the harvest were in demand, as the "Great Leap" rapidly destroyed agriculture (the superlative was correct; it was just the direction that was off). In fact, vernalization does not increase yields, since the additional opportunities for mold, and the breakage that happens when the sprouted seeds are turned during winter, offset any gains: the net result is a waste of labor and an opportunity cost.

Nevertheless, sensing the changing political winds, Nikolai Vavilov took Lysenko up—probably in the hope of humoring and containing him within a harmless fiefdom of the Soviet agricultural establishment, one cannot be sure. Political dexterity was a requirement for survival under state-directed research. But this proved to be a fatal underestimation of Lysenko's talents. The Ukrainian used his invitation to the 1929 All-Union Congress of Genetics in Leningrad to rapidly broaden his scope. Next he claimed to have discovered that vernalization altered the very nature of the wheat, transmuting the winter wheat variety into summer wheat. Moreover, there were, according to Lysenko, actually no varieties as such: all wheat was one variety, merely with different levels of "winterism" that could be altered at will. But soon, "winter wheat" was quietly dropped, and Lysenko was vernalizing spring varieties instead, at his new lab at the Odessa Institute of Selection and Genetics. He pressed his methods on collective farms, and though, given the chaotic state of Soviet agriculture, it is likely that only a few acres were actually planted using his methods, he kept claiming success, touting unverified questionnaires. Yields were always up by staggering amounts.

Nobody checked Lysenko's figures, and so he got bolder. Armed with research money and his own journal, *Vernalization*, he ranged from corn to sorghum and even fruit trees, all of which could benefit from the miracle of vernalization. By now Lysenko was emerging as a hero of agriculture, a shock worker who could get results where establishment figures offered only laborious experiments and decades of tedious data collection and comparison. Vavilov, unsure how this protégé could be contained, gingerly offered more praise. Lysenko, barely educated, was eventually elected a member of the prestigious Academy and given a whole institute to wield.

The appearance of Isak Prezent, a specialist in the Marxist philosophy of biology, added a new ideological dimension. Prezent became Lysenko's Marxist-Leninist spirit guide. It was his idea to name this new kind of agricultural science "Michurinism," after the homegrown, unorthodox but green-fingered plant-grafter, Ivan Michurin

(1855–1935). This term was useful when contrasting the new science with the decadent ideas imported from abroad. In 1935 Lysenko denounced the "saboteur kulaks" of scientific research—for they were to be found there too—deriding the adherents of "bourgeois genetics" who were mentally crippled by capitalist agriculture. Someone in the audience stood up and applauded, "Bravo, Comrade Lysenko! Bravo!" It was Joseph Stalin himself.

Lysenko was now a made man. Ideas multiplied. He began to advocate cross-pollination regardless of variety, a process he referred to as "love-marriage," as opposed to the "arranged marriages" of bourgeois geneticists, who labored under the illusion that varieties with known properties ought to be preserved. Since he had abolished the concept of variety, this all made sense. Here Lysenko encountered his first real opposition, as he began to endanger the patiently collected library of plant varieties that Vavilov had devoted his life to assembling. But Vavilov's careful criticism of this talk now backfired. He was demoted to vice president of the Lenin All-Union Academy of Agricultural Sciences (VASKhNIL) and attacked by Molotov for wasting time on his "useless collection of seeds." By political decree, Lysenko was elected to VASKhNIL, which was followed by the first of many Order of Lenin awards.

In December 1936, criticism of Lysenko's "theories" and violent counterattacks on genetics burst out at a special VASKhNIL conference. Leading geneticists, including Hermann Muller, pointed out that Lysenko had no experimental data, and that controlled data quietly collected by others positively disproved his claims about transmutation of varieties. Lysenko and Prezent were cordially invited to inspect slides of chromosomes that had been assembled for their benefit. They gave the slides a cursory look but were not impressed. The losers by this were their critics. Within the next four years, more than eighteen of Vavilov's senior men would fall. At the same time, geneticists and agronomists who were not publically involved in the Lysenko controversy were rounded up, too. The 7th International Congress of Genetics, due to have been held in Moscow in 1937, was suddenly "postponed" after years of preparation—in reality, it had to be moved to Edinburgh and was only held in 1939. Solomon Levit and Israel Agol were among those accused of being "Menshevizing idealists," "Nazis," and "Trotskyites," guilty of "racist falsification of biology."[35] Then Agol was arrested by the NKVD on December 8, 1936.

News of these arrests, the "postponement," and general persecution quickly reached the West and the press. It was feared that Vavilov himself might have been a victim. This part was premature. Vavilov was instructed to publically deny the rumors, which he did in a telegram to the *New York Times*.[36] Shrewdly, none other than J. B. S. Haldane himself was chosen as the addressee of an additional "open letter" to Western scientists in 1937, assuring them all was well. Vavilov was among the signatories. The letter included the ringing assurance that "In the USSR scientists have the right to make public their scientific views completely freely, and arrests on the basis of scientific opinions are completely impossible and contradict the whole spirit of the Soviet Socialist Constitution."[37] But the arrest of Agol as a "wrecker" had long been confirmed.

It is hard to say whether Haldane really took these assurances at face value; he certainly acted in public as though he did. After Vavilov was openly attacked by Lysenko and Prezent in the Soviet press, concern grew again. Haldane then wrote a letter to *Nature* that demonstrated the kind of arguments he would deploy in the future whenever these questions were raised.[38] They rely on a shift we will call *equivalence-by-naming*, whereby things are set equal merely by classifying them under the same general term, in this case "criticism."

> The attacks of Lysenko on Vavilov and other Russian geneticists reported in *Nature* of August 21 are not wholly dissimilar to Dr. H. Dingle's attack on Prof. E. A. Milne in a recent issue of this journal. Vavilov was accused of being anti-Darwinian, Milne of going back to Aristotle, in neither case perhaps with full justification. If these attacks have led to a curtailment of Vavilov's work,

the situation of genetics in the Soviet Union is indeed serious. If not, hard words break no bones, and the outlook for genetics in Moscow is at any rate no worse than in London, where I understand that the only department of genetics in the University is shortly to come to an end.

Haldane blithely ignored the arrest of Agol, who had disappeared into the meat grinder. Half a century later it was revealed that Agol was shot on March 8, 1937. We will return to Haldane's treatment of his case later. As it happened, Haldane had undisclosed private information about Vavilov and the general situation in the Soviet Union, personally conveyed to him by Hermann Muller. Charlotte Haldane later recalled,

> Professor Muller left the Soviet Union in 1937,[39] and, it may have been in that year or around that time, that he visited us in London. He expressed acute disillusionment and dissatisfaction with the changing scientific line in the Soviet Union, and described to us, although with a certain degree of reticence, the peculiar developments in Soviet genetics; the gradual unfolding of the Party attack on Vavilov, and the rise to power in genetical circles of one, I Present. I do not now recollect whether at that time he mentioned the name of Lysenko.[40]

Muller had thought better of his earlier resolution (quoted in chapter 3) not to disabuse Haldane. He may have been prompted by Vavilov's personal request to transmit a message to another Soviet geneticist, Nikolay Timofeev-Ressovsky, then in Berlin, that he would be in danger if he returned—Timofeev-Ressovsky took the advice.[41]

Koltsov had taken a public stance against Lysenko all along and was under fierce attack from 1937 onward. At an unusual meeting of VASKhNIL that year, he was accused of "fascist, racist conceptions" contrary to the "social-class line" of the country.[42] His work in the eugenics movement of the 1920s was repeatedly raised against him. He was, apparently, a "fascist" before fascism existed. Always defiant, he refused to back down and dismissed the charges. But they were repeated in the state-controlled newspapers, a sure signal from above.

In July 1937, Levit was removed for good as director of his institute, which was shuttered in September. On January 10, 1938, he was arrested by the NKVD. On May 17, he was convicted of the usual espionage charges. On May 29, he was shot. That year Lysenko was appointed president of VASKhNIL. His accession had been simplified by the expedient of shooting several competitors the previous year, including Aleksandr Bondarenko, D. S. Margolin, and A. I. Muralov. His immediate predecessor, G. K. Meister, was also arrested and shot as an "enemy of the people." Throughout the rest of 1938 and 1939, attacks on Vavilov in the official organs were amplified. The NKVD had already placed a number of informers in his institute by the early 1930s. Minor trade union members and graduate students were now used to intimidate and humiliate him. Vavilov's theories were considered bourgeois and harmful, while Lysenko had it right. Denunciations poured in. Lysenko demanded that research at Vavilov's institute be reorganized to serve the country and packed it with his own appointees.

In early 1939, Koltsov, along with Lev Berg, was nominated to the prestigious Academy of Sciences. A letter protesting his nomination soon appeared in *Pravda*. One of the authors was none other than Professor A. N. Bach from the Institute for Biochemistry, familiar to us from Haldane's 1928 visit. The letter denounced Koltsov as a "pseudo-scientist" who did not belong in the academy. Bach was an Old Bolshevik and was on the Central Committee of the Party, so the situation was now grave. He was quickly appointed head of a commission to investigate Koltsov's institute. Koltsov again dismissed complaints about his ideological deviations with contempt, but was removed as director. Dependably, Prezent renewed attacks in the press. By December 1940, Koltsov was dead, supposedly of a heart attack. His wife committed suicide the same day. A Russian biochemist has since revealed, as many long suspected, that Koltsov was poisoned by the NKVD.[43]

Vavilov, too, was under a barrage in 1939, after fighting back against Lysenko in the press with a spirited defense of genetics. Lysenko, describing "Mendelist" genetics as nothing more than metaphysics, was not impressed. VASKhNIL stepped into the fray, subjecting Vavilov to a cross-examination when he submitted the (usually routine) report from his institute. They were not satisfied that he was properly Marxist in his approach. In his defense, Vavilov invoked . . . J. B. S. Haldane!

> *Lukyanenko:* Couldn't you learn from Marx? Maybe you hurried through his works, satisfied yourself with broad generalities, but 'a word is not a sparrow on the wing'.
>
> *Vavilov:* A book by Haldane came out recently. He is a very interesting figure, a member of the British Communist Party, a prominent geneticist, biochemist, and philosopher. This Haldane has written an interesting book entitled *Marxism and Science*, in which he tried . . .
>
> *Lukyanenko:* He was criticized.
>
> *Vavilov:* The bourgeois press criticized him, of course, but he is so talented that he was admired even while being criticized. He showed that the dialectic has to be used skillfully. He says Marxism is applicable in the study of evolution, in history, wherever many sciences intersect; when we deal with complex matters, Marxism can be prescient, as when Engels foresaw by fifty years many of today's discoveries. I must say that I am a great devotee of Marxist literature, not only ours, but foreign as well. There, too, attempts are being made at applying Marxism.
>
> *Lukyanenko:* Marxism is the only science. Darwinism is, after all, only a part; Marx, Engels, and Lenin gave us the real theory of cognition. And so when I hear Darwinism mentioned without Marxism, it may seem correct, on the one hand, but it turns out to be quite otherwise.
>
> *Vavilov:* I studied Marx four or five times, and I am prepared to go on studying him. Let me conclude by saying that the staff of the institute consists basically of highly qualified specialists, pro-

ductive workers; and we ask the academy and you, Comrade Lukyanenko, to help our collective create conditions for good work. As for these tiresome labels, we ought to get rid of them.[44]

Haldane's Marxist biology wasn't good enough for VASKhNIL. In a strange conjunction of events, that same year Haldane solicited an article from Vavilov for the new Marxist journal that JBS had helped to start, *The Modern Quarterly*. Vavilov promised a long scientific paper on the "Origins of Cultivated Plants," but confessed that he was "not much of a Marxist," hastily adding that he did find it "useful" and had enjoyed Haldane's books on the subject.[45] Marxism-as-a-hobby versus Marxism-whether-you-like-it-or-not. Vavilov was never able to send the article. The NKVD dossier was nearly complete.

Many have debated whether the turn against genetics was a necessary consequence of Marxism.[46] As there are numberless varieties of Marxism, it is not *impossible* to reconcile the two, but it must be difficult. The Bolsheviks fell for Lamarck early and often. Stalin, an amateur gardener, was known to be a Lamarckian himself. Khrushchev followed him, and so too did Brezhnev, to some extent.[47] Internationally the same pattern emerged, and spread throughout left-wing thought, which remains hostile to genetic explanations outside of very limited spheres. Dissenters were few. Looked at from another angle, the question is beside the point. Marxism within the Communist Party and its international fronts meant exactly what Moscow said it meant. It could change overnight. As a signal of group membership, there are real advantages to requiring belief in an obviously false or absurd idea. Otherwise one might believe it merely because it is true.

5. WAR ON ONE FRONT

In 1938, Haldane published a book on air-raid precautions, titled simply *A.R.P.*, issued by the Left Book Club. This must be the least-read of all Haldane's books, even though it is aimed at a general audience. It repays close attention.

Air-raid shelters had been a Haldane theme since at least 1935, and he applied his experiences in Spain, where up until early 1938 he had witnessed Nationalist bombing attacks on the Republican-held areas. Although the Republicans had their own bombers, supplied and flown by the Soviets, he never mentions them. Some of the material in *A.R.P.* addresses gas attacks and contains practical advice about shelters, but the book's scope is far broader.

Haldane argued that the proper precaution against air raids on Britain was to build an enormous system of underground tunnels, at least 1,000 miles of them beneath London alone, to a depth of 60 feet. Unemployed miners would be happy to dig them. He called this his "two year plan" to provide "total protection" against air raids—one assumes a five-year plan was too long. By his own estimate, the cost would have been 180 million pounds, fully one-third of the amount then being spent on all rearmament by Britain. The cost could be made up to some degree by scrapping bombers and building only fighters to resist raids. An international moratorium on building bombers could be negotiated with other powers, knowing that the country's defenses were impenetrable. "It would no longer be possible to frighten the German people into supporting an extreme form of militarism by the threat of bombardment by French, Russian, or British aeroplanes."[1] Haldane had been making this argument at least since 1934. "If we had protection of this kind [air-raid shelters], we could afford to reduce our expenditure on ships and airplanes, and people in other countries would be less afraid of us."[2] As an added bonus, people could take shelter in the tunnels when their own government bombed them. For those who might be skeptical about the chances of the British government bombing its own citizens, he said that a man in his London club had proposed doing exactly that to get rid of socialists.

Of the many possible objections to his plan, each of which Haldane tried to counter in turn, "bureaucracy" stands out. Specialists might attempt to resist his ambitious proposal by producing recommendations elaborately designed to frustrate it, choking it with complex considerations (motivated by spite?). He claimed that he had once written a report during the First World War that was designed to do exactly that, and succeeded. "This form of activity is called wrecking (or rather the Russian word for wrecking) in the Soviet Union. Having done it myself here I am therefore the more willing to believe that others have done it there."[3]

Applying science to problems is therefore a double-edged sword. Too much science could get you shot, as in the case of the engineer Peter Palchinsky, who had the temerity to point out that the massive hydroelectric scheme in the Dnieper River basin would not necessarily pay off, due to the sluggish and variable flow of the river—a Soviet hydrologist later confirmed this, calculating that simply burning the hay grown in the massive area that the dam flooded would have generated more energy.[4] An unnecessary complexity! For those disinclined to believe that Haldane would go that far, consider his thoughts on those building owners who might deny the public access to their basements. "I can imagine very well what would happen in Madrid if the owner of a good cellar refused to allow its use as a public shelter. It would be suggested to him that the lamp-posts had not been in use since 1936, and that in spite of the shortage of rope he would constitute a suitable decoration for one. This would of course be another Red Atrocity. Actually no such case has arisen. A certain degree of public spirit is taken for granted in Madrid."[5]

Haldane stuck to this ARP theme for years, taking every opportunity to embarrass authorities at public meetings around the country, by arguing that the aboveground "Anderson" shelters they provided were inadequate compared to his own deeply buried "Haldane" shelter design. His arguments were usually spiced up with assurances that the capitalist government didn't care about the poor anyway. He very narrowly avoided a libel suit after claiming that the cement manufacturers were profiteering on the construction of shelters. He abandoned the campaign only after the Soviet Union was attacked by Hitler in 1941, and sullenly retreated to the position that aboveground shelters would have to do after all.

There is not much relevant science on offer in *A.R.P.* to support Haldane's "better defence by burrowing" theories. He had to argue that air raids were serious enough to shelter from, but not so serious that feasibly built tunnels would not be good enough. He could offer only his personal conclusions based on the bombing in Spain that he had seen, though he had no real data beyond his own authority, and nothing that could be critically examined. Later in 1938, when he was not there, Nationalist bombing became more effective, and he was forced to add an appendix to his book to try to counter that development. Maginot-like tunnels represented a vast fixed investment, an all-or-nothing bet against offensive weapons and tactics. Haldane's expertise in biochemistry, gas defenses, and other bits and pieces of mathematical modeling did not in the end convey all that much about global military strategy.

But it is not a given that Haldane took his own arguments seriously. Douglas Hyde, who was then a member of the Communist Party, recalled after his defection that the ARP campaign offered many useful opportunities to the Party, which was publically against "imperialist wars" only until the USSR was attacked. "The shelter campaign had everything to it from our point of view, since it had the appearance of being a crusade for greater safety for the common people, whilst at the same time it spread alarm about the provisions already made by the authorities. Moreover, it gave us an opportunity to use many of our crypto-communists on public activity. It was led by our scientist-members, among whom was Dr. Nunn May, later imprisoned for passing on atomic secrets to the U.S.S.R."[6]

On August 23, 1939, Foreign Ministers Von Ribbentrop and Molotov signed a ten-year non-aggression pact between Germany and the USSR, publically reversing the previously advertised undying enmity between the two countries. The public terms stipulated that neither party would attack the other or join up with those hostile to it, and would lead to economic trade and scientific cooperation between the two parties. A secret protocol to the Nazi-Soviet Pact arranged for the division of Poland between Germany and Russia, and was later modified, in part implicitly, to cede the Baltic States to the USSR along with Bessarabia in eastern Romania. Just over a week after the Pact was signed, Germany invaded Poland from the west to claim its portion. Two weeks later, so as not to appear too hasty, the Soviets invaded from the east to claim theirs.[7]

In later years, it would become convenient for some to explain away the Nazi-Soviet Pact as just a device used by Stalin to buy more time, in the knowledge that a German invasion of the USSR was looming. The evidence shows otherwise. It was Stalin's hope that Germany would become embroiled in a long fight to the death with Britain and France, while the USSR watched from the sidelines. The Pact secured Hitler's eastern front when he turned to invade Belgium and France, and thus allowed the Second World War to be launched. The Pact also ensured that the USSR shared some of the resulting spoils. Hitler's successes and the quick course of the war in Western Europe surprised Stalin as much as everyone else, showing that he had miscalculated badly. During the twenty months that the Pact arrangement lasted, the Soviets shipped critical raw materials to the Germans, including much-needed oil, and cooperated with them on a number of other levels—for instance, by sending them German communists, who had been sheltering in the USSR, to deal with. Anti-German books and films were removed from circulation.[8] Within Germany itself, the Pact also led to a dramatic decline of Communist Party activities. There is no reason to believe that Stalin was expecting an imminent invasion of the USSR by the Germans. When that did come, he was not prepared for it, mentally or militarily.

At the time, the Pact proved difficult for many communists and their fellow travelers to swallow, since they had expended a great deal of energy arguing that "fascists" were the undying enemies of socialists everywhere—that had been the basis of their extensive propaganda campaign over the Spanish Civil War. Now it was no longer clear what the correct Party line was. Charlotte Haldane was with Hans Kahle in London when the news broke, an "intimate, personal, friendship" having developed. "For him, of course, as a German Communist, the shock was personally greater than for me. We dutifully spent the week-end working out the new Soviet line. No doubt we were slow in the uptake, politically weak, owing to our respective bourgeois backgrounds."[9]

The British Communist Party completely misread Moscow's intent. Harry Pollitt rushed out a pamphlet arguing for a two-front war, one front being Germany, the other being fascism at home. Moscow rapidly shot Pollitt down. There was to be no anti-German campaign now, the pamphlet had to be withdrawn, and Pollitt was promptly replaced as Party leader by Rajani Palme Dutt, a suppler thinker by far. Now it was back to a "War on One Front," against Western bourgeois democracy. The majority of the Party accepted this correction without public protest. As Charlotte confessed, "party discipline and loyalty prevailed" for both herself and Hans. But Douglas Hyde—who, like Charlotte, later defected from the Party—recalled that things became difficult to explain to the working class. "I was sharing a flat with a leading communist shop steward, one of the Party's most active factory members. I learned the news from the Sunday papers, took them home and showed him the headlines: 'Russia Invades Poland'. On the face of things, for any communist, the news was devastating. His reactions were typically violent. 'Bugger Uncle Joe, bugger Molotov, bugger the whole bloody lot of them,' he cursed."[10]

None of this stymied the far more powerful mental machinery of JBS, who was able to adopt the new Party line at once. In a letter to the *New Statesman*, where harsh words had been said about the Pact, Haldane patiently explained that it was not a new Soviet policy at all, just an application of the old one! "I cannot understand how an intelligent person can find its policy in any way inconsistent. On the contrary it appears to be almost fantastically consistent."[11] Soviet policy was "based on an objection to two things, capitalism and wars," since one led to the other. Capitalism needed wars to solve the problem of unemployment, but the Soviet Union had no unemployment problem, so war was not in its interests—"the Soviet Union is the only nation which can make no economic gains from war." There would be no point in conquering other countries, since Soviet "production per man and their population are both increasing faster than those of any other country." As for the Soviet invasion of Poland—a curious admission, given the foregoing proofs—Haldane was confident that the Byelo-Russian and Ukrainian-speaking parts would have voted to join the USSR anyway. In the Polish-speaking parts, Stalin, being a "specialist in the problem of nationalities," would know how to accommodate national sentiment, even though the Poles "have not always respected that of other peoples." This accommodation, he was willing to concede, "will be the most searching test of Soviet policy." Those Poles who didn't like their new borders—likely "only a small part of the occupied territory"—would "very probably be given back to a free Poland." The Soviets would not deport Poles into forced labor camps, and if "the land is given to the peasants, as seems likely, the Russians will soon have millions of friends." The end result could be a "reconstituted" Poland that would be "ill-disposed to the Third Reich, though not to a peaceful Germany."

The last qualification is significant, as Haldane dismisses the idea that Stalin ought to prefer Britain and France to Germany because of their more liberal treatment of trade unions, etc. "Do you mean Britain or the British Empire, France or the French Empire? I would sooner be a Jew in Berlin than a Kaffir in Johannesburg or a negro in French Equatorial Africa. If the Czechs are treated as an inferior race, do Indians or Annamites enjoy complete equality?" Indeed, the Soviets were sitting out the war on behalf of the "coloured peoples." "The British and French people may prepare to fight their battles under the leadership of the men of Munich with the firm resolve to preserve their rule over the coloured peoples of their own empires. But if so they can hardly complain that the Soviet Union remains neutral in the struggle, and occupies itself in stemming Hitler's advance and abolishing feudalism in Eastern Europe." Haldane closed with the issue of the Baltic States annexed by the USSR. There he was confident that the inhabitants "welcome Soviet protection." A "leading Lett," not a communist, with whom he had dinner in London, assured him so. Stalin, since he was a Georgian, would never impose Russification. And anyway, the harbor in Ventspils was too big for Latvian needs.

At this point one might be forgiven for speculating that this letter was not really written by Haldane at all, but rather by, say, a cunning intelligence service, and planted in the *New Statesman* to make him look craven to his contemporaries and odious to posterity. However, a handwritten copy can be found in his archive of papers at University College London. That is the text used here—a full transcript with erasures appears in Appendix 2.

Leonard Woolf, Virginia Woolf's mild-mannered husband, had bolted from the fellow-traveling crowd by then. Haldane flew into a rage when Victor Gollancz's previously dependable Left Book Club issued Woolf's *Barbarians at the Gate*, with its equation of Soviets to Nazis, in late 1939. In a review for the *Daily Worker*, Haldane hoped that the venerable Left Book Club would stop issuing titles with such deplorable attitudes.[12] A vain hope, since even the stalwart Victor

Gollancz now found it impossible to defend the Pact.[13] As events unfolded, Haldane had his work cut out.[14] That same day, November 30, 1939, Finland was invaded by the USSR, which was surprisingly unhampered by its lack of a domestic unemployment problem and its advertised gains in productivity. The invasion was very unpopular with the British public. Douglas Hyde remembered,

> In our demonstrations we had sung the Soviet Airmen's song: 'We drop them leaflets whilst we bomb their houses,' but when the raids began on Helsinki we saw just how optimistic and unreal those words had become. . . . Sellers of the *Daily Worker*, women as well as men, were spat upon and assaulted on the streets; canvassing, they had doors slammed in their faces, even chamberpots emptied on their heads from upstairs windows. Often, out selling papers or pamphlets, we would have housewives shouting vituperations at us until we disappeared from their street. It was a testing time, and most Party members rose to it and gloried in their martyrdom. There were, of course, the weak ones, and the Party was glad enough to be purged of them.[15]

Haldane's own faith was not of the weak kind, but a series of embarrassing defeats with heavy losses for the Red Army, at the hands of the tiny but well-organized Finnish army, was hard to deal with. Explanations involving "timber" were forthcoming.

Speaking to a meeting in Kingsway Hall in December, held to celebrate the anniversary of the Russian Revolution, Haldane told a meeting of 1,000 people that the Soviet Union was only protecting itself by "fighting fascism" in Finland. The timber exporters and financiers who governed Finland were to blame for the war, not to mention the Finnish leader Carl Mannerheim, in reality the tool of the timber-men.[16] In January 1940, Haldane evolved this theme at the Usher Hall in Edinburgh, identifying the culprits as nickel mine investors and paper and cellulose manufacturers; he called on the workers to "agitate for peace" before millions were killed. At St. Andrews later that month, he pinned the blame for the whole war on "the capitalists" and requested the working class to neutralize this by appealing to their fellow workers in Germany over Hitler's head—here he was straying a little from the new, post-Pollitt, Party line. He assured a meeting at Liverpool Grove in Walworth Road on January 21 that he was opposed to the current war and ready to go to prison for his beliefs—in December 1939 he had told a Peace Council in Edinburgh that he had joined up in the First World War only because he had been deceived.[17] On March 5, 1940, the *Daily Worker* reported that Haldane had signed a petition that warned of the "grave and imminent danger" that Britain and France might attack the Soviet Union. The point was that invariably "the British ruling class have allied themselves with German militarism to crush democracy when it suited them."[18] In parallel, Haldane continued to develop his ARP line of talk, first introduced in 1938, which was designed to show that the government didn't really care about the working class and was happy to sacrifice them to Nazi bombs.

The culmination of Communist Party efforts to influence the public along these defeatist lines was the People's Convention of January 1941, which they organized at a discreet distance. By design, the Convention ignored ways in which the war might actually be won, and aimed instead to stir up discontent about its likely beneficiaries, effects on the working classes, and general conditions in the country. As the Blitz raged on outside the convention hall, the speakers inside called for higher wages, more and deeper bomb shelters, "friendship" with the USSR, nationalization of banks, ending of the partition in Ireland, independence for India, and so on. Douglas Hyde summed it up in retrospect:

> The campaign for the People's Convention was, I suppose, the most effective and, from the Government's point of view, potentially dangerous thing we had done. As a communist tactic it was perfect. It united a reasonably wide and varied number of people and organisations on limited, short-term political and economic demands which deliberately did not include direct opposition to the war. Yet had those demands been won, they would

have crippled the war effort—as they were intended to do. And the campaign for those demands was unhelpful in the extreme and was calculated to weaken public morale. Into the campaign were drawn people and organisations far removed from Marxism; pacifists whom we had for years fought and derided became our allies; non-communist musicians, literary people, university professors and other well-known figures were prevailed upon to add their names to the hard core of Left-wing trade union leaders whom we had managed to entangle. The deep longing for peace which everyone instinctively felt was carefully worked on for revolutionary-defeatist ends.[19]

The Labour Party and the Trades Union Congress banned attendance in advance and quickly expelled anyone who participated in it. It was said that more than 2,200 delegates attended. The press in the USSR was enthusiastic. Haldane was instrumental in organizing the convention and tried to rope in as many intellectuals as possible, with a premium placed on those not openly communist. J. B. Priestley received an appeal to attend. Getting Michael Redgrave, an old friend from "Chatty's addled salon" at Cambridge, was easier. Others who attended included D. N. Pritt, Indira Nehru (later Indira Gandhi), Olaf Stapledon, Beatrix Lehmann, and Hewlett Johnson, the "Red Dean" of Canterbury. Leaflets were issued explaining that "society women" were being shampooed with rationed eggs, and that canteen food was not up to standard. Housewives, it was said, would no longer put up with conditions.[20]

The only concrete result of this attempted undermining of public morale was the banning of the *Daily Worker* ten days later. Nominally, the People's Convention continued to meet until May 5, 1942. Haldane was absent then, but still on the committee, which seems to have disposed of its funds to other front organizations and dissolved. Once the USSR was invaded in July 1941, and the Party immediately switched its line back to winning the war, the Convention had lost its usefulness. In the meantime, Haldane worked hard to get the *Daily*

Worker unbanned; he eventually succeeded in early 1942, but only after pledging to support the war effort from then on.

Haldane had been wholly absorbed into the Party, in much the same way that other cults envelop and isolate their members, supplanting their families and friends. Almost all of the people that he had voluntary dealings with were Party members. Even his secretary had to be a member, vetted both for political reliability and to his own standard, a requirement that gave the Party some trouble (the position regularly fell vacant). Charlotte had been replaced by Dr. Helen Spurway, who was also within the Party. Constantly producing and assimilating corrupting propaganda added to this sympatric isolation. The extent to which all this screened Haldane and his new companion from reality can be judged from Spurway's reaction to the invasion of the Soviet Union on June 22, 1941. After hearing the news at her parents' house, she immediately set off for the Rothamsted Research Station to find JBS, expecting him to be locked up at once when the British government joined up with the Nazis to destroy the USSR.[21]

6. IVOR MONTAGU
AND THE X GROUP

In 1985 a former MI5 man named Peter Wright caused a sensation by publishing his reminiscences in a book he called *Spycatcher*. Hoping to evade the Official Secrets Act, he published it abroad, in Australia. The British government forlornly tried to suppress it even there, but only managed to get it more publicity; the book was subsequently published in the United States anyway. Buried within its pages, but unmentioned in its index, were startling claims about the mysterious VENONA program. Decoded intercepts from the Soviets apparently showed that "J. B. S. Haldane . . . who was working in the Admiralty's submarine experimental station at Haslar, researching into deep diving techniques, was supplying details of the programs to the CPGB, who were passing it on to the GRU in London." Moreover "the Honourable Owen[1] Montagu . . . was used by the Russians to collect political intelligence in the Labour Party, and to a lesser degree the CPGB."[2]

At least, these claims ought to have startled someone, but if they did nothing much appeared in print—a conclusion that can now be made with the full confidence of Google Books and digitally searchable newspaper archives.[3] The trouble was that, while Wright made a great many claims, not all of them proved accurate. Here his memory was faulty in at least one respect: it was Ivor Montagu, not his brother

Ewen; the VENONA intercepts certainly showed that Ivor headed an espionage ring called the X Group.

Soviet embassies abroad communicated back to Moscow through diplomatic pouches, but during the war this process was extremely slow. More urgent material was transmitted either by commercial telegraph, which was the practice in the Washington Embassy, or by radio broadcasts, the method used by the London Embassy with British consent. The communications were encrypted using a system called the one-time pad that was highly secure in theory. Even though the communications were recorded by the U.S., British, and other governments, they could not be read without a significant investment in cryptanalysis. Before 1942, the only serious attempt to make sense of the Soviet code was made by the Finns, with some limited success.

Despite this initial lack of interest, there were intelligence riches to be had in the intercepts. The Soviets ran an extensive espionage system, with networks run out of their embassies through "illegal residents," by both the NKVD/KGB and the GRU (military intelligence, with its own network). These networks received extensive assistance from the Communist parties operating in each jurisdiction, and had penetrated deep into the state institutions in both the United States and the UK, as well as elsewhere. They were greatly assisted by politically sympathetic intellectuals, who rose rapidly and exerted pervasive influence in state circles. By 1942, having seen the benefits of British success in cracking the German Ultra code, the United States turned its attention to the trove of messages they were accumulating from their Soviet ally, whose intentions they had come to distrust (at least in military circles).

The Soviet code was a two-step process. First, characters were encoded as groups into numbers, which was something like a dictionary lookup for most common words, but a little harder for unusual ones. Then the resulting groups of numbers were encrypted by using a one-time pad of random numbers, which were added to them. Decryption involved subtraction of these numbers and then reversing the code lookups. That required both sides to work off the same sheet from the pad, which had a serial number to allow it to be referenced in the message and matched. The pads had to be created in the Soviet Union and then distributed abroad for use. If it was used as intended, the system was highly secure, especially given the lack of computing power at the time. The weakness proved to be a manufacturing deficiency. As with so much else in the Soviet state, a shortage of pads developed; to make up production quotas, sheets were reused, though they were spread over multiple pads to make it harder to find duplicate uses in intercepted traffic from individual stations.

The US program to decrypt the intercepts went on for many years, under various code names, eventually becoming known as VENONA. The Finns had supplied useful initial groupings of cables by patient cryptanalysis, and eventually the duplicate uses were detected by collating cables from multiple international sources. A partially burned codebook—for the first stage of translating characters to letters—had been captured by the Finns, though inexplicably other codebooks captured from German territory in 1944 were returned to the Russians at the request of the White House. No copies were kept! Nevertheless, by 1947 the first messages from the Washington Embassy were being decoded, and soon the emergence of the first mechanical computers allowed the pace to pick up. These messages immediately paid off, showing that the US Manhattan Project and the State Department had been penetrated. Though cover names were commonly used in the cables, detective work easily identified Klaus Fuchs, Julius and Ethel Rosenberg, Alger Hiss, and others. Analysis went on until as late as 1980. There were several hundred cover names in play, even though only a small percentage of the recorded traffic was ever successfully decoded. Many of the cover names remain unidentified today.

For most of its existence, the Allies treated VENONA as highly confidential, so much so that the CIA only heard about it in 1952, and Truman was kept in the dark for three years. This secrecy was not irrational, given that the highest levels of government had been compromised. As it turned out, Stalin knew about the program long before Truman, since the intelligence services of both the United States and the UK had been penetrated, too—by William Weisband (1908–1967) in the United States and Kim Philby in MI5. Regardless, the program was only publically disclosed in 1995, and was never used in the trials of the handful of those who were charged with crimes. Documents released from it have become one of the most important resources for historians of communism, along with the Soviet archives that briefly opened up to foreigners in the 1990s.

Messages from the London Embassy took much longer to crack than the Washington traffic. Intercepts had only been recorded up until April 1942, apparently owing to transfer of resources and attention elsewhere. It was as late as June 1945 before interceptions began again. The first messages to yield information were from September 1945; these showed that MI5 had been compromised, without fingering precisely how. The Washington traffic and other intelligence eventually disclosed the Cambridge Five—Guy Burgess, Kim Philby, Anthony Blunt, Donald Maclean, and John Cairncross—as well as a host of others. Nevertheless GCHQ kept working on the London messages. In the early 1960s, Swedish intelligence shared their own rich store of intercepts from the transmissions between the Stockholm Embassy and Soviet Military Intelligence, covering March 1940 to April 1942, and September 1945 to March 1947. Duplicate sheets from the one-time pads had been used, and suddenly some of the London messages from 1940 onward began to yield. They revealed "Gruppa Iks," the X Group.

Peter Wright's revelations resurfaced in 1999 with the publication of *Venona* by "Nigel West," the pen name of Rupert Allason.[4] This was the first full treatment of the British VENONA intercepts, complementing the landmark study of the American intercepts by John Haynes and Harvey Klehr.[5] But the London Venona intercepts then available were redacted—as we will see in detail below—and West was forced to do some guessing, bolstered by other sources of information, to try to penetrate the numerous cover names used in the messages. Two of the people cover-named were evidently members of an espionage cell referred to as the X Group. The first, called INTELLIGENTSIA, seemed to be the chief liaison between the group and the illegal resident in the Soviet Embassy; the second, NOBILITY, was a senior member of the group superior to INTELLIGENTSIA. It seems that West followed Peter Wright's lead given in *Spycatcher*, along with other circumstantial information—possibly including inside information from contacts in MI5—and identified these two as J. B. S. Haldane and Ivor Montagu, respectively. It was a shrewd guess, but it was not quite accurate. INTELLIGENTSIA was Montagu, not Haldane, and, though it is still uncertain who NOBILITY was, Haldane was certainly a member of the X Group and is known to have passed classified information through it to the Soviets.

West's identification of Haldane as INTELLIGENTSIA proved to be influential, most likely because of the authority lent by the use of the VENONA intercepts, which had swept away decades of obfuscation about the instrumental role of Western Communist parties in the global Soviet spying operation. In 2005, Andrew Brown repeated West's claims in his notable biography of J. D. Bernal, the communist crystallographer and contemporary of Haldane.[6] Others followed suit.[7] It is even possible that this confusion will always live on somewhere, forever finding new habitats, since Haldane is a much more interesting and consequential figure for playing INTELLIGENTSIA than Montagu, who is now remembered mostly for his table-tennis exploits. But with the release of the unredacted VENONA intercepts to the National Archives of the UK, the facts are available.

Eighteen VENONA messages mention either INTELLIGENTSIA, the X Group, or its other members, NOBILITY, BARON, and RESERVIST. Complete unredacted transcripts of the most important messages can be found in Appendix 4. The first mention is dated July 25, 1940, and introduces the X Group to GRU in Moscow. The illegal GRU resident at the London Embassy, code-named BARCh in the messages and identified by MI5 as Simon Davidovich Kremer,[8] acted as the coordinator of the group's activities. Kremer, the secretary to the military attaché in the embassy, liaised regularly with Moscow for advice on the most promising lines of enquiry to follow. On this occasion he reported new activities initiated by an organization he called the X Group.

> I have met representatives of the X GROUP [GRUPPA IKS]. This is IVOR MONTAGU [MONTEGYu] (brother of Lord MONTAGU), the well-known local communist, journalist and lecturer. He has [1 group unidentified[9]] contacts through his influential relatives. He reported that he had been detailed to organise work with me, but that he had not yet obtained a single contact. I came to an agreement with him about the work and pointed out the importance of speed.

Since the messages were thought to be fully secure, Kremer felt free to introduce real names and other identifying information. Although cover names were routinely used for contacts, agents, and other persons of interest, they were not used as a security measure per se, and in early practice they even conveyed meaningful information, modulated by the Russian sense of humor, which offers a sporting chance for identification from the cover names alone. Kremer at once introduced a cover name for Montagu: "He (INTELLIGENTSIA [INTELLIGENTsIYa]) reported the following . . ."—going on to repeat a long list of gossip peddled to him by Montagu. This settles the identification of INTELLIGENTSIA as Montagu rather than Haldane, which is recorded in the annotations made to the messages by either GCHQ or MI5.

Montagu's gossip included attitudes and alliances in the Cabinet and British political parties, and their mood regarding Hitler and Germans in general, who are cover-named SAUSAGE-DEALERS by Kremer. Although INTELLIGENTSIA was able to pass on some information about some reshuffling of military brass from a parliamentary secret session, Kremer was not satisfied, and the remark "he had not yet obtained a single contact" shows this. Subsequent messages showed mounting frustration on Kremer's part about Montagu, who took some time to get into his stride as a GRU agent. Kremer took this frustration up with the X Group itself.

From Kremer's description, Montagu had been put in touch with him *by* the X Group—"he had been detailed to organise work with me"—and later messages show Kremer complaining to other members of the group about Montagu. It is still not clear exactly who the members of this group were or where they were insinuated. As we shall see, they probably constituted an underground cell within the Communist Party of Great Britain, but MI5 never made any definite identification. They could not have been Montagu's own creation, since they preceded him and could have replaced him. This distinction is made clear in the next message, from August 16, 1940, which shows that Montagu was trying but still failing to impress his handler—despite promising classified information from none other than his old friend from Cambridge, J. B. S. Haldane.

> INTELLIGENTSIA [INTELLIGENTsIYa] has not yet found the people in the military [C% finance department] [VOENNYJ FINANSOVYJ OTDEL]. He has been given the address of one officer but he has not found him yet. He has promised to deliver documentary material [MATERIAL] from Professor HALDANE who is working on an Admiralty [MORSKOE MINISTERSTVO] assignment concerned with submarines and their operation. I have taken the opportunity of pointing out to the X GROUP [GRUPPA IKS] that we need a man of a different calibre and one who is bolder than INTELLIGENTSIA.

It took Montagu a few weeks to get this report from Haldane, and it was September 6, 1940, before Kremer was able to report back to Moscow that "INTELLIGENTSIA [INTELLIGENTsIYa] has handed over a copy of Professor HALDANE's report to the Admiralty on his experience relating to the length of time a man can stay underwater." Still, Kremer was not satisfied that he had been given the right man by the X Group, since Montagu had not taken up leads that he was given for a new contact, who would later be cover-named RESIDENT.

> However he does not deny the main point that for a month he has not been in touch with the British Army colonel picked out [VYDELENNYJ] for work with us although the latter does come to

> LONDON. I have told X GROUP [GRUPPA IKS] via NOBILITY [ZNAT'] to give us someone else because of this. INTELLIGENTSIA lives in the provinces and it is difficult to contact him.[10]

Who was NOBILITY, evidently a senior member of the X Group? It is tempting to assign this name to Haldane himself—he was not really nobility, but the Russians could easily have taken him to be such, given his uncle's career—but this suggestion is made improbable, though not impossible, by the construction of this report. Haldane figures in it explicitly under his own name, and is not linked in the text to NOBILITY, who seems to be a separate individual. Certainly, MI5 never made an identification of NOBILITY in their annotations.[11]

Haldane's research into midget submarines and the physiological challenges posed to a crew by the need to breathe oxygen under pressure was considered important by the Admiralty, which sponsored it, and the specific findings were considered secret and classified—though Haldane was publically known to have worked on the broader problem. At the time, the Soviet Union was an ally of Nazi Germany, but even after they joined the Allies when the Germans invaded them in 1941, the work remained classified. We can be sure that Haldane was conscious of this, since MI5 had queried him about the spread of information from this research in the past, when he employed communist volunteers from Spain for some experiments. Moreover, in 1946 the British physicist Alan Nunn May was revealed to have been passing information to the Soviets about atomic research. Haldane covered the case in the *Daily Worker* and, employing a characteristic inversion, promptly blamed the UK government for not sharing such secrets with the Soviets in the first place. "I can speak from personal experience. I was engaged in research on underwater operations. The work in progress was shown to naval officers of several navies, including American and Dutch, but not to Soviet naval representatives, though it would probably have saved the lives of Soviet sailors."[12] Except that it was, by Haldane himself.

Montagu was eventually able to allay Kremer's fears that the

X Group had sent the wrong man. On October 2, 1940, Kremer reported that he had obtained information from the (apparently broadly connected) X Group about the breaking of a Soviet code.

> INTELLIGENTSIA [INTELLIGENTsIYa] has reported that the X GROUP [GRUPPA IKS] has reported to him that a girl working in a government establishment noticed in one document that the British had broken [RASKRYLY] some Soviet code or other and apparently she noticed in a/the document the following [words:] "Soviet Embassy in Germany". I stated that this was a matter of exceptional importance and he should put to the group the question of developing this report [further].

This was most likely a mistake on the part of the "girl working in a government establishment" for early results from the Ultra project, which worked on decryption of German Enigma messages. The British were not focused on Soviet codes at the time (unless this early success has remained concealed). The X Group seems to have straightened this out soon enough. By December 20, 1940, NOBILITY had "passed material [12 groups unrecovered] [which has been] received from a member of the CORPORATION [KORPORATsIYa] working on technical work in intelligence departments [ORGAN]." Here the "CORPORATION" was the Communist Party of Great Britain. A fragmentary message from April 3, 1941, demonstrates that Ultra had been compromised by another X Group member, BARON.

> This information originates solely from BARON. [Its provenance is] well . . . known to you, the intercept [B% bearing] the designation ENIGMA [ZAGADKA] . . . The information about . . . the intention of the SAUSAGE-DEALERS [KOLBASNIKI] . . . on our part. What . . . It is well known that . . . but . . . the government of the COLONY[13] [KOLONIYa] . . . and this delicate sources's possible connection with Military Intelligence and ~~one~~ can [B. take it] that his information is [B. fully deserving of attention] and therefore I consider that it would be a profound error to take his . . . BARON's facts prove that . . . the COLONISTS'[14] [KOLONISTSKIJ] . . . that . . .

Judging from the other information that he supplied, BARON appears to have been connected with Czech military intelligence, but he has never been positively identified. Montagu had extensive contacts with the Czech exiles in London during the war, as we will later see in connection with the show trial and execution of Otto Katz. The leaking of Ultra secrets to a country allied to Germany at the time was a critical security compromise. The intelligence services of the Pact countries are known to have cooperated. In that respect, the X Group was of considerable use, supplying ready intelligence of British military countermeasures to German weapons. On October 11, 1940, Kremer informed Moscow that "INTELLIGENTSIA [INTELLIGENTsIYa] confirms that the British really do render delayed-action bombs safe by freezing the bombs' exploder mechanism." Better still, Montagu reported by October 15, 1940, "as a result of a conversation with an officer of the Air Ministry," that the British believed that the German bombing raids had a new method for accurate targeting: "the SAUSAGE-DEALERS proceed towards the target along a radio beam." This was the *Knickebein* "bent-leg" system, which identified bombing targets by the spot at which two radio beams, transmitted from the Continent, intersected over the UK. There was a frenzied search by the British government for effective countermeasures at the time.

Even though only a modest number of the VENONA communications from London to Moscow were successfully decoded, and many were lost forever simply because the government had stopped recording them at the time, it is obvious that the X Group had a wide reach and developed important contacts, especially regarding the application of science to military problems. Here Haldane was an invaluable asset; he worked on a number of government projects and enjoyed the sort of trust that a Fellow of the Royal Society and the Weldon Professor of Biometry at University College London could expect to get. Haldane considered himself primarily a scientific link-man, or "cooperator," as he put it, using his broad reach across disciplines to make connections others might not see. It was this sort of influence that had helped to get Hans Kahle back into the UK from internment in Canada.

After they had met in Spain, Haldane stayed in touch with Hans Kahle and other veterans of the International Brigade. After representations by the MP "Red Ellen" Wilkinson and Clement Attlee, Kahle arrived in the UK in February 1939, supposedly to write a book about the war and then go on to the United States or Cuba. He came via Paris, which he had reached by December 1938, and where he had met up with the versatile Soviet agent Otto Katz. While he was in England, Kahle remained in touch with Katz via Tom Driberg (who was later a Labour MP and a double agent, of uncertain loyalty, for MI5 and the KGB). Colonel Hans offered to teach British soldiers about modern warfare later that year, but was turned down when MI5 was informed he was organizing an OGPU spy ring, along with Jürgen Kuczynski (who was interned in February 1940). The French complained that Kahle was sending money to communists there, and an MI5 source reported that he kept company with John Heartfield (Helmut Franz Josef Herzfeld) and Kuczynski. The Soviet defector General Walter Krivitsky had identified Heartfield as an OGPU agent. Other sources reported that Kahle was involved in underground propaganda activities with the CPGB. Charlotte Haldane put him up for a while in her house, describing him as "a typical German," a type she generally found it difficult to like, but *he* was "tall well-built, with an ugly face, but great charm of manner." Later she claimed that he had become a communist only when Hitler came to power, a story which he may have told her.[15]

According to Haldane, he approached Kahle in the spring of 1940 to see if the German would like to be a "rabbit" for his dangerous physiological experiments related to submarine decompression.[16] Kahle had agreed, no doubt attracted by the chance to observe British military preparations up close. He was taking part in experiments conducted at the Siebe Gorman works (then in Lambeth[17]) when one day that summer he disappeared. After inquiries were made, it turned out he had been arrested as a foreign national of an enemy country and was to be deported. Haldane's interventions on his behalf failed, even after Wilkinson was pulled in to help. Kahle was deported to Canada in the summer of 1940 and held in Internment Operations Camp L near Quebec City.[18] There he organized a communist cell, recruiting the physicist Klaus Fuchs and the economist H. W. Arndt, among others. Correspondence from Hemingway to Kahle during this internment survives and shows that Hemingway idolized him. Hemingway wished, in a letter to his old friend, that Kahle was "commanding the 45th Division and that I had a small job on your staff," recalling "when we were walking through the shelling as happy as children because we were the same age and re-living our boyhood in that great state of invulnerability that all we old soldiers have instead of the state of grace."[19]

Haldane had intermittently, over many years, worked on the physiological problems posed by high pressure and decompression after diving. His father, John Scott Haldane, had pioneered the field in 1908, inventing the staged method of achieving decompression. JBS had helped him with these experiments when still a boy. Prior to the current war, JBS had also assisted in the official inquiry into the HMS *Thetis* submarine disaster of 1939, conducting experiments with the help of human "rabbits." In that regard he had started a curious practice that persisted throughout the war. He wanted exclusively communist volunteers. For the *Thetis* inquiry, he used only himself and returned International Brigade members—Bill Alexander, Patrick Duff, George Ives, and Donald Renton. Ives had been trained as a radio operator in Moscow in January 1936. Renton and Alexander had been political commissars in the brigades, and were thus hardened Comintern men.[20]

The experiments being conducted in 1940 were part of the preparations for the design and construction of a new class of 16-meter three-man attack submarines, the X-class or "midget" submarines. The midget submarines were still in the conceptual stage and needed an escape chamber with a watertight compartment, which had been mocked up. After initial testing (presumably at Siebe Gorman) the mock-up was tested at Portsmouth in the torpedo-testing tank, which was 60 feet deep. Testing there seems to have been conducted by Haldane himself and navy personnel only: Captain Godfrey Herbert, Commander C.H. Varley, and Colonel Millis Rowland Jefferis.[21] Colonel Jefferis was in charge of the top-secret MD1 unit known as "Winston Churchill's Toyshop," which would soon develop the limpet mine. Commander Cromwell Hanford Varley (1890–1949), a retired veteran of the First World War and a talented inventor, would go on to design the prototype X-3 sub, assisted by a Commander Bell. Limpet mines would be deployed from the midget subs during the war. The X-3 prototype craft was developed in a yard near Portsmouth.

The testing crew used here may have been different because of a lack of security clearance for the current crop of Haldane's (perhaps exclusively) communist helpers. Those included Kahle (before his deportation), Edwin Martin Case (from Cambridge days), Helen Spurway, Elizabeth Jermyn, and, later, Juan Negrín, the exiled prime minister of Republican Spain. The experiments were harrowing and often produced hallucinations and severe convulsions, with Haldane and Spurway the main victims of these. JBS badly injured his spinal cord in one episode and carried the injury for the rest of his life.[22]

The X-craft midget subs proved to be a partial success. They inflicted serious damage on the German battleship *Tirpitz* as part of Operation Source, by slipping under defenses and laying mines with timed detonators under the ship. They were also used to sink the floating dock at Bergen in April 1944, and formed part of the surveying preparations for the D-Day invasion of Normandy. Several of the submarines were scuttled during missions to prevent their discovery. However, Haldane's role in this work was curtailed in 1943 when the Admiralty decided that the security risk he posed, given his cell of communist assistants, was no longer tolerable.[23] But Haldane worked on several other projects for the military during the war, including a study of how stick-bombing techniques affected the distribution of bomb hits.[24]

Kahle returned to the UK from Canada on January 22, 1941, after he was again offered as a "rabbit" for Haldane's submarine research, at Haldane's repeated initiative. The Admiralty had put in a word.[25] Haldane had also given MI5 a written undertaking that Kahle was not a security risk. While Charlotte Haldane was abroad in the Soviet Union for the latter half of 1941, Kahle occupied her apartment in Hampstead.[26] His fellow internee and recruit Klaus Fuchs had also returned to the UK, finding a position in the British contribution to the Manhattan Project. During the remainder of the war, Kahle worked officially as a military correspondent for the *Daily Worker*, *Labour Monthly*, and the American *Life* magazine. He frequently addressed public meetings, duly reported by the major newspapers, and wrote pamphlets extolling Stalin's virtues as a military commander.[27] MI5 dutifully added copies of these to his file.

Kahle kept up regular contact with Haldane throughout this period, addressing him familiarly as "JBS" and closing his letters "fraternally," requesting, among other things, contributions by JBS to pamphlets on the "scientific" aspects of race and anti-Semitism.[28] Kahle's active assistance at this time in conducting Soviet espionage was later confirmed by Ursula Kuczynski (sister of Jürgen), who had been the resident Soviet handler of Klaus Fuchs and Allan Foote from her home in Oxford.[29] Kahle supplied Ursula with what she called "useful reports" and undoubtedly enabled contact between the Kuczynski espionage clan and Klaus Fuchs. Ernest Hemingway, fraternal comrade to both Kahle and Haldane in Spain and later, had also tried to sign up with the "neighbours," i.e., the KGB, in 1941. Despite repeated contact and receipt of a code name ("Argo") he was never able to supply them with anything interesting, and the KGB considered him a failure.[30] Fuchs was more successful and was convicted in 1950 of passing atomic secrets from the Manhattan Project to the Russians.[31]

Kahle went back home to the Soviet Occupation Zone of Germany (later East Germany) in February 1946, undertaking not to return to the UK. MI5 were finally paying close attention to him and noticed that he was escorted to Germany by Bill Rust of the CPGB.

Kahle was appointed "Police President" of Mecklenburg by March 1946. By the following year, he directed the Information Service of the Police with the assistance of Soviet MVD officers, but died while undergoing a stomach operation on September 8, 1947. MI5 had information that shortly before his death he was in the running for head of the Stasi. In 1966, East Germany issued a postage stamp in his honor.

In Haldane's later career, the subject of scientific espionage often cropped up in public, with the uncovering of atomic spies like Alan Nunn May, Klaus Fuchs, and Bruno Pontecorvo. Haldane retained his security clearance until 1950 and proved to be remarkably brazen. Questions had been asked about this clearance in Parliament in 1948—"In view of the fact that the Prime Minister has announced his purge of the Civil Service, is he not aware that purges, to be effective, must apply to people like Professor Haldane and the B.B.C.?"—but Attlee's undertaking to remove known communists from security-sensitive work had not been applied to Haldane—"I am aware that Professor Haldane is working on two committees of the Medical Research Council. The Government's attitude towards the employment of Communists upon work which is vital to the security of the State was announced in my statement of 15th March, to which I have nothing to add."[32] This was the policy that action would only be taken against communists who posed a definite security risk because they had access to secret government information, construed narrowly; merely being a communist was not cause enough for exclusion from government work.

When Fuchs was arrested in February 1950, belated action was taken. In March, Clement Attlee contacted MI5 and was advised that Haldane was a security risk.[33] The Admiralty refused to pass on classified information to the committees Haldane worked on, and he was asked to resign. A protest followed—"the Board of Admiralty has been misinformed concerning me"—and a challenge—"If the Board has any other reason to suppose that I would divulge secret information imparted to me, I should be interested to learn it."[34]

P.F.45762/B.1.a./MBT. 22nd October, 1946.

Dear Mr. Philby,

You may be interested to know that Professor J.B.S. HALDANE who was mentioned in my letter S.F.492/F.2.a(b)/MBT of 12.7.46, has received an invitation from the Institute of Philosophical Studies, via Montanelli 4, Rome, to attend an international congress on Philosophy to be held in Rome on the 15th - 20th November, 1946. We do not know yet whether Professor HALDANE has accepted the invitation.

Whilst there is every reason to suppose that this congress will be a perfectly reputable gathering, it has occurred to me that if HALDANE goes he might possibly be asked by the British Communist Party to make contact in his spare time with the Italian Communist Party, and you may therefore like to have a record of his visit.

As you are already aware, Professor HALDANE is a convinced Marxist and an active and influential member of the British Party.

Yours sincerely,

MBT.

M.B. Towndrow.

H.A.R. Philby Esq., O.B.E.,
S.I.S.

MBT/EDA

Note from M. B. Towndrow to H. A. R. "Kim" Philby about Haldane, October 22, 1946. From National Archives, KV 2-1832.

In his resignation letter to the Admiralty, Haldane dropped nonspecific hints about other scientists who had been careless with information and were probable "sources of leakage." By November 1951, he added more details in public. In September 1950, the nuclear physicist Bruno Pontecorvo (1913–1993) had fled to the Soviet Union, never to return; he had become alarmed after being questioned by MI5, who had finally realized after a tip-off from a Swedish source that Pontecorvo was a communist. The "Brothers Pontecorvo" had arrived in the UK in 1938 as communist refugees from Mussolini. Bruno was a nuclear physicist who had studied under Enrico Fermi and would later work on the British contribution to the Manhattan Project at Montreal; Guido Pontecorvo (1907–1999) was a geneticist who had worked with Hermann Muller in Edinburgh; Paolo the engineer worked on radar during the war; and Gillo, the youngest, became a radical filmmaker. Haldane knew Guido Pontecorvo well and, through him, Bruno.

At a public meeting on "Science for Peace" held at Holborn in London on November 21, 1951, Haldane argued that Bruno Pontecorvo was the victim in this affair. An MI5 source in the audience reported that, according to Haldane, "there was no evidence that he had breached any law of this country. The newspapers had, however, made him the spy of spies, and if he ever returned to this country, which even Professor Haldane doubted, he could make a million pounds out of Press libel actions." A second source reported hearing that "he (HALDANE) would not be surprised at all to learn that the United Kingdom had secret agents in the USSR trying to spy out Russian scientific achievements. Conversely, it was quite possible that the USSR had a few such persons in the United Kingdom (Laughter)." Working the audience, he said that the problem with Alan Nunn May was that he had been so "stupid" as to believe Churchill's "promises" in 1942 to share intelligence with the Russians, and so deserved to be in jail. "He assured the audience from personal observation and experiences over a period of years of association with classified government projects that the procedures and investigative methods of government agencies such as MI5 (in this country and elsewhere) are not really capable of uncovering really clever people."[35]

Revelations were again hinted at. "The people who are really responsible for so-called breaches of security are people in high places who whisper things to their friends to impress the latter. He illustrated this point with alleged examples from his own experiences during World War II, but did not specifically name the scientists who 'had confided in him' things which they should have retained." The first MI5 source was more expansive: "in one instance a high ranking Government Official had informed him, although aware of his political opinions, of the development of radar, another when he had been informed of top secret Cabinet decisions, and one man, who only because he wanted to swank, had told him of the Government expenditure in a special electronic sphere nuclear fission." Haldane, he reported, then hung out more bait. "If there was a representative of MI5 in the audience who wanted these three names and addresses, Professor Haldane would be pleased to give him them after the meeting."[36]

This promised information would perhaps have placed MI5 on a level footing with the X Group. Curiously, MI5, who as we have seen really did have agents in the audience, were greatly interested in Haldane's promise to reveal his sources and for over a week debated internally whether they ought to approach him for help. In the end, they thought better of it.[37] Meanwhile, Pontecorvo pursued atomic physics in the Soviet Union until his death there in 1993; for many years he enjoyed support from scientists and others in the West, who resolutely maintained the fiction that he had gone there merely to escape persecution.[38]

Some additional hints about X Group–like underground activities were dropped by Charlotte Haldane after she defected from the Communist Party in 1942. In her confessional *Truth Will Out* (1949), she recalled that Haldane was involved in a curious sub-group within the Party that she was not allowed access to.

The Communist Party ... organized an A.R.P. [Air Raid Precautions] Bureau, of which J.B.S. was the head. For some reason unknown to me, I was not invited to be a member of this Bureau, although it used to meet in my house. One afternoon, feeling slightly unwell, I was resting in the front sitting-room. The Bureau was meeting in the back sitting room behind closed folding doors. I could not help overhearing its deliberations. At one point it was clearly in a difficulty. Knowing the answer to the problem under discussion, I knocked at the door, asked if I might be allowed to speak, and gave the comrades the information they lacked. The only reward I received for my unsolicited help was a terrific wigging, afterwards, from my spouse, who accused me of 'eavesdropping' on a secret Party meeting.[39]

Was this the X Group or some other clandestine operation? It could hardly have been about the humdrum ARP, which Haldane thundered about from every platform across the land, and it is hard to understand why Charlotte would have been excluded from "secret" meetings unless there were far-reaching security implications. The Party were not happy with her disclosures. MI5 recorded a conversation at the King Street premises in which Harry Pollitt and JBS debated what they could do about it.

> PETER KERRIGAN heard talking to HARRY POLLITT about something which had been published. HARRY heard to call out "Come in, JACK". Told V. to take the easy chair. Previous visitor (SAM) left almost immediately. HARRY then turned to JACK (? HALDANE) and asked him what he could do. V. did not think HARRY could do very much. . . . Comrades now appeared to be discussing some article (probably written by CHARLOTTE HALDANE). HARRY very indignant about this, and said that had it not been for the fact that HALDANE was in the Party, he would have advised libel action, because the passages which related to him (HARRY) were just awful lies. He, however, thought that when everything came out it would be a "pale glory"; everything would be against them.[40]

These rumblings drew the attention of MI5, but only briefly. A note in J. B. S. Haldane's file records that "Charlotte HALDANE has not been interviewed by us, nor have we hitherto thought it worthwhile to make an approach to her, since any information not already publicly disclosed by her is necessarily some years out of date."[41] A cursory review of her book shows that she left a great deal out and was well connected to a network of which many members were probably still active. Basic intelligence work would have led to an interview at the very least. But clarification about whether this was ever followed up on awaits the release of Charlotte Haldane's own MI5 file, which is still embargoed.

MI5 never got very far analyzing the X Group after its exposure in the 1960s. "We have not established what the X Group represents. It is not the Communist Party as such, but it is probably some fraction or undercover group of the C.P. Moscow obviously visualised it as a source of military intelligence but it is difficult to trace the connection between Ivor Montagu (whose interests were largely in Film Production, Jewish affairs, International Table Tennis etc.), a Colonel in the R[oyal] A[rtillery], a girl in a Government Department and NOBILITY, a journalist."[42] On the contrary, MI5's bulging Personal File on Montagu was packed with leads still relevant in the 1960s when this note was written and he was still alive and active—very little of it concerns Ping-Pong. Little wonder, then, that the KGB was dumbfounded by the internal collapse of the Soviet Union decades later—*the World was going our way!*[43]

7. THE FATE OF VAVILOV

By the time that Germany invaded the Soviet Union on June 22, 1941, Charlotte had been supplanted in Haldane's affections by his pupil Dr. Helen Spurway, but they remained officially married. Though she had not worked as a journalist since her marriage, she managed to land an assignment to Russia as a correspondent for the *Daily Sketch*, hoping that she might get a personal interview with Stalin. She left Liverpool on August 16, 1941, on the *Llanstephan Castle*, which formed part of the Dervish convoy to Archangel via Iceland, arriving on August 31.[1] But before she left, JBS asked her to make some urgent enquiries on his behalf. The first concerned unpaid book royalties he was owed for his publications in the Soviet Union. The second was about his old friend, the plant geneticist Nikolai Vavilov, whose fate Haldane was privately worried about. He asked Charlotte to try to contact Vavilov when she was in the USSR, or find out what had happened to him. Vavilov had been expected to attend the International Congress of Genetics in Edinburgh in 1939 but failed to turn up. Officially, Vavilov communicated that research in the USSR proceeded normally, but privately he wrote to a Western colleague at the conference that the opposite was true. This letter had been forwarded to Haldane, who was therefore privately aware that there was trouble, at least by this date but probably even prior to that.[2]

After her arrival in Moscow in September 1941, Charlotte duly made inquiries about Vavilov. Definite information proved hard to find, until she learned to read the signals that had long ruled Soviet life. She asked to be put in touch with Lydia Bach, a daughter of the director of the Institute for Biochemistry in Moscow, Professor Aleksei Nicolaevich Bach, an acquaintance from her 1928 trip. The Bachs were Old Bolsheviks. Lydia, with multiple books to her name, was well integrated into the Soviet establishment. It was more than a month before she arrived unannounced at Charlotte's hotel room. Charlotte later guessed that her request had been delayed by the required NKVD approval.

> As soon as I met Lydia Bach, and we began to talk of old friends, I asked after Vavilov. During our 1928 visit to Moscow there had never been the slightest difficulty with regard to meeting individual scientists and many other Russians, even of the former bourgeoisie. In reply to my present inquiry, however, Miss Bach, who had in the meantime become a member of the Party, and secretary of the biological section of the Academy of Sciences, raised her eyebrows and remarked thoughtfully: 'Vavilov? Vavilov? I do not remember what he is doing now. One has not heard of him for a very long time'. . . . Even so dull a scholar as I then was learned fairly rapidly when not to press a point.[3]

The trip proved disillusioning for Charlotte. The publishing rubles due to JBS led her on a long tour through the Soviet bureaucracy, only to be told, eventually, that the organization owing the money had supposedly gone bankrupt. She discovered that Soviet air-raid precautions were rudimentary and far worse than those in England so bitterly complained about by her own Communist Party. The authorities in Moscow did not care. And the Russians were surprised to be told by her that the Germans had bombed London. The censors had made the Blitz a nonevent during the Nazi-Soviet Pact years. Then she saw processions of forced laborers in miserable condition. The NKVD arrived to arrest a man who had vomited in her presence. She was told he would probably be shot. Throughout her journey, the NKVD continually intruded. She was refused access to factories, and staged displays of "captured German equipment" insulted her intelligence. Evacuees were starved to death through official indifference. Religion was now openly permitted. The sight of an evacuee with a baby fallen victim to starvation proved the breaking point for her former faith. Her "masochistic devotion" to the Communist Party was over, leaving her with "a deep and strong sense of guilt" for deserting it, mixed with "pure rage."[4]

On her return to London in late November 1941, Charlotte met with JBS to convey the news about Vavilov and to share her disillusionment. "My reports caused him, undoubtedly, considerable surprise and mental uneasiness." Soon, she heard from Bill Rust, the head of the Communist Party, who had been tipped off by JBS. Rust concluded that she had been "got at" and, when told that she planned to write a book about her trip, asked if she would accept "Party collaboration" in writing it. She refused.[5]

Charlotte's trip and the information she gave about Vavilov are curious, given the campaign that Haldane subsequently conducted against reports that Vavilov was dead. These had been circulating since 1940, when Vavilov had completely dropped out of sight; even a casual observer of the Russian scene would suspect that he had been "disappeared" deliberately. Concerned scientists in the UK had him elected to the Royal Society in 1942, in his absence, in the hope that this foreign prestige might protect him from further harm if he were still alive. Among the twenty proposers on the membership nomination, one stands out: J. B. S. Haldane.[6]

Writing in the *Labour Monthly* later in 1942, Haldane mentioned Vavilov again, much as he had often done before, as an example of a great Soviet scientist.[7] No reference was made to the rumors that he was missing. But by 1944 these could no longer be ignored. Haldane dissimulated: Vavilov's institute "was cut down to some extent in the years before the war, largely because the best varieties had been selected, and partly because Lysenko's invention of vernalization rendered many of them less valuable." Of course, in reality it had been closed down in 1940 as a nest of "Mendelist-Morganist" wreckers. Moreover, "Vavilov was shot about once a year in the American press, though he continued to communicate papers to the Academy at least up to 1942."[8] But what of Charlotte's specific information about that?

Haldane reassured his readers with information that he certainly did not have. "Vavilov still directs research on a vast scale. So far from having been muzzled for his alleged anti-Darwinian views he communicated seventeen papers on genetical topics to the Moscow Academy of Sciences between January 1st and April 10th of 1940." A later footnote adds that "Vavilov's name is now less prominent, but up till June 1941 the output of genetical work showed no sign of abatement."[9] Previously Vavilov had supposedly been heard from in 1942, but now the date is 1941. More diversions followed, along with an early use of the *what-about-racism* tactic. "Anyone who studies the record . . . will certainly realise that thought on scientific topics is pretty free in Moscow. . . . I could wish that those of my European and American colleagues who have taken up the cudgels on behalf of Vavilov, who is not incapable of self-defense, would transfer some of their energies to an attack on [doctrines of racial inequality]."[10]

After 1945, more specific reports of Vavilov's demise required a change in tactics. Cyril Darlington and Sydney Cross Harland had published an obituary of Vavilov in *Nature* that year. "The circumstances are not precisely known, but the time was after December 1941 and the place probably Saratov."[11] From now on, Haldane's previous statements about Vavilov's active direction of research during the war, on a "vast scale," vanish without explanation, and instead his fate is neatly assimilated by sleight of hand into those of "other refugees" from Leningrad. "It is widely believed that plant genetics in the Soviet Union came to an end with the death of the geneticist Vavilov, who had done very fine work on the origin and geographical distribution of cultivated plants. I do not know how

Vavilov died, nor do I know how tens of thousands of other refugees from Leningrad died. As his institute at Detzkoë Seloe became a battlefield throughout the siege of Leningrad, it is not surprising that no research has been done there in recent years."[12]

The story changed again in 1948. "You may have been told that Vavilov, a famous Russian plant breeder, died in prison. His research station outside Leningrad became a battlefield in 1941, and according to a very anti-Lysenko article in the *Journal of Heredity* he appears to have died at Magadan in the Arctic in 1942 while breeding frost-resistant plants."[13] None of this was true either, though one has to admire the detail about "frost-resistant plants" and puzzle at Vavilov's demotion from a giant of modern science, in Haldane's previous estimation, to a "famous plant breeder."

The truth was that Vavilov was arrested by the NKVD in 1940 and tortured to try to produce a confession to a farrago of fantastic crimes. By then the ritual of confession was standard practice. The NKVD file on him stretched back to 1931 and was packed full of denunciations by suborned coworkers and planted agents, kept handy until the time was right. His numerous trips abroad and extensive correspondence were useful for introducing foreign governments into the story.

Nikolai Vavilov, photographed in the Gulag, 1942.

Great care had been taken to arrest him out of sight, on a plant-collecting trip to the newly conquered Carpathian mountains (a bonus from the Nazi-Soviet Pact). Vavilov endured four hundred interrogations, totaling nearly 1,700 hours over a period of eleven months, after his arrest on August 6, 1940. A single interrogation could last twelve or thirteen hours at a stretch. The longer the session, the fewer the written records left behind. The notorious thirty-three-year-old interrogator Aleksandr Khvat was assigned to his case.

He usually got prisoners to confess to something, and Vavilov was no exception. A cellmate, the artist Grigori Fillipovsky, was there to witness the aftermath of some of these brutalities.

> When Fillipovsky was thrust into the cell, he noticed immediately a strange figure among the prisoners who were lying, sitting, and standing around—an elderly man lying on a bunk with his swollen legs raised up. It was Academician Vavilov. He had only recently returned from a night-long interrogation during which the interrogator had kept him standing for more than ten hours. His face was swollen, there were bags beneath his eyes indicating heart trouble, and the soles of his feet were swollen and looked to Fillipovsky to be huge and gray in color. Every night Vavilov was taken off for questioning. At dawn a warder would drag him back and throw him down at the cell door. Vavilov was no longer able to stand and had to crawl on all fours to his place on the bunk. Once there his neighbors would somehow remove his boots from his swollen feet and he would lie still on his back in his strange position for several hours.[14]

Vavilov told Fillipovsky that interrogations invariably began: "Who are you?," "I am academician Vavilov," "You're a load of shit and not an academician."

Khvat had worn Vavilov down to destruction, physically and mentally, by the summer of 1941; but he did not get exactly what he was after. Vavilov refused to confess directly to espionage. After the invasion by the Germans, Khvat had to rush to completion. Interrogations with dead people were invented for the files. A troika convicted Vavilov of espionage regardless, identifying him as the member of an organization of "right wingers," who had carried on widespread agricultural "wrecking" and transmitted state secrets to White Russian émigrés abroad. The eminent geneticist was sentenced to be shot.

Vavilov's appeals were turned down, and two of his colleagues, who were tried at the same time as part of his nefarious ring, were immediately shot. Unexpectedly, a last-minute appeal for clemency, sent to Lavrenti Beria (head of the NKVD) from the appointed place of execution, Butyrskaya prison in Moscow, succeeded. (Recall that at this time Charlotte Haldane was still in Moscow.) Vavilov was then transferred to Saratov, as the Germans threatened to overrun Moscow. Like the other prisoners, he was forced to kneel for six hours at the Kursk railway station on October 16, 1941, to wait for a train in the thawing snow, excoriated by the gawking citizens of Moscow. The train journey took two weeks, in appalling conditions. Vavilov, unlike many, made it there alive, only to be crammed into a cell for the condemned, to wait for word from Beria about his disposal.

By the end of 1941, Vavilov had scurvy. In 1942, the Royal Society election almost paid off, after diplomatic intervention by the British government, and on July 4 Vavilov's sentence was commuted by Beria to twenty years of hard labor in the Gulag. Or perhaps, as Vadim Birstein speculated, Beria intended Vavilov for work in the "special projects" unit run by Pavel Sudoplatov.[15] Either way, it was too late. Vavilov would die at Saratov of starvation on January 26, 1943. Meanwhile Vavilov's colleague Georgii Dmitrievich Karpechenko, another host of Haldane on his 1928 tour, had been arrested in 1941, tortured, and forced to sign a confession. He defiantly retracted the confession at trial. Karpechenko was executed on July 28, 1941.

These finer details could not have been known to Haldane and only emerged when Vavilov was posthumously "rehabilitated" more than a decade later in the 1950s, and the entire NKVD/KGB file was eventually released to Vavilov's descendants. But Haldane had been privately concerned, enough so to ask Charlotte to make inquiries. Her report back could only have been interpreted one way, when combined with Vavilov's dried-up publication record and newspaper silence. Haldane's own sister, Naomi, later suggested that "he might, possibly, after the last genetics conference, have been able to do something to help Vavilov."[16] What Haldane had really done was offer the Soviets the equivalent of covering fire, before, during, and after this episode.

In confidence, Haldane used different arguments. In a letter to

Bertrand Russell at around this time, he claimed that Vavilov's case was equivalent to that of the surgeon Sir Victor Horsley.[17] During the First World War, Haldane had served with Horsley in Mesopotamia. Horsley had gone there after falling out with his military superiors, only to perish. Haldane's implication was that Horsley had been, in effect, eliminated by the British authorities, just as Vavilov had been by Stalin. But Haldane was being dishonest. Horsley—a gifted but eccentric, quarrelsome, and fanatical teetotaler—apparently believed that teetotalism would protect him from the sun. He promptly died of heatstroke in the ordinary course of his duties. He was not ill-treated in any way and served in Mesopotamia during the war as a surgeon in the usual manner.[18]

The dissimulations continued as late as 1948, artfully incorporating plausible counterfactuals; they might have gone on even longer had Haldane delivered on a promise he had made to the Royal Society. It is customary for Fellows of the Royal Society to receive detailed obituaries in the form of biographical notices, summarizing their accomplishments and publications. Haldane was asked by the Society to write Vavilov's notice in 1945, and accepted on condition that he be sent a bibliography of Vavilov's publications—one assumes this was supplied. Years passed. Repeated inquiries produced no result.[19] Ultimately Vavilov's biographical notice did appear, seven years later, in 1952. But it was written by the geneticist Sydney Cross Harland.

And so it was that Vavilov, who was wrongly accused of espionage, met his death, while Haldane, who was guilty of espionage, got off scot-free.

Several Soviet scientists risked their lives to intervene on Vavilov's behalf. The academician Dmitry Pryanishnikov (1865–1948) sought and obtained a personal interview with Beria to protest. When shown Vavilov's supposed confessions by Beria, he refused to believe them unless told so directly by Vavilov. Once Vavilov had disappeared into the meat grinder, Pryanishnikov stubbornly kept up appeals and audiences with Beria. Once he heard the rumors that Vavilov was dead, he pestered Beria about Karpechenko, not knowing that he, too, had been murdered. Pryanishnikov died in 1948, before the consequences of his appeals could catch up with him. The botanist Leonid Ipatevich Govorov (1885–1941) had also issued inquiries about Vavilov. A trip to Moscow in 1941 to appeal to Stalin found only closed doors, and Govorov was immediately arrested by the NKVD when he returned home. He was shot on July 27, 1941.[20]

In his study of Lysenkoism, David Joravsky counted at least 105 "repressed specialists" (agronomists, biologists, philosophers of science, and physicists) where "repression" meant anything from public execution after a nominal trial to disappearance and execution without trial or publicity; or to internal exile, often leading to death by overwork, starvation, or disease; or, more rarely, external exile.[21] This toll is, he conceded, greatly underestimated. Joravsky necessarily counted only the more visible cases, and did not count those who merely had their careers ruined. As we have seen, this process was not an attack on just plant genetics and Vavilov; it warmed up with social biology, proceeded to human genetics, and only got to plant genetics last.

In 1987, Aleksandr Khvat, comfortably retired with four children, was interviewed by a journalist. She asked him if he had ever felt compassion for Vavilov. Khvat laughed out loud.[22] But the Vavilov case was just one of many. Lina Stern, for example, had followed a different trajectory.

The Haldanes met the Jewish-Latvian physiologist Lina Stern on their tour of the Soviet Union in 1928, and spent a lot of time with her, staying in the apartment above hers in the Arbat district of Moscow—"modern though shabby," because each expected the other to keep it up.[23] Stern, who never married, made her scientific name through her pioneering work on what she called the "blood-brain barrier," which she formulated in the early 1920s when she was still in Switzerland. When she arrived in the Soviet Union in 1925, at the invitation of Professor A. N. Bach, she was already forty-eight years old, with more than a hundred publications to her name. She was left-wing by sympathy, but later confided that she was not political by nature. Only in 1939 did she officially join the Communist Party, a useful thing to do in the circumstances. Bach had long been a member and sat on the Central Committee.

Stern had a reputation for bluntness and was not overawed by JBS. A worn sports jacket and gray flannel trousers with a soft collar and an old wrinkled raincoat might pass muster in Cambridge, but not in her laboratory, where she didn't want him setting a bad example to her students. Nor did she hold off lecturing her guests, who had spent a lot of time in the pubs of Fitzrovia, about the evils of alcohol as they shared dinner in her comparatively luxurious two-room apartment with maid—the privileges enjoyed by senior academics. In the years after this visit, JBS would keep in touch with Stern's work on neurophysiology and other areas. As we shall see later, Haldane also contributed introductions to Soviet propaganda films about Stern that were adapted and distributed during the war by his old comrade Ivor Montagu.

When Charlotte revisited the USSR in 1941, she tried to look up Stern, whom she seems to have admired for a mixture of political and feminist reasons. By then, Stern had become the first woman to be elected to the Soviet Academy of Sciences, a body similar in intent to the Royal Society. She was also a professing communist, though she had balked at the Nazi-Soviet Pact: when assured that it was a marriage of convenience, she wondered what the offspring would look like. She had her own institute in Moscow. Stalin had even awarded her the rare privilege of a personal car. But now she had been evacuated to Alma-Ata (present-day Almaty), where she would remain until 1943, so Charlotte was not able to see her.

A letter to Stern from JBS survives from the following year, dated February 1942. He offers to send her reprints of articles he had published connected to physiological problems posed by escape from submarines. Also mentioned is "unpublished data which the Admiralty would undoubtedly allow me to give to your Naval attaché in London" in case she was interested in it. He hopes that, wherever she was, she was "200%."[24] It is likely, then, that JBS was in regular contact with Stern. This is important in light of what followed.

After the German invasion of Russia, Stalin had created "Anti-Fascist Committees" to tap international contacts possessed by people living in the USSR. One of these was the Jewish Anti-Fascist Committee, which Stern joined when it was founded in 1942. After the war, this body steadily fell out of favor because Stalin was suspicious of any independent centers of power. Increasing hostility to Jews, called "cosmopolitans," was bound up with this change. By 1947, Stern's scientific theories suddenly came under withering attack in the papers, a promise of trouble to follow. She was refused leave to travel to Switzerland for an academic conference. The following year, Stalin had the leader of the Jewish Anti-Fascist Committee, Solomon Mikhoels, quietly assassinated; it was disguised to look like a car accident. Stern also seems to have attended and spoken at the 1948 meeting of VASKhNIL, which was concerned with Lysenko's theories, and which we will cover in greater detail later. She was indelicate enough to be critical of Lysenko.[25]

At the Joint Scientific Session of the Academy of Sciences and Moscow Society of Physiology, held on October 5, 1948, Stern was once again under fire for unscientific, oversimplified theorizing. Especially troublesome was her known opposition to the theories of Ivan Pavlov. Her frequent trips abroad in past years were suddenly highly suspicious. By the end of the year, her institute was closed down on the pretext that it was being moved to Leningrad.

Late one night in January 1949, the Secret Police (then calling itself the MGB) arrived at Stern's apartment to escort her to what they said was an urgent meeting with their head, Lavrenti Beria. Three years of detention, conveyor-belt interrogation, and physical torture followed. She was seventy-one at the time of her arrest. Though she was made to stand upright for twenty-hour stretches in tiny punishment cells, it was hard to get much out of her. The endlessly barked questions only confused her. Screaming obscenities at her didn't work, since she spoke bad Russian and couldn't decode the slang used. Why, she asked, did they keep shouting about her mother, the one word she recognized?[26] Years passed. She wrote an appeal to Viktor Abakumov, the minister of state security, without realizing that he, too, had been arrested in 1951 and was being held in the same building.[27]

Stern was one of fifteen people from the Jewish Anti-Fascist Committee put on trial in 1952 for treason. They were "unmasked" as heads of a national organization aimed at subverting the Soviet Union. The sentences had all been decided beforehand, once confessions had been duly obtained. Thirteen were shot, one died in prison, and one was given a reduced sentence of ten years' exile to Kazakhstan. Lina Stern was the lucky one. The families of the executed were exiled to Siberia; being related to a "criminal" was a criminal offense in its own right. The trials and executions were kept secret at the time, for obscure reasons. Some have guessed that Stern was spared because she had worked on ways to prolong life, which she connected to cell metabolism, and Stalin hoped he might be able to use her research. It is a story so implausible that it may even be true.

These trials were to be the opening salvo in a new pogrom that Stalin was laying plans for. By then he had already launched the Doctors' Plot, fantasizing that Jewish-cosmopolitan doctors were trying to murder the Soviet leadership. In order to halt the resulting pogrom, the Jews were, so the plan went, to be deported to Siberia—like the Volga Germans and the Crimean Tartars before them. Stalin's death in 1953 terminated all this, and Stern was permitted to return to Moscow in 1954, "rehabilitated." Incredibly, she resumed her career, and lived to see the year 1968.

Haldane would soon have discovered by word of mouth through communist and scientific circles that Stern's institute had ceased to exist and that she had disappeared. W. W. Gordon, in *Soviet Studies*, fully covered the 1950 Pavlov conference in Moscow at which Lina Stern and other critics of Pavlov, such as Leon Orbeli, came in for strident, and telling, criticism.[28] Gordon sent a copy of his report to Haldane, who calmly replied that "I think both Stern and Orbeli needed criticism. They have probably got too much and too late. I am sorry to hear Konorski, who seems excellent, got it in the neck."[29]

8. EXPERIMENTS IN THE REVIVAL OF ORGANISMS

Creating propaganda for Stalin had a much broader scope than fraternal journalism. In July 1942, Ivor Montagu wrote to his X-Group recruit through the auspices of the Soviet War News Film Agency, a part of the press department of the Soviet Embassy that Montagu was now working for.[1] Haldane was asked to appear in and supply commentary for some documentary films about Soviet science. Montagu actually listed three variant films to be cut from a larger project: a popular film, *Soviet Science Snipits* [sic]; a short film, featuring Lina Stern, containing material excised from the first; and one with more bracing material, called *Experiments in Bringing the Dead to Life* ("if we dare," Montagu wrote). It is not clear what happened to the first two projects, but Haldane's contribution to the third, eventually released as *Experiments in the Revival of Organisms*, remains extant today.[2] It raises a number of questions about Haldane's scientific probity.

The film concerned experiments purportedly made by the Soviet scientist Sergey Bryukhonenko (1890–1960), in which disembodied organs—severed heads of dogs on platters, human lungs—were kept alive and functional using a machine called the "autojektor," which circulated and oxygenated blood. More ambitiously, deceased animals could, it was said, be brought back to life by the autojektor fifteen minutes after death, after having had all their blood drained and then reintroduced. The main footage, to which Haldane added English commentary and a short preface, had been produced for Russian consumption in 1940, based on a script written by Bryukhonenko himself.

Haldane, an enormous rumple-suited bald head with glasses, delivered his commentary in a plummy accent, awkwardly reading from a script.

> I should like to tell you that I have seen some of the experiments shown in this film actually carried out at the All Russian Physiological Congress. As you can imagine, technique is everything. Besides such work as you are about to see, Bryukhonenko shares the credit for the methods of human blood transfusion, which were first developed in the Soviet Union, and are now practiced in this country, which have saved so many lives during the war.

The All Russian Physiological Congress had been held in Moscow in 1928 when Haldane had paid his first and only visit to the Soviet Union. Although he reported on multiple occasions about the science he had seen there, he seems to have made no mention of Bryukhonenko at that time.

Over some stock footage of women in a laboratory, Haldane went on. "What enables Soviet scientists to solve this problem? Long ago, science established a fact that animal organisms and tissues, isolated

from the whole organism, can be maintained in a living state, but in order to achieve this, special artificial conditions must be created." A dog's heart is shown, attached to tubes, hanging on a stand. "Isolated organs can be brought to life

EXPERIMENTS IN THE REVIVAL OF ORGANISMS 135

even though they've been removed from the animal's corpse some time after death. There's a dog's heart. It can function as well in artificial conditions as in a living organism, and for this purpose, blood is introduced into the cardiac vessels. The isolated heart beats just as it did a few hours previously in the living dog." A dish holding expanding and contracting lungs appears, with the now-familiar tubes. "Bellows distend the lungs and fill them with air. The venous blood is forced into the lungs by the action of the pump. The dark venous blood passes through this tube [shown attached to lungs]. In the lungs, it takes up oxygen, and becomes arterial blood. The isolated lungs breathe, producing the same chemical changes as in the living animal."

The film moves on from these isolated organs to severed heads of dogs, dinner-plattered but living on. "An animal's head can also live in the isolated condition. Here is the plan of the experiment." Helpful cartoons show the general organization of circulation from the autojektor. "The arterial pump takes arterial blood from the reservoir to the head, while the venous pump drains out the venous blood. The blood is arterialized in the reservoir, where there is a steady flow of oxygen. The artificial blood circulation ensures the metabolism necessary for the life of the head."

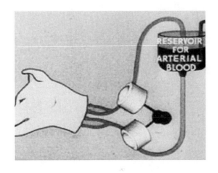

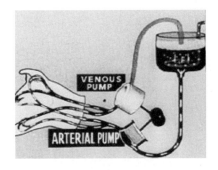

The severed dog's head then appears on a platter. "The isolated head lives on for hours, and reacts to external stimuli." Then the eye is poked with a probe, and the dog blinks. The nose is tickled with a feather repeatedly and the dog's head grimaces, opening and closing its mouth. It blinks

again. Citric acid is applied to its mouth. It licks its lips several times, and moves its head noticeably. As it does so, its throat region expands as if it is breathing. "The isolated head even reacts to light"—a spotlight is shown and is played on the head—"and to sound"—a hammer is tapped on the table, the ears twitch, and the head moves distinctly *away* from the noise.

Then we shift from merely reviving individual body parts to reviving whole organisms, à la Lazarus. "The revival of individual organs enables scientists to proceed to experiment on reviving the whole organism. The revival of the whole organism can be achieved with the help of an apparatus called the autojektor." The device looks exactly as one would expect an apparatus in the laboratory of a deranged scientist to look.

A cartoon shows the circulation of blood. "The autojektor carries out the functions of the heart and lungs. As we know, the heart, by its rhythmical contractions, supplies the body with arterial blood rich in oxygen. After losing its oxygen, the blood returns to the heart through the veins. From there, it flows into the lungs, whence with a fresh supply of oxygen it returns to the heart, and thence flows into the arteries of the organism."

Haldane next describes the arterial pump. "The autojektor works on the same principle. The apparatus includes a system of pumps, for supplying blood and drawing it off. The arterial pump supplies the organism with arterial blood. When the blood has given up its oxygen, the venous pump draws it back into the reservoir. Here, just as in the lungs, it is enriched with oxygen, and returns into the organism. The blood passing into the arteries of the body ensures the necessary metabolism. In this way, the autojektor can perform the work of both the heart and lungs." We are shown a bed with an autojektor next to it.

"We begin the experiment of revival. The experiment is carried out on a dog." The unfortunate dog is shown, lying muzzled on a bed. "A substance which prevents clotting is introduced into the animal's blood. The dog is under an anesthetic. It doesn't feel pain. No interruption of the animal's normal functions has yet occurred." "The dog reacts to touch"—the eye is poked with a probe—"its pupils are normal. . . . A special apparatus, a kymograph, registers the breathing and the function of the dog's heart. The pulse and breathing are normal." A graph of instrument recordings continuously traced on paper is shown. "The experiment begins. All the blood is drained off through the carotid artery." Blood is shown draining into a flask. "The heart has stopped. This is one of the animal's last gasps. This is a final blip. The dog is dead."

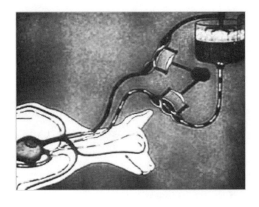

Now a stopwatch is started. "Without operative interference, death would be final, as the disintegration of the body cells would gradually set in. The autojektor is being attached before starting the revival. The arterial pump is connected with the artery. The venous pump is connected with the vein. Ten minutes have elapsed since the animal died. The blood removed from the animal is pumped back into its vessels by the autojektor.

"The autojektor ensures a normal blood circulation in the organism, replacing the action of the dead heart and lungs. The artificial heart circulation gradually induces the heart to start beating again. The heart's action begins to be normal. . . . The first sign"—the dog's head and upper body moves. "The respiration is gradually restored. The dog breathes more normally and evenly. The animal's condition approaches normal. We can now disconnect the autojektor, and leave the organism of the dog to maintain life with its own resources."

Heartbeat sound effects ensue, and then portentous music. "The dog soon shakes off the effects of the anesthetic." The dog is shown lying on a cot. "The dog is still weak and can't move." We switch to triumphant music.

A few days later. "After ten to twelve days, the dog returns to its normal state." A nurse is shown walking a dog on a leash. "After the experiment, the dogs can live for years, they grow, they put on weight, and have families. For a number of years now, three dogs have been under observation in the Voronezh Medical Institute, after being revived by artificial blood circulation. This dog 'Bunny' [shown] was revived in 1939, after having been dead for eight minutes. 'Black ears' [shown] is the offspring of revived parents. She herself was revived in 1939, after eleven and a half minutes of death. 'Naida' [shown] was revived in 1938 after fifteen minutes of death." (So the Lamarckian transmission to offspring of the effects of death and reanimation could be ruled out, although Haldane does not explicitly make this connection.)

"These experiments on the revival of dogs have shown that the process has no harmful effect on the animal organism. The question of the revival of animals is one of the most interesting problems in physiology today. Experiments on revival have added to the valuable store of our knowledge of experimental medicine." *The End.*

The film was screened in London before the end of 1942[3] and then in New York the following year. *Time* magazine reported that a thousand scientists had attended the New York screening, and that the "scientific audience thought this work might move many supposed biological impossibilities into the realm of the possible."[4] The *Time* report had something of Haldane's own style—that is, it read like a press release from the Soviet press agency. "Red science is a vast, centrally directed enterprise, with the U.S.S.R. Academy of Sciences at the top and hundreds of institutes working on assigned problems. No scientific frontier is neglected." Great achievements were duly listed, including "a method of planting winter wheat (in unplowed stubble) that enables it to withstand Siberian temperatures of 40 below zero." (In fact, this method was a failure and led to heavy crop losses, but it took the Soviets fifteen years to penetrate the misleading nonsense surrounding it.[5]) The imprimatur of J. B. S. Haldane, F.R.S., surely helped to persuade these audiences.

Bryukhonenko was born in Kozlov (present-day Michurinsk), near the city of Tambov, on April 30, 1890. After studying medicine at the University of Moscow and qualifying in 1914, he served as a doctor in the First World War. Apparently he invented the autojektor machine, for artificially circulating the blood via mechanical roller pumps, in 1926, after discovering that the drug suramin could be used as an anticoagulant. By 1928 he had applied for a Soviet patent for his "Device for Artificial Circulation."[6] The patent application contained helpful diagrams showing his experiments on dogs.[7]

This was really an early heart-lung machine. As Haldane indicated, Bryukhonenko demonstrated some of his experiments at the Third Congress of Physiologists of the U.S.S.R. on June 1, 1928—but made rather more modest claims, showing nothing more than some lingering reflex reactions.[8] He showed no resuscitation experiments then. In 1960, Probert and Melrose, writing in *The British Medical Journal*, summarized his research as follows. "In his total perfusions Brukhonenko stopped inflating the lungs of the experimental animal and excluded its heart from the circulation by mechanical means. In three such experiments when he discontinued the perfusion and restarted pulmonary inflation he succeeded in restoring a temporarily effective heart-beat. He also observed that cooling of blood in the extracorporeal circuit reduced the temperature of the perfused animal, and to overcome this surrounded some of the tubing with warm water."[9]

After this respectable start, Bryukhonenko fell victim to the relentless escalation of Soviet expectations, driven by their use of scientific claims as propaganda. In 1936 he was given his own Institute of Experimental Physiology and Therapy to work (strictly materialist) miracles within. Soon *Nature* was reporting that Bryukhonenko had progressed to "reanimatology," more ambitious even than merely keeping severed heads functioning—insofar as repeating "facts" presented by the Soviet Union Year Book Press Service can be considered reporting.[10] Corpses could now be raised from the dead after as long as half an hour of induced death, and *restored to a fully functioning state*—that is, not brain-dead. After the war, these sorts of claims faded from plain sight without ever totally vanishing from the Soviet journals, where "reanimatology" resurfaced from time to time to set ever-extending records for the times claimed between death and reinvigoration with total recovery.[11]

By 1995, as the country was being reintegrated into the international scientific establishment, one author was exasperated enough by this idiosyncratic survival to lay down the known facts explicitly.

> Clinical experience confirms the fact that the time taken for clinical death to occur does not exceed 5–6 min in adults and 7–8 min in children. At low temperatures, these times may be extended a little. Information in the literature concerning the complete revival in ordinary temperature conditions, after 20–30 min. clinical death, cannot be considered proven. It only disorientates the doctors and does harm to the development of clinical resuscitation.[12]

Bryukhonenko had tapped into a popular theme in Soviet science-fiction writing in the 1920s. In 1925 Aleksandr Beliaev (1884–1942) had enthralled his audience with his novella *The Head of Professor Dowell*, which was serialized in *The Worker's Gazette*.[13] Set in the United States, the story had the unfortunate Professor Dowell suffer decapitation followed by disembodied reanimation in a laboratory, where the talking head eventually reveals *whodunit*, but only after many adventures. Inevitably, Bryukhonenko's work became intermingled with this material through the ordinary mechanisms of journalism, and perhaps because, as Nikolai Krementsov has ventured, the Soviet scene then was "at one and the same time permeated by omnipresent death and by high hopes for the future." By 1929 the theme had circulated broadly enough for George Bernard Shaw to try to get in on the ground floor, volunteering his own head for the new procedure. "I am greatly tempted to have my head cut off so that I may continue to dictate plays and books independently of any illness, without having to dress and undress or eat or do anything at all but to produce masterpieces of dramatic art and literature."[14] The disembodiment theme was a very old one, in both literary and scientific terms.[15]

Faking photographs was an active area of interest in the Soviet Union, starting with the Revolution—many of the iconic photographs from the period are doctored to grossly inflate crowds.[16] This accelerated with the un-personing of Old Bolsheviks, who had to be retrospectively removed from the company of Stalin wherever they were unfortunate enough to have been photographed with or near him.[17] Documentary film was a capital-intensive extension of this process, with the advantage of much greater verisimilitude. Moving pictures, giving the impression of motion, suggest truth in an immediate way that still photographs cannot. Improving the facts—using heads of concealed live dogs—is easier and more convincing than inventing everything wholesale. Ivor Montagu had himself learned from Eisenstein that the film is not in the footage but rather in the editing. Bryukhonenko's research was ideal material for this. Those motivated enough would find a prior media trail, presentations at scientific congresses and other suggestions that the work was real, as long as they didn't inquire too closely about the way it had crept from blood circulation to reanimated cadavers. Haldane proved he was willing to collude.

9. IT IS YOUR PARTY DUTY, COMRADE!

Haldane had known Barnet "Doggy Woggy" Woolf since his early days in Cambridge at the Dunn Biochemistry Laboratory. There was also the Communist Party, of which they were both members—Woolf openly, Haldane secretly until he came out in 1942. Woolf had left Cambridge by the mid-1930s for a brief spell as a lyricist in the Unity Theatre, but broke his femur and underwent multiple operations, which left him with a lifelong limp. By 1943, he was back in the academic world at the Department of Zoology at the University of Birmingham, under Lancelot Hogben, a zoologist with mathematical leanings. Hogben, a science popularizer in his own right, was on friendly terms with Haldane. Though a declared socialist and a fellow traveler, he was not a Communist Party member.

At Birmingham, Woolf had conducted statistical inquiries into the incidence of child mortality in the north of England, believing that it could be causally tied to socioeconomic conditions there. In late 1943, he had written up the first of two papers on the results of this research and submitted it through Hogben to Haldane for possible publication in the prestigious statistical journal *Biometrika* (founded by Francis Galton and Karl Pearson, and then edited by Pearson's son, Egon).[1] Haldane was a member of the editorial board. Woolf's paper also included some stringent criticism of previous statistical work. Hogben wondered if *Biometrika* would be more likely than other journals to have the necessary paper stock, then in short supply. He hoped that Haldane would not approach the paper "with the absent-mindedness of a professor whose mathematical right hand does not know what his political left hand is doing." Woolf needed "all the backing and encouragement he can get from you and me."

Within a week, Haldane replied with a cursory and blunt review, pointing out several typos and other slips in the paper and objecting to what he thought were statistical inadequacies, requiring "drastic revision at certain points."[2] Woolf had not, he argued, properly considered the effect of skew distributions (one of Karl Pearson's favorite topics) when working out correlations, or made other due-diligence checks for normality. Woolf had also failed to cite relevant work recently done by others, and Haldane cautioned him that "If you attack other statisticians you want to avoid a vitreous domicile." JBS also suggested that the *Journal of Hygiene* might be a more suitable venue for the paper.

The fury from Woolf that followed gives a rare glimpse, behind the scenes, of the extent to which left-wing politics had crept into scientific practice by this date, and the ways in which the personal networks established by political activities and shared sympathies could be exploited. Hogben replied first, politely regretting the errors pointed out by Haldane, but suggesting that they could easily be fixed, and reiterating that the article was "of first rate social importance" and more original than Haldane had supposed. Then once more he appealed directly to Haldane's political sensibilities: "I am a simple Englishman, albeit an old-fashioned social democrat, apt to assume that when you address audiences about the social orientation of scientific research in the Soviet Union, you mean what you say, and am therefore anxious to promote socially oriented research in this country." In response to complaints by Haldane that he was too busy to assist Woolf, Hogben chided, "you are not the only person who is distracted by a variety of responsibilities." He thought that *Biometrika* would confer more prestige than the *Journal of Hygiene*, and the question remained about the shortage of paper.[3]

Haldane didn't budge.[4] A few weeks later, Woolf lost patience with his old mentor in a ten-page letter that combines contempt for Haldane and modern statistical practice with mordant wit that would not have disgraced Karl Kraus, reinforced with invocations of Communist Party duties. "It pleases you to use your position to make reckless and false accusations of incompetence and ignorance against those who are younger and lower-paid than yourself. You take a perverse pride in this so-called 'plain speaking' to those you judge too dependent on your favour to answer you back. Let's see if you can take a little of what you so lavishly hand out."[5]

Woolf then proceeded to pour scorn on Karl Pearson, R. A. Fisher, W. S. Gossett, and modern statistical practice in general—and therefore Haldane in particular—confiding that he would have to reinvigorate the subject himself. Fisher's foundational *Statistical Methods for Research Workers*[6] he dismissed as a "masterpiece of muddle." Gosset's t-test was "out of date since 1929." While Haldane might "swallow the idealist hocus-pocus behind this approach to experimental data," "I as a Marxist do not." "My generation has rightly cleansed its habitations of antimacassars and Pearson's skew curves, of partial correlation coefficients and aspidistras. You tenderly rescue them from the dustbin of history." Worse, "all our statistical journals are dominated by two rigid prejudiced unbending autocrats who are determined to defend to the last ditch the established orthodox dogma. I refer of course to R. A. Fisher and J. B. S. Haldane." The finer technical details of Woolf's critique are beyond our scope here, though we may note that the statistical field has yet to take the slightest notice of it.

Woolf dismissed the errors pointed out by Haldane as either unimportant or mistaken criticism—"has Perfection blundered? Has the Pope, speaking ex cathedra, been found fallible?"—and jibed that he easily found similar errors of a similar order in Haldane's own book *Causes of Evolution* (1932): "I think you might consider including 'to give up sneering' among your New Year resolutions." The prior work mentioned by Haldane he waved away as not relevant enough to merit space: "I see no reason, out of all this mass, for singling out the two Wright studies for mention, just because they are the only two you happen to have heard of." And anyway, "practically all the previous work has employed faulty or inconclusive statistical techniques." Haldane was "thoroughly wrong and un-Marxist," having let himself become worked up into a "frothy frenzy of self-complacent superiority." We can only marvel at the abuse on offer, as for example:

> You ... have a most powerful mind. It can, and does, harbour three or four obsessions at one and the same time. This gives your conversation and writings a certain element of unexpectedness, because, although one can usually predict exactly what you will say on every occasion for several months ahead, one cannot be sure of the order in which you will say them ... Reliability is a rare and praiseworthy quality. In every speech and letter and almost every article you have composed for months past, you have contrived to convey the astonishing information that, unlike anybody else in Britain, you are on urgent war work. It was with considerable anxiety that I noticed this was not specifically mentioned in your first letter. But you did not keep me in suspense too long.

More damning still was the reiterated suggestion that Haldane was not much of a Marxist after all. "It is a pity you use up so much Marxism in your *Daily Worker* articles and week-end speeches that you have none to spare for statistics. Otherwise you might, like me, be watching with interest and excitement the emergence of a new and genuinely dialectical materialist theory of statistics out of the needs of practice." There is accidental bathos in much of this. "As a materialist, you must be aware that publication at the present time depends on a very tangible factor—paper supply." Then Woolf gets down to more serious business.

> The next thing I want to ask is whether you will sponsor my paper for *Biometrika*, subject to any outstanding differences of

opinion about its composition being settled by mutual agreement between us? I am not asking this as a personal favour.

I am not asking, I am demanding. I am putting it to you as a Party duty. It is not only important that my work shall be published, it is important that it be published quickly. It is politically important that it shall be out before the Parliamentary debates on the Government Beveridge proposals.... The political importance of such investigations is obvious....

... I am demanding, as a political duty, that you sponsor my paper, take or send it to Pearson, and press for early publication. And I demand at any rate a definite reply to this request.

Woolf drew to a close with a sententious attack on Haldane's contemporary reputation, and some gratuitous advice for his mentor:

May I end by saying that I did not enjoy writing this letter.... I have written it as a Party duty.... It is bad for the Party that its No. 1 scientist should so behave as to get himself laughed at by those who have no direct dealings with him, and loathed by those who do. It is bad for the Party that a member of its central committee should get the reputation of being always ready to do down or discourage a young worker, that he should publicly humiliate young men specifically gathered to do him honour, that he should give on public platforms exhibitions of indiscipline, egotism and contempt for his audience. It is bad also that he should be called a back number, as I have heard you called more than once, and that people should be discussing when he will retire.... Where is the old J.B.S. of Daedalus, of the Tyrol—hell, I'm no good at preaching. Why don't you get a grip on yourself, stop antagonising those who like you most, realise your potentialities to the Party and to science?

Haldane had a notoriously short temper, but by the time he replied to his former protégé (after offering a preliminary deflection through his secretary) he seems to have mastered it enough to only beg off replying in detail, because he disagreed with so many points that he did not have the time, just as he did not have time to meet Woolf personally.[7] There is no way of knowing whether he accepted Woolf's strictures about "Party duty" or considered himself under orders, but Woolf's paper was finally published in 1945 in the *Journal of Hygiene*, just as Haldane had suggested.[8] Nevertheless, Haldane always took his Party duty seriously.

Haldane played an active role in the contribution of the Communist Party of Great Britain to the Cold War, on top of campaigning extensively for them in elections and running the editorial board of the *Daily Worker*. JBS explained to his readers that "We do not take orders from Moscow as, for example, some members of the late Conservative government took orders from New York and Berlin when they were directors of firms belonging to international cartels."[9] The Party viewed his scientific work and active journalism as an extension of his work for them, but it was concerned about the demands made on his time. An intercepted note from Emile Burns to "Dear Sam" underscored this. "As for Haldane's other commitments—do not raise this one. An article for *Nature* is Party work of a very high order—I only wish more comrades realised this & acted accordingly."[10]

Haldane was often present at street meetings, trying to drum up interest for the Party. Conditions could be difficult, and the public fickle. An MI5 report records one such meeting in late 1946, held in "torrential rain" and attended by "about 30 patrons of the East Street Sunday morning market, sheltering in nearby doorways." Jimmy Bent opened the meeting, speaking through a loudspeaker attached to a van, but disappointed the damp and dripping audience by announcing that Haldane could not make it because of the bad weather. The van then sped away. A few minutes later it sped back, bearing Haldane, who, it was said, had arrived unexpectedly. JBS boomed through the loudspeaker that police had been discriminating against the Communist Party by arresting their leaders, rather than people like Oswald Mosley. "He said that in 1938 he was present at a communist meeting which was broken up by the fascists. He was himself attacked and took refuge in the doorway of a public house. Some policemen arrived but made no effort to protect him, being more concerned with safeguarding property in the public house, the window of which had been smashed."[11]

At another meeting a few weeks later, regarding the International Policy of the Communist Party, he anticipated that it would be at least five years before they were in power, but suggested that until then the Labour government should emulate the foreign policy of the Soviet Union, who were "working for peace" and were trying "not to interfere with internal policies of other countries." This was proved by the dissolution of the Communist Third International. The problem was that past foreign policies of the UK had made the Soviets suspicious. "We want a socialist policy and not an imperialist policy."[12]

Early the next year, in January 1947, he was in New York City to speak at the "Lenin Memorial Meeting" held at Madison Square Gardens—MI5 received a helpful report from a U.S. government source. There was a crowd of 15,000 people, and a backdrop of Lenin silhouettes in black against a red and white background, with the hammer and sickle arranged in a cross with the stars and stripes. Copies of a book on Lenin could be purchased from "comrades of both sexes" in the crowd. As the fifth speaker, Haldane was a figure of some importance, and received heavier applause than most.

"In his speech Haldane stated that the Soviet Union, given thirty years of peace, would by 1975 demonstrate a higher standard of living for its people than any other nation under capitalistic economy. He stated that Lenin was a philosopher whom no scientist could neglect." It was possible that, if Britain and France got the atom bomb, a small group within them might hold those nations to ransom. Lenin, he believed, would have regarded it dangerous to one's own citizens to possess atomic bombs.[13] A convoluted argument ranked Lenin with George Washington: "one could not say that a non-Russian person was pro-Russian if he revered and loved Lenin for his contributions to mankind, any more than one could label a Latin American as pro-American if he admired George Washington for his contributions."[14]

As the Iron Curtain imposed by the Soviets rapidly isolated the occupied countries of Eastern Europe, Haldane was a dependable participant in "Peace" congresses and science get-togethers organized there. "The 'iron curtain' is round capitalist science, not Soviet science," he assured his readers. Before the war was even over, he had been confident that German reconstruction would work out well. "Fortunately the Soviet union, which has suffered far more from German aggression than Britain, let alone the U.S.A. will have a large share in determining the policy adopted."[15] Some idea of his hopes for this process followed: "Germany will only cease to be aggressive when its ruling class is wiped out. I do not mean that they should all be massacred, though I hope that those who are actually responsible for murders will be killed. I mean they should cease to exist as a class. . . . This could be accomplished in several ways, of which I should prefer the method of the Russian Revolution adopted."[16] As for fears for the political freedoms of the residents there (whose countries would all be taken over by Soviet-imposed regimes), the "plain fact is that over most of the world such parliaments as survive are at least as subservient to Big Business as is the Supreme Soviet in Moscow to the Communist Party."[17]

Later in 1947, Haldane attended a conference in Czechoslovakia, then under a communist-dominated regime bolstered by the NKVD. Broadcasting on a government radio station there, he enthused about what he had found. "I think the average Czech is better off than the average Englishman. I also think he is slightly happier than the average Englishman and a good deal happier than the average American." Even though everything had been collectivized, there was "plenty of room for the little man who wants to work for his own profit, provided he fits into the general plan of production." Touring the coal mines, he discovered that they had completely solved the problem of dust. Every miner lived in a beautiful village and had his own vegetable garden, and even a cow. The only drawback Haldane could find was that pipe tobacco was hard to get, at least for foreigners—and that the latter might not finish their dinners, given that food was so plentiful.[18] The communist coup of the following year, during which the prime minister fatally "fell" from his window, and remaining non-communists in the government were purged, went unmentioned on Haldane's lecture circuit.

Wroclaw in Poland hosted the World Congress of Intellectuals for Peace in late August 1948. There were some six hundred delegates in total, dominated by fellow travelers ranging from Bertolt Brecht to Pablo Picasso, Hewlett Johnson to Ivor Montagu, and Irène Joliot-Curie to J. B. S. Haldane. The congress was funded by the USSR, which repeatedly throughout the Cold War was to harness international intellectuals to promote "peace"—at all times the main aim of the Eastern Bloc, or so they said. Haldane's contribution to the congress was unswerving, as he warned those present that American imperialism was the main threat of war in the world. But he worried that the Soviets were not releasing enough information about the superior living conditions in the USSR, which would help win the propaganda war in the West.[19] American doctors who did research on bacteriological warfare were, he said, traitors to the human race like the Nazi doctors in Auschwitz. The Politburo was pleased with the results of the Wroclaw congress and immediately voted a large budget to fund the next event, held at Paris in April 1949.

In the meantime Haldane moved on to Hungary, where the International Medical Congress was being held at Budapest on September 11, 1948. This congress was another front for the Soviet-backed "Peace" movement; it passed a resolution that the "fight against common diseases must be linked with the battle for lasting peace," urging all to "despise the traitors of science and the medical profession who sell themselves and serve the warmongers." Addressing the delegates, Haldane was grateful for the "opportunity to see the rhythm and élan with which Hungary has carried out the work of reconstruction and surmounted all difficulties caused by the war." In a subsequent lecture, he explained that racial arguments should not be used "to prevent any human being from enjoying the benefits of higher culture," for to do so was a "most revolting injustice." Before he left, he informed the press that he and his wife Helen had wandered about the city without a guide and had been delighted to see "an air of democratic luxury" everywhere. Visitors could see for themselves that the "Iron Curtain" did not exist.[20] The Haldanes then returned to London via Prague.

Pairing talk of "peace" with proclamations about race had long been a staple of anti-Western propaganda, and Haldane frequently returned to this theme on other occasions—though note his careful restriction to indefensible *uses* of race. In his novel *The First Circle*, Alexander Solzhenitsyn explained how this technique was used in the unlikeliest ways in the USSR. American visitors to a Gulag camp, including "Mrs. R.," are treated to a spotless example of prisoner rehabilitation and shown dormitories of respectable prisoners, tolerably spruced up and putting on flesh after being hastily fed and clothed shortly before the visit. Sensing a travesty, the Potemkin prisoners protest excitedly in Russian. When the visitors ask their concierges to translate, their conductors are quick to explain that the prisoners are protesting about the maltreatment of Negros in the United States.[21] Similarly, Soviet cartoons portrayed menacing troikas on the march: a hooded Klansman with lynching rope and a club, a gangster policeman, and a white-coated geneticist gleefully brandishing a hypodermic needle and calipers.

Did Haldane really believe this Eastern Bloc fantasy? There is solid evidence that he did not. An intercepted telephone call from the previous year (October 13, 1947) recorded some Communist Party members sharing their concerns over anti-Russian remarks that

Haldane had made at a Party meeting held at Muswell Hill. Haldane had complained about the restrictions imposed on science and research behind the Iron Curtain. Although he believed that the Americans had started this, it was "very bad" that the Russians had reciprocated; Britain was left to be squashed between the two parties. Mary Jones wanted something to be done about Haldane, while Kitty Cornforth suggested that Mary could take that up directly with Haldane himself.[22]

The UK now abounded with organizations that had been organized along the same lines as the Society for Cultural Relations with the USSR, to promote "friendly relations" with Stalin's dominions, old and new. Haldane served on many of these, including the British Soviet Society and the British Roumanian Friendship Association. Likewise, he lent his presence to gatherings like the International Conference for Aid to Democratic Greece held in Paris on April 11, 1948.[23] This was organized by the Communist Party of Greece, then engaged in a long civil war aimed at taking the country into the Soviet sphere, which it finally lost in 1949. If there were front organizations he did not join, it could only have been through lack of opportunity. He was invariably joined on these by his old comrade INTELLIGENTSIA—Ivor Montagu.

Montagu appears to have kept active throughout this period, but he received a shock when developments in Eastern Europe progressed inexorably to that intermediate stage of full socialism, show trials and rampant terror. His concern was personal rather than altruistic, as he had long been associated with some of the accused, a dangerous position to be in. In Czechoslovakia, the number-two communist Rudolf Slansky was arrested in late 1951, along with many of his associates, and charged with the usual nefarious plots. Included among the co-accused, who were mostly Jewish, was none other than Otto Katz, an old associate of Montagu dating back to the 1930s and perhaps even earlier. Katz had helped Montagu extensively with his propaganda documentaries, including those he made about the

Spanish Civil War. This was the same Otto Katz visited by Hans Kahle en route to England after the Spanish Civil War. Now back in Czechoslovakia under the name Andre Simone, Katz had been arrested.

In late October 1952, Montagu was recorded discussing the Slansky/Katz case with his fellow Party members. Montagu was worried that if Katz was going to be "scrubbed" he might implicate his old friend, Ivor the foreigner, who was looking forward to being decorated by the Bulgarians. There was also the delicate matter that Montagu had had extensive dealings with Otto's old boss Willi Münzenberg, who had been "liquidated" in the late 1930s. The transcript gives important insight into the extent that Party members had real knowledge of events in Eastern Europe at the time.[24]

> **14.05.** Visitor comes in saying he came to see the International Department but there is no-one there.
>
> BOB[25] enquires for his health and then asks if all the people are home now.
>
> No, BASIN?? ph[26] should have come yesterday, but the rest are waiting until November. Relates a long and complicated story of muddled arrangements involving several cables.
>
> BOB asks if it was a successful conference.
>
> Visitor's immediate reply inaudible. He then says he has brought a lot of stuff for the International Department. Visitor says he wanted a word with BOB because of the arrest of OTTO KATZ, connected with the Reichstag Fire. Then Visitor (believe he is Ivor Montagu) outlines KATZ' history for BOB. He tells BOB that when he went through Prague between 8 and 15 he rang KATZ up, as he was in the habit of doing when in Prague to arrange for them to lunch together. His wife answered and said he would be sorry to have missed him but he (KATZ) was away for ten days. A day later a girl employed in the Ministry of Information called into the World Council and said as MONTAGU knew KATZ she thought she should tell him he'd been arrested. MONTAGU left a message with ROY to tell HARRY when he went through as HARRY knew him well. MONTAGU is a bit worried in case KATZ is 'scrubbed'. He (KATZ)

has many times been near people who have turned out bad ones, e.g. WILLI MUNZENBERG. If on the one hand KATZ is all right IVOR does not want to stop seeing him; on the other hand if he has done anything wrong he does not think he should conceal the fact that he used to go and see him. They were in the habit of exchanging propaganda material. Asks BOB what he should do about it; should he keep his mouth shut and say nothing or if that is the wrong thing to do.

BOB doesn't think it is easy to say. Adds 'If you talk to them here at the Embassy they don't know a damn thing'.

IVOR was wondering if he ought to say anything to the Soviet comrade in charge of . . . (GOULAEV? phon[27]) or to (HODINNA-SPURNY phon). He did not say a word when he came through this time, but thought he would ask BOB. Goes on to say one irony is that on the 29th the Bulgarians are going to decorate 'us all'. And of course it is he who ought to be decorated. Laughs.

BOB replies they must hope there is nothing wrong with the chap; of course everyone who has been in that bloody country is under suspicion.

IVOR says KATZ spoke to him about that once or twice, IVOR thinks extremely ably. A joke was made of what they called the English disease that it was a trial for comrades and a test of character because those who hadn't made a good job of this didn't realise the inevitability of it—on the other hand those who were real comrades understood the necessity of it and took it well. KATZ gave as an instance of this the young fellow (Paul) EISLER who married the daughter of the man who owns the News Chronicle, Layton[28] (Jean LAYTON). He'd been private secretary to GOTLOV[29] and on the basis of the precautions against everybody who had been outside they'd removed him from his post and put him in a factory and how well she'd taken it. She said she was delighted with it because she said it was the first time her husband had ever worker[30] on regular hours.[31] IVOR says KATZ was putting himself in the context of that of course from his clearing out of (RUDE PRAVO[32] phon.) IVOR thinks 'they' are under heavier tension than any of the others.

BOB explains that is because they are not such experienced people.

> IVOR goes on they have such a vulnerable frontier with the Americans concentrating on it.
>
> BOB answers you judge that from them here. Says, a bit sadly, our friend KAVAN[33] is also in jail.

Apparently Katz—who knew the conveyor-belt interrogation routine only too well—immediately signed a full confession and was promptly executed.

Haldane, for one, was certain that Slansky and his co-accused were guilty and had got the punishment they deserved, or so he told his incredulous colleague Peter Medawar.[34] Logically, JBS would have been bound to see Montagu hang too, if Comrade Ivor had been so unlucky. But Montagu survived this scare and went on to upgrade his Bulgarian medal to the Lenin Peace Prize (1959) and to holiday often in Mongolia. The inevitable travelogue featured homely yurts, a beaming Ivor, and blissful communal farmworkers.[35] As late as the 1970s, he was to be found writing scathing dismissals of the notion that Stalin had in 1940 ordered the massacre at Katyn of 20,000 Polish officers, and lashing out at the "Solzhenitsyn Industry."[36] The first installment of his wistful autobiography *Younger Son* (1970) made no mention of his days in the X Group. Death without benefit of reanimatology in 1983 meant that the second installment was left unfinished, but those who have consulted the manuscript say that it, too, makes no mention of INTELLIGENTSIA, BARON, or NOBILITY.[37] To save some blushes, he was never prosecuted.

As for Haldane, he never let go of Stalin. The Society for Cultural Relations with the USSR passed him a letter of condolence to co-sign when Stalin, "one of the great men of world history," died in 1953. He was glad to do so. A letter to a friend written during his final days in India shows that he did not even accept Khrushchev's renunciation of Stalin in his secret speech of 1956, which had flushed so many of his old comrades in the Party out of the fold. "I certainly don't go all the way with Khrushchev. As you know, I disagreed, during Stalin's lifetime, with some of his actions. But I thought, and think, that he was a very great man who did a very good job. And as I did not denounce him then, I am not going to do so now."[38]

If Haldane ever criticized Stalin, as he claimed, in any important respect, in any known forum, no trace has ever surfaced. The treasure lies in the opposite direction—Haldane was always drawn to violence. Khrushchev's revelations provoked only regret: the inconvenient details were now generally, undeniably, known.

10. LYSENKO AND LAMARXISM

Since Lysenko's victory over Vavilov and the rest of the mainstream geneticists in the Soviet Union, his reach had steadily expanded. Those of his opponents who were not shot, imprisoned, or otherwise done away with were forced to disperse through the sprawling reaches of the state-science complex of research institutes and universities. By 1948, the dispersed biologists had regained enough confidence to mount a fresh challenge to Lysenko's increasingly obscure theories, provoking what may be thought of as the second major crisis of genetics in the USSR. The reaction of the Communist Party to this put Haldane's own communist loyalties into an ideological wind tunnel where no deviation of form could be concealed.

Haldane had been one of Lysenko's earliest and most ardent supporters in the West. "A Great Soviet Biologist," a *Daily Worker* article from the late 1930s, was entirely devoted to the Ukrainian's scientific advances.[1] It is an important early statement of Haldane's approach to the matter, and shows his style of popularizing science. Since it has been overlooked in the literature and is now rather hard to find, its text is given in full here.

> Most of the prominent scientists of the Soviet Union are over forty. So they were at least partly educated under Tsarism, and many had also studied abroad. The younger men and women,

who hardly remember Tsarism and take Socialism for granted, have seldom done enough to achieve international reputation. But Lysenko is an exception. He is the son of a peasant, and only thirty-nine years old; and so far as I know, his first work was published in 1928 in the Caucasian Republic of Azerbaijan.

He has played a great part in the improvement of Soviet agriculture, and although some biologists doubt his theories, there is no question that his practical methods work. One of the main lines of crop improvement in the U.S.S.R. has been the selection of the best races of plant and the production of new ones by crossing. Here Vaviloff did great things for his country.

And, of course, the development of the electrical and mineral industries has assured a supply of chemical fertilisers. But Lysenko has studied, not so much the plant or its environment, but the relation between the two. Here is the problem. Everyone knows that seeds do not always germinate the moment they are sown. For, of course, wild plants generally sow their seeds in autumn, while they come up in spring.

Again, annual plants flower in their first year of growth, whilst others wait much longer. Thus if you try to grow a tulip from seed, instead of from a bulb, you will have to wait for at least seven years. Sometimes we want a plant to flower in its first year. For example, maize often fails to do so in England, and is killed by frost before it is any use. Sometimes we do not want it to flower. For example mangolds which flower in their first year use up the material stored in their roots to make flowers and seed. This is called "bolting," and if you look at a field of mangolds in early autumn you will generally see a few bolters.

Now Lysenko distinguishes sharply between development and growth. The seedling does not grow during the winter. It looks no different in March to what it did in October. But it has undergone an internal development which enables it to flower at the proper time. Similarly, all mangold and maize plants grow. But only some undergo the internal development which is needed to make them flower.

A very great deal of work has been done on the conditions for plant growth. But Lysenko was the pioneer in working out the conditions for development in seeds, while Garner and Allard in America did the same for the later stages in development. Their work has been extended by Lysenko and other Soviet biologists, such as Razumov and Liubimenko.[2]

Lysenko's main practical problem was this. The summer is often so hot and dry in the Ukraine that the wheat plants may be damaged or killed unless they form their ears before the end of June. So rapid development is essential. Now, in England all kinds of wheat can be sown safely in the autumn, and the seedlings are not damaged by the winter frosts.

But this is not so in Canada and Russia, where the winters are very cold. The hardy but generally slow-growing forms which can be sown in autumn are called "winter wheats." The more delicate forms, which must be sown in spring, in some places as late as May, are called "spring wheats." If a winter wheat is sown in spring it may not produce ears at all, or may do so very late in the season. So many kinds of wheat which are useful in other countries are no good in the Ukraine.

Lysenko set out to treat the wheat seeds, before sowing, so that they could be sown in April, and yet get off the mark with a flying start, so to speak, and flower in June. The method, which is called *yarovizatzia*, or vernalisation (from the Latin *ver*, spring), differs for different wheats, but is as follows for some varieties:

The wheat is watered and kept at about 50° F. for twenty-four hours until a few seeds begin to sprout. Then it is spread out about 6 inches deep on the granary floor and the door and windows opened at night till the temperature falls to about 1° above freezing-point. The granary is shut in the day-time to keep it cool, and the seed stirred every day for a fortnight to a month, when it is ready for sowing.

As an example, one kind of wheat from Azerbaijan, if sown in the ordinary way, formed ears so late when sown at Odessa that it only gave 8 per cent. of the yield of a local wheat. When vernalised it ripened three weeks earlier, and gave a yield of 41 per cent. above the local variety. Of course, this technique, which requires a thermometer, good ventilation and careful weighing and measuring, is beyond the resources of an individual peasant, but quite easy on a big collective farm.

By 1937, 22 million acres were sown with vernalised crops. For the method does not apply to wheat alone. Maize must be kept in darkness for a fortnight at about 70° F. before sowing. In the case of potatoes, the aim is not, of course, to encourage seed formation, but that of tubers, and the treatment is quite different. Instead of keeping them in the dark, they are threaded on string, and hung up in a greenhouse at 60–70° F., exposed to sunlight by day, and electric light at night. The idea of continuous lighting was due to Garner and Allard, but Lysenko and Dolgushin showed that it could be applied to the seed potatoes in an economical way.[3]

Of course, we are only at the beginning of an understanding of the change which occurs in seeds and potatoes during vernalisation. These are being studied by biochemists in the Soviet Union. When they are worked out, still greater improvements will be possible. Lysenko is also tackling problems in plant breeding. Here he is engaged in a lively controversy with some of the older workers. This has not been very fully reported in English, but I hope that he may be able to attend the International Genetical Congress at Edinburgh next August to describe his work.[1]

Lysenko is not only an Academician, but a Deputy in the Supreme Soviet. He believes in a flying start for boys and girls as well as wheat and potatoes. "In our Soviet Union," he said "people are not born. Organisms are born, but people are made here—tractor-drivers, motor mechanics, academicians, scientists. I am one of these men who were made, not born. And one feels more than happy to be in such an environment."

[1] Unfortunately this was not possible.[4]

Haldane uncritically accepted all of Lysenko's claims about the benefits of "vernalization," including the imaginary harvest yield increases that his colleagues in the USSR had been so skeptical of, before those critics were swept aside and under. The general reader would not know that the practical matters of crop production were very far from Haldane's remit, and that his career was built rather on the theoretical analysis of experimental data about genetic traits collected by others (leaving aside his physiological self-experimentation).

On the more delicate matter of Lysenko's claim to have transmuted varieties at will—the "problems in plant breeding" that Lysenko was "tackling," and which had led to a "lively controversy with some of the older workers," even if many of the participants were no longer that "lively" themselves—Haldane employed a dodge that would also prove useful in the future: *waiting for more data but promise to report back then.* We will return to more examples of this later. Is it possible that Haldane did not grasp the full implication of Lysenko's boast that "Organisms are born, but people are made here"?[5] That was of a piece with Lysenko's abolition of varieties, now applied to humans.

The supposed bounty of vernalization and the emergence of new varieties through its application periodically cropped up in subsequent years. In July 1940, Lysenko's speech to the 1939 VASKhNIL conference was translated and published in the West, drawing critical attention. Haldane responded that the "capitalist press" and "Hitler's friends" were blackening the name of Soviet science.[6] They could not see that Lysenko was "world-famous" as "the inventor of vernalization" and practitioner of "vegetative hybridisation." He assured his readers that in the West, unlike the USSR, "our leaders despise genetics." In 1944, Haldane was confident that "Lysenko's work is being applied, not only to speeding up the development of existing varieties of wheat, but to making new varieties . . . vernalization and the breeding of new early varieties have made it possible to grow wheat in northern regions of the union where summer is short. This will undoubtedly save many lives."[7] Recall also his odd argument, quoted previously, that the destruction of Vavilov's world-famous collection of plant varieties was not much of a loss since "Lysenko's invention of vernalization rendered many of them less valuable than they were before."[8]

Addressing general attacks on genetics, now too widely reported to be ignored, Haldane preferred to create diversions.

> Two first-rate Russian geneticists have refused to return to their country and are occupying positions elsewhere, Dobzhansky[9] in Pasadena and Timofeeff-Ressovsky[10] in Berlin. In the Soviet Union Tsetverikov,[11] Agol, and Levit have lost their posts. Agol is alleged to have been imprisoned, or even executed . . . Dobzhansky and Timofeeff-Ressovsky got good jobs abroad, as dozens of British scientists have done in the last twenty years without any suggestion that British science is persecuted. Tsetverikov was a serious loss to research. The other two dismissed workers had not done work of great originality. But several good British geneticists have recently lost their posts, one for marrying a Chinese wife, another for trying to expose corruption in an institute, and a third for disproving one of his professor's pet theories. Similar events have occurred in America.[12]

His technique relies on the fact that no more than prima facie plausibility is required. Close examination by the general reader is unlikely, but shows that the comparisons are hollow and devious. British scientists may be "compelled" to seek positions abroad, but not through fear of prison sentences or execution, which is what kept people like Timofeev-Ressovsky in Berlin at Vavilov's suggestion, which had been relayed through Hermann Muller. The British geneticist who supposedly "lost" his "post" for "marrying a Chinese wife" was Sydney Cross Harland, well known to Haldane.[13] He was perfectly aware that Harland had actually arranged his own dismissal from the firm that employed him by suing them for libel, after taking care to accept a job in South America beforehand, a story that

Harland recounts in his autobiography; it had nothing to do with the fact that his wife was Chinese, even though that caused social difficulties in color-conscious Trinidad, where Harland was based at the time.[14] But this would not be the last time that Haldane deployed the Harland story.

The comfort to be gained from the reassurance that Agol and Levit "had not done work of great originality" is obscure when it is remembered that Agol and Levit were *shot*, and that Haldane had met them both. Recall his casual description of the arbitrary execution of his fellow detainee during the Spanish Civil War, when Bethune's entire Blood Transfusion Unit had been under arrest. However, there is another loose end here.

Hermann Muller had worked with Levit in Moscow shortly before Levit's opportunity for originality had been foreclosed. Later he wrote a friend that he had forwarded a paper on mathematical genetics and heterosis, written by one of Levit's students, to Haldane for possible publication.[15] The paper vanished, and Muller suspected that Haldane had discreetly dropped it for fear of embarrassing the regime.[16] The student in question, Vladimir Pavlovich Efroimson (1908–1989), had been sent to the Gulag. In fact, he did *two* terms of hard labor: first from 1932 to 1935, after he was arrested when Levit's institute was first purged; and again, after criticizing Lysenko, from 1948 to 1955. Somehow he survived both horrors. However, his vanished paper was not on heterosis per se but rather on the estimation of genetic mutation rates.[17] It had been written just before Efroimson's first spell in the Gulag, around 1932. Haldane published his own calculation of mutation rates in 1935.[18]

Did Haldane squash Efroimson's paper, as Muller suspected, because of Efroimson's embarrassing Gulag imprisonment? There are too many uncertainties to tell: exactly when he received it, if at all; what his personal politics were like at the time of receipt; whether there were other grounds for rejecting it; or whether Muller's antipathy to Haldane at the time he wrote the letter clouded his judgment. However, Levit himself thought Efroimson's paper was more general than Haldane's and worth publishing.[19] Efroimson went on to author several books, including *The Genetics of Genius,* which was published posthumously in 1998. It was possibly a work of some originality, as Efroimson was an early sociobiologist and could not publish material like that in the USSR.

Returning to Lysenko, Haldane volunteered himself for correction. "I have little doubt that when I taught genetics (owing to the war I no longer do so) I made a number of misleading statements. I should be a better teacher if these were pointed out in a public debate to which I could reply." There was, Haldane averred, equivalent criticism of rival theories in journals in the West—equivalence through naming—but, incredibly, "such attacks are not hot news in New York or London, because the publics of those cities are much less interested in genetics than is that of Moscow."[20]

This led Haldane to wonder if perhaps, as the result of Hermann Muller's persuasion, Soviet geneticists really had spent too much time on "such questions as locating genes in chromosomes, rather than in finding out how they act in the development of an individual, or arise and spread during evolution of a species." "It may be that under the stimulus of so brilliant a teacher as Muller, an unduly large fraction of the younger Soviet geneticists had occupied themselves with formal genetics." The stuff was being taught the wrong way. "Some of Lysenko's points are, I think, valid against genetics as often taught, rather than against the theories held by competent geneticists."[21] In truth that is not what Lysenko meant at all, since he had no time for the "Menshevizing idealism" of genes as a concept, nor did he have any desire to be taught about them. But it all looked a lot more reasonable when it was dressed up by Haldane.

The specific question of the Lysenkoist transformation of varieties proved a lot stickier for Haldane to handle, especially in more technical forums where hard questions might be asked. It was contrary to orthodox genetics, which had banished Lamarck's inheritance through use, and required stochastic changes to the sex or germ cells in order to alter the nature of offspring. Lysenko, having no use for "Mendelist-Morganist" concepts, offered no coherent theory, apart from repeatedly invoking "development." Here Haldane did his best over the years to hedge, holding out hope for an orthodox interpretation that would vindicate Lysenko if things were just formulated the right way.

Thus, one of Lysenko's more offbeat claims was that he could actually produce new varieties by *grafting*, so that the results bred true. He had supposedly done this with potatoes, which are in the nightshade or *Solanaceae* family. In his 1940 lectures on genetics, Haldane hedged with a hopeful tautology. "If Lysenko is correct, there is no sharp line in the *Solanaceae* between viruses causing disease and graft-transmissible factors responsible for morphological characters."[22] This neatly ignores the question of whether Lysenko had offered *any proof at all* for viral-style transmission of heredity in potatoes, which would indeed be an extraordinary discovery. Good reasons for believing in an effect are advisable before speculating about its causes.

Haldane pursued much the same type of hopeful interpretation, but now with transfer to other anomalous cases, in an article devoted to Lysenko in the Marxist journal *Science and Society*. The article was provoked by reports printed in the previous number of that journal from the 1939 Moscow Conference on Genetics and Selection, including large chunks of Lysenko's speech. Haldane submits that Lysenko often "goes too far" in his criticisms of genetics but works hard to hold out for "some" cases where Lysenko's non-genetic transmission might work, arguing that they were not impossible. This fudges the obvious fact that Lysenko's claims were *overwhelmingly unlikely* given extant knowledge, and that peculiar instances of heritability in extremely rare instances would actually make no material difference to the probability *of the instances Lysenko really was claiming*, like the creation of new varieties by grafting fruit trees. We will consider one of Haldane's attention transfers in detail because it cropped up several times in his writings on the subject.

I think that nine times out of ten Lysenko is wrong, that is to say that you cannot improve a breed of animals by improving its food. But there are cases where this is possible, and they may be common enough to make Lysenko's principle of great practical value. The clearest of such cases was discovered at Bar Harbor, Maine, by Little's group of workers on mouse genetics and has been specially studied by Bittner. For many years they had kept different pure lines of mice. Each line had a characteristic liability to mammary cancer in females. In one line 90 per cent of all females who did not die of some other cause before the age of two years would develop this disease, in another line only 5 per cent. The members of the immune line were no more likely to develop it if they were caged for months with the susceptible line. The liability seemed to be hereditary. But it turned out that if the young of the susceptible line were separated from their mothers at birth and suckled by immune females they were much less likely to become cancerous. And this partial immunity is handed on to their children.

Nothing of the kind has been discovered for other forms of cancer. And I believe it to be a rarer phenomenon than Lysenko supposes. But it is futile to deny its existence and to regard Lysenko's assertion of its possibility as in any way unscientific.[23]

This is a transfer of attention to an experimental anomaly never referenced by Lysenko, abstracting "Lysenko's principle" away from his own putative examples and neatly sidestepping the question of whether he had conducted an experiment at all, or whether, say, spring wheat *really* can be transformed into true-breeding winter wheat by vernalization—but look over there! Haldane knew that the most probable explanation of the Little-Bittner anomaly was that there was transmission of a carcinogenic virus through suckling. This was strongly suspected at the time, as the cancer switched tendencies in both groups; the previously susceptible also benefited from being suckled by the non-susceptible. Bittner himself attributed this to a "milk factor" that we now know to be the MMTV virus.[24] Haldane makes no mention of the viral explanation here, as it would ruin the effect of introducing the anomaly. But elsewhere he *does* mention exactly this possibility: in his 1940 lectures on genetics mentioned above! "The epoch making work of the Bar Harbor group has shown that 'spontaneous' cancer not only has a genetical basis, but at least in one case depends on an agent, perhaps a virus, which can be transmitted by the milk."[25]

In 1944, criticizing an article by Cyril Darlington in *Nature*, Haldane shifted the burden onto the doubters by splicing in an imputed fallacy, complaining that "Lysenko and his colleagues in the U.S.S.R. have reported a large number of cases in which characters have been transmitted from one plant variety to another by grafting" but that Darlington had ignored his work. "It may be that Lysenko, with the enthusiasm of a pioneer, has criticized Mendelian conceptions unjustifiably, as the biometric school did in Great Britain. But this is scarcely a sufficient reason for ignoring his work, particularly as some of his publications antedate that cited by Dr. Darlington."[26] But rather than making the dubious inference imputed by Haldane, might Darlington not have ignored that "work" because it provided no substantial data in the form of controlled experiments beyond the claimed results?

At other times, Haldane enlisted what might be called the *argument from overcorrection*. "There can be little doubt that some Soviet plant breeders had underestimated the importance of environment. Lyssenko,[27] who has produced most conspicuous results by changing the environments of plants, has probably swung the pendulum too far in the other direction, but I do not doubt that it will settle down."[28]

By now this ongoing defense of Lysenko by shifts and contrivances was exasperating many of Haldane's colleagues at home. A *Daily Worker* reader, Professor F. G. Gregory from the Research Institute of Plant Physiology at Imperial College, was moved to write to him in August 1947 to complain. "Lysenko has spoken a great deal of nonsense on vernalisation, just as he has about genetics, and I really cannot see why you should give all his utterances your official blessing. I have never had the opportunity of talking with you about vernalisation . . . but should you care to hear from me what the position with regard to vernalisation really is I would only be too pleased to discuss the matter with you."[29] What Gregory meant was that experience, including his own, had shown that vernalization was not worth the trouble in practice.

At home, Lysenko was coming under steadily increasing pressure. The chaos of the war and the deep layers of deceit that the Soviets called *tufta* could disguise the variance between his theories and the way things worked in the real world, but only for so long. His promised novel varieties, to be delivered in as little as two or three years, did not materialize, and vernalized spring wheat (he was vernalizing almost everything by now) simply didn't work in Siberia. The dire state of agriculture was worsening, not improving, after the war. The technical specialists and the regions of the research establishment not under his direct control began to notice, and by 1947 open challenges to his theories were emerging. This sort of tension—an *internal contradiction,* in Marxese—would periodically recur throughout Lysenko's career until his final eclipse, and could only be contained by political action. If Lysenkoism didn't work as a scientific theory, it would be made to work as a political theory. Those who had the bad manners to notice that the promised results never materialized simply had to stop noticing. Crucially, Lysenko still had the backing of the politicians, and of Stalin in particular.

An article in a "bourgeois" journal in the West provided the impetus. The geneticist L. C. Dunn had written a glowing recollection of Soviet science, based mainly on his visit to the USSR back in 1927, to inform the readers of *Science* about its state there in 1944.[30] Dunn was enthusiastic about "the control and organization of science by and for the whole community," and the "vigorous science" that could "survive in a socialist state" despite naysayers. This provoked a rejoinder by Karl Sax, who pointed out the widely reported persecution of Vavilov and orthodox genetics, which Dunn had passed over in silence.[31] News of this exchange filtered through to the USSR, and in 1945 a reply to Sax appeared in *Science*, written by Anton R. Zhebrak (1901–1965), who was at the Timiryazev Agricultural Academy of Moscow. Zhebrak assured the world that "the science of genetics is making progress in the USSR," pointing out that there were a "number of important genetics laboratories" that were "doing good work." This despite Lysenko's "naïve and purely speculative" criticism, which he implied was not taken seriously. The pill was sugared with the observation that Lysenko had nevertheless made "a number of practical suggestions which have been of great value."[32]

Back home in the USSR, this response was not taken kindly by Lysenko's allies. Two years later Zhebrak was unceremoniously slapped down for "unpatriotic acts" in the state organ *Pravda*: first by Prezent,[33] and then by the agronomist Professor Ivan Laptev and others.[34] For a while Zhebrak went into hiding. He was dismissed from the presidency of the Byelorussian Academy, but otherwise escaped harm. This encouraged the latent opposition.

The controversy was followed with interest in the West. During the war, the British geneticist Kenneth Mather had kept up scrutiny of Lysenko in *Nature*.[35] Lysenko's book *Heredity and Its Variability* (1943) had also been translated into English by the exiled Russian geneticist Theodosius Dobzhansky and circulated in manuscript by late 1945 for critical review, to increase interest in the West—this translation was eventually published the following year. Lysenko had sent the Russian edition of the book to Haldane in 1944, with a cheerful cover letter.[36]

Dear Professor Haldane!

I have read with highest interest your book 'New Paths in Genetics', which according to my request has been translated in Russian and is now available in manuscript.

I am sending to you my book on Heredity and its variation,

which in many points differs essentially from the orthodox Mendel-Morganistic genetics.

Sincerely Yours,

President of the Lenin Academy of Agricultural Sciences
acad. T. D. Lysenko

Haldane replied that he could not read the language, adding somewhat obscurely that he would "perhaps" agree with some of Lysenko's views, and tactfully asking for experimental evidence.[37]

Dear Professor Lysenko,

Many thanks for your book. Unfortunately my knowledge of Russian is very poor, and at the present time it is difficult to find translators who also know enough biology for the purpose. I should like to add that the book of mine to which you refer so kindly was written in 1941, and that as a result, if I wrote it again today, I should modify my views in several respects, perhaps bringing them nearer your own. Unfortunately it is extremely difficult in this country to obtain accounts of the actual experimental results on which your views are based, and I should be very grateful if I could obtain some reprints giving accounts of the facts in question (on genetics), as well as the general conclusions based on these facts.

During the war I have done little genetics, as I have been engaged on problems of human physiology arising in connection with the war. However, I hope shortly to begin full-time work on genetics again, and it would be of great value to me to have access to your reprints.

Yours sincerely,

J. B. S. Haldane

Now Haldane was offered Dobzhansky's English translation of *Heredity and Its Variation* to review, by a group of geneticists based in America: Kurt Stern, Salvador Luria, Hermann Muller, L. C. Dunn, Milislav Demerec, and Dobzhansky himself.[38] Haldane refused outright, causing Muller to break off relations with him.[39]

A well-informed article in the *Nineteenth Century* by Haldane's former colleague Cyril Darlington, "The Retreat from Science in Soviet Russia," proved the most provoking. Darlington, unlike Haldane, had taken the trouble to read the English translation of *Heredity and Its Variation*. He summarized the controversy in detail, and Lysenko's role since its inception, along with an account of what was then known about Vavilov's arrest, trial, and disappearance into the Gulag, as well as the general persecution of geneticists that accompanied and preceded Vavilov's demise. After an accurate précis of Lysenko's blossoming inspirations, Darlington concluded that "It may be difficult in such a man to distinguish between the enthusiasm of the charlatan and the frenzy of the fanatic."[40]

Laptev made sure to single out Darlington for especial opprobrium as a foreign troublemaker, but Darlington had provoked even stronger reactions in the UK from the small group of geneticists in the fold of the British Communist Party. One of these, Angus Bateman (1919–1996), wrote a hurried letter to Haldane, asking him if something could not be done about Darlington.[41]

6th Nov.

Dear Haldane

I suppose you have read Darlington's latest in the '19th Century and After'. He qualifies for membership of the Committee on un-American activities! A full dozen Russian geneticists who have died or faded from public view in recent years are assumed to be the victims of oriental despot Lysenko. He treats all other existing geneticists as worthless. A week or so ago I heard him denounce Zhebrak as 'another gangster'. Now that news reaches us that Zhebrak has been denounced in Pravda (presumably by Lysenko) he has been elevated to martyrdom. Do you not think it undesirable that this anti-Soviet tripe should be pedaled by a leading British cyto-geneticist without his being openly disowned by his colleagues? I have been thinking whether it would not be a good plan for a resolution to be tabled at the next Genetical Society meeting. It might state that the members of the G.S.[42] while not wishing to defend Lysenko's views on genetics are entirely opposed to attempts to insinuate without any evidence that it is regarded as a political crime to be an orthodox geneticist in the USSR—or words to that effect.

I could not associate myself openly with such a resolution. You will know that it would be more than my job is worth, and I have a wife and child to keep. Perhaps you would be prepared to move it. One can't help feeling that if C.D.D.[43] believed all he wrote and was really interested in the welfare of Russian science he would desist from these anti-Soviet tirades which can only inflame Lysenko's antagonism to anyone who professes similar genetical beliefs to Darlington and strengthen his case in the eyes of the Soviet public.

Darlington is notorious for scathing criticism of his colleagues behind their backs. I feel that he can indulge in his slanderous attacks on Russians not merely with impunity but is at the same time hailed as a bulwark of Western civilization and gets as much publicity as he could wish.

Now I have got that off my chest I want to come to the point of my letter. Of course, if the G.S. is of such a nature that the resolution would be rejected by an overwhelming majority it were best not moved. . . .

No such resolution seems to have been introduced at the Genetical Society. Bateman would go on the following year to do influential work on sexual selection among fruit flies that would no doubt have made Lysenko even angrier.

In 1948 the anti-Lysenkoists found a more powerful champion in the form of Yury Zhdanov, who had married Stalin's daughter and was a son of the Politburo member Andrei Zhdanov. Yury had studied genetics himself, and evidently felt that his political connections would shield him when he launched the first serious public attack on Lysenko on April 10, 1948. In this he did no more than notice what others had noticed, and point out that spurious appeals to Marxism and anti-fascism would not change the facts on the ground or in the lab. Lysenko, who had a well-developed sense for the ramifications of this kind of criticism, promptly wrote a letter to his patron Stalin, resigning his presidency of VASKhNIL—an unprecedented step.

Stalin had a vested interest in Lysenko, who had eagerly agreed to develop a miraculous "branched wheat" that would vastly increase yields. Since this had been Stalin's own suggestion, and he had even supplied Lysenko with supposed "branched wheat" seed that had been sent to him by peasants from his native Georgia, Stalin depended on Lysenko's success. The "branched wheat" was a long-standing myth, but Lysenko compensated for this in the same way that he always had, by simply reporting large harvests using it, and claiming victory. A Politburo meeting was called. Stalin lit into critics of Lysenko, and Zhdanov in particular. They had "insulted" and "defamed" Lysenko. Emboldened, his protégé now sought complete demolition of the foreign-influenced "Menshevizing idealists."

A session of VASKhNIL was arranged in August 1948, packed in advance with new members elected by Lysenko himself. His opening "report" to the congress was privately approved and annotated by Stalin.[44] Stalin did not notice Lysenko's bad Russian, but struck out his more outrageous claims, such as "any science is based on class," which Stalin marked up "Ha, ha, ha . . . And mathematics? And Darwin?"[45] The "report" merely redoubled political phrase-mongering and attacks on the motives of Lysenko's critics. The speeches that followed were assigned to Lysenko's supporters, who called for police action against his critics, to "cleanse" the institutions they infested. Western geneticists and statisticians were excoriated, including Francis Galton, Karl Pearson, R. A. Fisher, and Sewall Wright. But J. B. S. Haldane was not mentioned. Raissa Berg later suggested that this was because Lysenko was counting on Haldane's support.[46]

Few participants dared to speak out against Lysenko, partly because most critics had been excluded in advance. Those who did were repeatedly shouted down; they later found that their contributions had been largely edited out of the published transcript. By the last day of the conference, *Pravda* had printed Zhdanov's personal repentance before Comrade Stalin: "I unquestionably committed a whole series of grave mistakes."[47] The congress officially adopted Lysenko's edicts, "On the Situation in Biological Science." The requisite purge of anti-Lysenkoists proceeded, enforced all over the country by Communist Party inquisitions, with obligatory public recantations. Within months this process would sweep around the world to "cleanse" orthodox "Marxist science" everywhere.

Haldane was behind the Iron Curtain at a conference in Budapest when the news about VASKhNIL broke; he had just come from the World Congress of Intellectuals for Peace at Wroclaw in Poland. *Time* magazine tracked him down there for commentary, but he pleaded a "lack of exact information." "Until he could be sure, said Haldane, that Lysenko's current theories are unscientific and that opponents had been punished for disagreeing, he would make no decision." But when he lectured on genetics in Prague in the coming winter, "I shall say just what I think . . . and if what I say does not agree with Lysenko, it's just too bad."[48] That promise was never put to the test, since Haldane ultimately canceled the Prague lectures. On his return home to the UK, he discovered that he was now required to fall in with the Party line on Lysenko, whether he liked it or not. Declaring him *good in parts*, as he had loyally done in the past, would no longer cut it.

Several prominent Western scientists resigned their "corresponding" memberships of the Soviet Academy in protest at these latest developments, including Hermann Muller. Haldane was a corresponding member too, having been elected in 1942. A few years earlier, in response to a claim made by H. G. Wells, Haldane had himself stated that "It is alleged that where, as in the Soviet Union, almost all research is paid for by the State, the free expression of scientific opinion is necessarily checked. . . . As a member of the Academy of Sciences of the U.S.S.R., I should have protested had this been proved correct." Having satisfied himself of the baselessness of the claim that any such suppression existed—his sources included, he stated, White Russian exiles—he had no need to.[49] Now he neither protested to the Academy nor resigned, and appears to have remained a member for the rest of his life.

Responding to growing publicity about the affair in newspaper editorials and journal articles, the BBC arranged a radio symposium on the Lysenko affair, which was first broadcast in the early evening of Tuesday, November 30, 1948.[50] A brilliant panel had been assembled: R. A. Fisher, the leading statistical geneticist in the world; Cyril Darlington, the cytogeneticist who had made fundamental contributions to the study of chromosomes; Sydney Cross Harland, a fellow of the Royal Society and pioneer agro-geneticist; and Haldane himself. Each contribution had been recorded separately, so that the element of an adversarial debate was removed and each had to address the question at hand, despite Haldane's protests to the producers.

Harland was put on first, perhaps because he was the only panelist who had personally met Lysenko, during a visit in 1933 that had been arranged by Vavilov. "In Odessa we went to see a young man called Trofim Lysenko, who Vavilov said was working on the vernalisation or treatment of seeds in order to secure earlier maturity or greater productivity. I interviewed Lysenko for nearly three hours. I found him nearly completely ignorant of the elementary principles of genetics and plant physiology." Moreover, "to talk to Lysenko was like trying to explain the differential calculus to a man who did not know his twelve times table." Worse, he was "what I should call a biological circle squarer." More ominously, Vavilov had confided to Harland that Lysenko was of an "angry species."

Darlington's contribution was characteristically crisp in its summary of the affair to date, and perceptive about its xenophobic aspects and Lysenko's attempts to improve Darwin on the same lines as Prince Kropotkin the cooperator. "Last year . . . Lysenko had asserted that there was no such thing as competition between members of the same species. Even pigs and potatoes, it seems, have the right communistic attitude; even cannibals co-operate in a friendly way with their victims." Darlington also dismissed the idea that Lysenko was a fluke. Rather, he was "the inevitable end of the long persecution of science in Russia," but less shocking than "the indifference to this persecution on the part of our own scientists"—an obvious reference to Haldane himself. He closed with a call to support the dissidents within Russian genetics.

R. A. Fisher was less inclined than either Harland or Darlington to write Lysenko off as a mere crank or yokel, feeling that Lysenko's "mind does not seem to work in either of these ways." Rather, Lysenko came across to Fisher as a more straightforward dogmatist and megalomaniac, unembarrassed by the need for any evidence, unconcerned that he might be refuted with ease by anyone who cared to graft tomatoes for a while, but happy to make up for this with political threats and invocations of Stalin. "The reward he is so eagerly grasping is power, power for himself, power to threaten, power to torture, power to kill." In the light of subsequent knowledge, this was an insightful remark.

Haldane was up last, and the most confrontational of the four. He disagreed with many of Lysenko's points but agreed with some. He held out for the possibility of heritable acquired characters in rare cases, which he speculated would not obey Mendel's laws. Lysenko might derive support from the fact that X-rays and other treatments can alter genes, gently sliding past the difficulty that those are not acquired characters, but rather random mutations. Lysenko's more unusual claims, he conceded, would be "very revolutionary" if they were true, but more evidence was required before minds could be made up. Still, he was "inclined to believe" in them, supposing that Lysenko had somehow altered germ cells on shoots by altering the temperature; perhaps others had not replicated those results because they had not done the experiments the right way. More reasons to believe could be derived from the fact that the Soviet government was backing Lysenko, and they had done well at fighting the Germans. He was ready to grant that Lysenko had been "quite unduly critical of earlier workers," but then R. A. Fisher himself had been critical of Mendel when he questioned the too-perfect ratios Mendel claimed in his sweet pea experiments, "a charge at least as serious as Lysenko's accusation of idealism." And there were some scientists in England who agreed with Lysenko. Regarding persecution of scientists in the USSR, he was "very sorry to hear" that N. P. Dubinin's laboratory had been closed, but the situation in London concerned him more, since there was "no regular practical course in plant genetics" to be had there. *Samizdat* humor suggests itself.

> *Comrade Haldane,* leaping energetically in the air and waving his hands: Comrades, comrades! Come quick, someone set my house on fire!
>
> *Comrade Dubinin,* bending double and rocking with laughter: Now, now *tovarich*! You know perfectly well there are thousands of people in Moscow who cannot even get a light when they need one!

Haldane went on to include the dissimulations about the fate of Vavilov that we have noted above. His performance was unconvincing, especially when measured next to the other participants. Where he bobbed and weaved, they took definite positions. The evasions he offered were much the same as his previous attempts to avoid commitment. Equating "criticism" by Fisher with "criticism" by Lysenko would not fool many, given that one involved hurling denunciations that led to the Saratov prison, while the other was stated coherently enough to roil a tea party at Trinity College, but not much more.

Curiously, Haldane's *waiting for more information* bluff was called by a few patient listeners. Four years later, one of these, a schoolteacher at Eltham, politely wrote to him, noting that Haldane had reserved judgment in 1948 during his BBC broadcast and wondering if he had by now digested the relevant "printed matter" so that his considered opinion could be given. Haldane replied that, alas, he was too busy at the moment.[51] Other requests elicited similar responses.

The Engels Society unanimously endorsed Haldane's BBC talk, but the Party was far from pleased with Haldane's performance. A meeting of the inner circle held on December 1 attracted the interest of MI5. Emile Burns and several others had met with Haldane about his views, and "EMIL said that for the first time he had had the impression that HALDANE was being dishonest in the expression of his views." Johnny Campbell said that "his own attitude on the whole controversy would be to defend the right of the Soviet Government to intervene." Emile went on to say that "he had felt HALDANE was being dishonest because instead of trying to see what there really was in LYSENKO's theory, he only tried to prove that it was illogical, contradictory and unscientific." He would have to have a talk with Haldane. James Klugmann (1912–1977) anticipated trouble for Haldane within the Party after his speech to "the enemies," leading to "rumours, discussions and coldness," and phrases like "the silly old fool."

Bill Rust tried to defend Haldane, who had "foolishly" agreed to a public debate, but agreed that they had to "do a certain job with HALDANE." Klugmann was worried that "the fact that a well-known Communist scientist like HALDANE should have differed from LYSENKO would give the reactionaries all over the world an opportunity to attack them," and thought that Haldane "because he was not a sufficiently good MARXIST" was "letting his own position and prestige interfere with the interests of the Party." Bill Rust countered that "HALDANE, although he disagreed with the line of the Party, had in his own way tried to defend the Party and the Soviet Union." Klugmann pointed to the French, who he said had "put up a terrific show" endorsing Lysenko. As well as neglecting the "terrific conception" of the collective farm, Haldane had "failed to bring out" most of all "the Party's basic conception that you can change nature, which to the thinking man is the greatest thing of all."

Klugmann emphasized this revealing point. "The great conception of man conquering nature and LYSENKO's theory in support of this, was not brought out. This was what they ought to be fighting on." The meeting concluded with a fond speculation that, provided the Party could get enough money for advertising on the BBC and in the papers, they ought to be able to sell enough copies of Lysenko's books to retire on.[52]

Moves were already being made within Party circles in December 1948 to establish the correct line on Lysenko and genetics. A working group led by Angus Bateman drew up a document "In Support of Lysenko" and circulated it among Party members.[53] The contents are all the more peculiar if read while keeping in mind the fact that Bateman was a working geneticist himself, and would even establish an enduring reputation in years to come.

To warm up, Bateman identified something called capitalist science as "a weapon in the conduct of imperialist war, fascist racialism, the fight against the working class, and the U.S.S.R." Under the sway of this wicked influence, biology had become "limited, pessimistic, and relatively sterile." By contrast, in the USSR science "had all the means for unlimited advance," since it could dispense with "the last limitations imposed by reactionary bourgeois theory." Included in the last was the theory pushed by Mendelians, which had wrongly separated changes in the cell from "changes in the mode of life of the organism" by insisting that genes changed only rarely by chance mutations, and could only be selected if already present. This was "a pessimistic view of the possibilities of directing evolution in a way valuable to man," given that it was very slow.

Mendelian pessimism was useful to "Imperialist exploiters" bent on "showing that colonial peoples have an inherent inferiority in consequence of their possession of a poor complement of genes." It was also useful closer to home, where it inured the working class to "exploitation with its consequent malnourishment, poverty, and ill-health" by telling them that this had "no effect on their children"—that is, was not *heritable*, a new form of ideological mystification to be sure! Mendelists were also said to be committed to the idea that "every member of a species is constantly engaged in a struggle against members of its own species for the limited supply of food," an idea devised only to "divide the workers in their struggle by making each the enemy of his fellows." But this struggle was "exceptional and of little evolutionary significance."

Fortunately, Lysenko had cleared all this up. "The definition of heredity given by Lysenko as 'the property of a living body to require definite conditions for its life, and to respond in a definite way to various conditions' is accurate and complete." There was no need to wait for genetic variation, for it could be produced on demand. "Changes in the heredity of an organism are made by influencing the germ cells it will later shed. Those may be influenced by fundamental changes in the function of the organism caused by its adaptation to its environment." To forestall the objection that mere use and disuse, or wear and tear, would be inherited in this way, Bateman decreed that only changes "which affect processes involved in the formation and life of the germ cells" would be inherited. (No explanation is given as to how the *character* of the experience determines *which cells it affects* to maintain this fortunate discrimination.)

The path forward had been shown by Lysenko. In particular, "physiological treatment of the organism (graft hybridisation, partial vernalisation, distant hybridisation etc.)" is crucial for deliberately creating useful variation on which selection could operate. "Marxist theory correctly rejects a dominating role for chance in any process," we are assured. Doing so banishes "philosophical idealism, the ideology of reaction." All this is proved by the well-known advances in Soviet agriculture achieved by Lysenko.

Haldane's marked-up copy of this document shows that he found some of it hard to swallow. His scattered annotations include sporadic "evidence?," "nonsense," and similar exclamations. He added footnotes that give references to literature contradicting some of the assertions made, or ask for proof, for example, "Evidence for yield per acre or per cow > England or Denmark" (later, a comrade would point out to him that official Soviet production figures demonstrated *precisely* that). He was not convinced that mutations are rare—they happen all the time. He disputed the correctness of the Marxist exegesis. "As a Leninist I don't believe real development can be imposed on animals or plants from outside. Can do so by introducing internal contradictions or making them more important. Ignorant, sectarian." Bateman's document seems to have stalled after this cold reception.

Next, the Party called a meeting of its members concerned with genetics, at which one of its senior leaders, Maurice Cornforth, was present. Some attendees were upset that the *Daily Worker* had published an account of the BBC debate that implied Haldane was fully in agreement with Lysenko. Cornforth supposed, amiably, that there must have been a typo. Haldane suggested that the Party might use the controversy to establish an independent line from Moscow, a deviation that was coldly received by Cornforth. Once it became clear that the geneticists would not pass a resolution definitely in favor of Lysenko, the meeting ended inconclusively. This did not prevent the *Daily Worker* from issuing an "educational" pamphlet on its own, fully supporting Lysenko, which was directly based on Angus Bateman's document.

One of the truculent geneticists in the orbit of the Party (almost certainly Lionel Sharples Penrose[54]) later recalled that Haldane had responded by privately distributing a reply to the *Daily Worker* pamphlet. In fact, this response was written up as a mildly conciliatory article by Haldane, "In Defence of Genetics," in the Summer 1949 issue of *The Modern Quarterly*.[55] The gist of this was that the faults attributed by Lysenkoists to geneticists were greatly exaggerated. Genes were neither immortal nor unalterable, since they could and did evolve by mutation, but were stable enough for heredity to operate. Geneticists didn't limit themselves to statistical distributions but also studied the biochemistry of heredity. They did not claim that all inheritance is chromosomal, but Soviet claims to show otherwise needed experimental verification. He conceded that Mendel may have used language that suggested idealism, and that British genetics needed more practical experience with farmers. "We are not infallible, but we certainly do not hold many of the opinions which are attributed to us." In the meantime, a meeting of the Engels Society at Marx House had been called in February 1949 to adjudicate the dispute, with thirty to forty scientists attending, including Haldane. But no consensus emerged from this debate, either.

An analyst at MI5 summarized the impasse as it stood. "It is known that HALDANE has refused to accept the Lysenko theory in toto on the grounds that such an acceptance would not only ruin his own scientific reputation but would also do irreparable damage to Communist Party recruitment amongst scientists. The Party appears to have decided that HALDANE is too valuable a Party member to be lost on a theoretical issue." However it had "censured" him, and, moreover:

> Further information from Top Secret sources shows that Maurice CORNFORTH had an interview with Harry POLLITT on 10.1.49, at which they discussed a report that HALDANE might be going to Czechoslovakia in February 1949. CORNFORTH had mentioned the matter to the Soviet people, who took a serious view of the proposed visit, as they thought it would do a great deal of harm in Czechoslovakia, where there is an ideological struggle in progress, should it become apparent that HALDANE, a Communist, disagreed on any point with the accepted Soviet line. It is not clear whether the British Communist Party is making any definite attempt to deter HALDANE from his trip to Czechoslovakia, but CORNFORTH suggested that if it proved impossible to stop HALDANE, a note of explanation should be sent to the Czech Party to explain the matter, in an attempt to minimize any possible harm HALDANE might do.[56]

As it happened, Haldane did not go on that planned tour of Czechoslovakia. But more trouble was on the way. Haldane's reluctance to sign up for the full Lysenkoist program did not please his fellow party members, or his leader, Harry Pollitt. The growing tensions within the Party interested MI5. In March 1949, an analyst there noted

> That J.B.S. HALDANE has been under a cloud in the Communist party was demonstrated at the St. Pancras bye election, when MAHON[57] invited him to occupy the platform at one of his meetings. HALDANE replied that he would like to have done this but as he was under criticism from certain members of the Party and was under orders from Harry POLLITT to undertake no Party work with the exception of his articles to the *Daily Worker*, he was unable to accept MAHON's invitation.[58]

In July 1949, Harry Pollitt made the invitation to visit Russia that began our story. Haldane was deft enough not to accept, suggesting that he knew exactly what might await him if he accepted. But if Comrade Haldane was too busy to go on holiday, others were happy to have the chance.

Many communist scientists were quite prepared to accept the entire party line on Lysenko, including the cheerfully pliant physicist and crystallographer John Desmond Bernal. Bernal was of Irish origin and had studied at Cambridge, spending time in the renowned Cavendish laboratory, and going on to lecture at Birkbeck College in London. Long steeped in Marxist rhetoric, he had helped Haldane to found the *Modern Quarterly*, frequently contributing articles. He had published a piece that summer alongside Haldane's, arguing that Lysenko merely had rough edges because he was a practical rather than a theoretical scientist. Socialism wanted practical men who were not afraid to break with tradition.[59] Lysenko's other virtue was his holistic view of the problem, integrating theory where appropriate. All of this was vindicated by his successes in vernalization and massively improving grain yields, which Bernal compared to the successes of the Red Army in the war. Reports about persecution of scientists were, he was confident, unfounded. These were flimsy arguments, to be sure, but critics would have to prove otherwise.

Along with J.G. Crowther, Hewlett Johnson, and Haldane's old friend Ivor Montagu—perceptively described by Bernal as "the greatest expert on the minor rodents of Eastern Europe"—Bernal visited the USSR in August 1949, where he spoke at yet another "Peace

Conference," and specifically visited Lysenko. Relating his experiences to the Society for Cultural Relations with the USSR back home in London,[60] Bernal said that Lysenko's methods reminded him of Darwin. He found Lysenko's demonstrations of graft hybridization convincing; he could not say what the mechanism was that enabled it, though it reminded him of animal embryology. Bernal was quite prepared to believe in the inheritance of acquired characters. Vernalization was especially convincing. The work showed "real genius" by a man who "got results" because he worked quickly. Bernal appreciated this all the more because he too had grown up on a farm, in Ireland. Lysenko also told him that he did not believe chromosomes were very important, and wasn't exactly sure what their functions were; opinions delicately relayed by Bernal in order to sound measured and reasonable.

Bernal went on to remark that criticism of Soviet science in the West was likely just prejudice and that he thought "planned science" far superior. When an audience member asked if Bernal could clear up what had happened to the colleagues of Lysenko who happened to disagree with him, he was assured that they were now tilling different fields—Dubinin, for example, was now working on insect pests. Crowther hastened to add that this only meant that, whereas Dubinin previously had something like twelve important jobs, he now had eleven.

By now Haldane and Bernal had fallen out dramatically—at least, for the moment—over a review that Bernal wrote for the *Daily Worker*.[61] A journalist, John Langdon Davies, had written a sharp account of the Lysenko affair, *Russia Puts the Clock Back*, which pursued Haldane's BBC evasions vigorously.[62] On October 6, Bernal sent a draft review to Haldane, laced with invective: "cheap journalism"; "hypocritical nonsense"; "dirty and low"; "old stories" which are "dressed up in lurid colours" about the "disappearance, execution and even imprisonment" of scientists, for which "there is no proof whatsoever"; and so on. Bernal also pledged unconditional endorsement of Lysenko's achievements.[63]

MI5 monitored a phone call on October 8, 1949, from Bernal to his own secretary, Anita Rimel, complaining that "he had just been rung up by HALDANE who was in a most violent fury that if his (BERNAL's) review appears he would resign from the *Daily Worker* and so on. Apparently because BERNAL had said something—he could not find what BERNAL had said because he hadn't got it with him. BERNAL said that he was in a very bad temper saying things like he wouldn't tolerate things of that sort. RIMEL inquired—what sort of thing. Bernal had asked him that, after which he got wilder and so BERNAL begged him not to be so childish about it, so he rung off."[64]

Bernal's secretary tried to contact the Party functionary Johnny Campbell for help, but failed, causing Bernal to conclude, "Well, after what we've heard you know, the whole situation may collapse I'm afraid."[65] Haldane resigned from the *Daily Worker* the same day, as he had done a great many times before over other disputes, but Harry Pollitt was able to change his mind over lunch. By the end of the month, better persuasion would be available in the form of Lysenko's deputy Ivan Yevdokimovich Glushchenko (1907–1987), who had been dispatched abroad to graft the new Michurinism onto old stock.

On October 27, 1949, Glushchenko addressed the British public through the always-obliging Society for Cultural Relations at Beaver Hall. An MI5 source reported that the audience became restive— Glushchenko spoke through an interpreter. Written questions were submitted, one of which asked if Glushchenko "took any account of the work of Professor HALDANE." Glushchenko replied that "he had adopted a correct position in many questions."[66] Julian Huxley was in the audience and reported in the *Spectator* that it was all "a flow of tendentious and unscientific nonsense."[67] But his old colleague from New College Oxford and fellow Etonian was not there.

The previous day Haldane had phoned the Soviet Embassy to arrange a private lunch with Glushchenko, which seems to have been held in Haldane's office at UCL. Haldane's colleague, the medical geneticist Lionel Sharples Penrose (1898–1972),[68] was also present at this meeting. According to MI5, "at this lunch considerable argument developed, and GLUSHCHENKO was left with the firm opinion that HALDANE was unconvinced and uncompromising on the subject of Lysenko."[69] Many years later, the evolutionary theorist John Maynard Smith—who at the time was a student under Haldane and a long-standing Communist Party member—remembered haltingly that "the Russians sent to Western Europe a hatchet man of Lysenko's, called Glushchenko, who called, among other people, on Haldane, to—he arrived with a couple of goons, it was awful. He arrived in the department. And he and the goons, I can't remember, I think and the goons, were actually closeted in Haldane's office for about three hours, trying to persuade Haldane how right Lysenko was. And Haldane was absolutely unapproachable for two or three days afterwards. He was clearly pretty shaken by what I take to be Glushchenko's ignorance."[70] Penrose found Glushchenko "curiously unsatisfactory," and asked him if there was any research in the USSR on medical genetics at that time. Glushchenko replied that as far as he knew there was none, but that he was a plant geneticist himself. Penrose was therefore astounded when Glushchenko openly attacked orthodox human genetics in public at Beaver Hall, with "a great diatribe" implying that he understood the field. All this impressed Penrose that Lysenkoism was no more than a "kind of mysticism."[71]

Valery Soyfer was told a suggestive story by Glushchenko himself, many years later in 1976. In the 1930s, when Glushchenko was preparing his dissertation, he wrote a letter to the U.S. Department of Agriculture to ask them about the characteristics of the Russian wheat that had been used to develop American varieties. He then inserted a Russian translation of their detailed reply into his thesis, as his own work. Seeing the technical summary, Lysenko exclaimed, "When did you become so smart?" Glushchenko confessed that he had simply copied the material from the USDA report. Lysenko was impressed by his resourcefulness.[72]

The Party was, in the end, unable to cow its reputable geneticists into unambiguous fealty to Lysenko's line, despite Angus Bateman's discussion group and Glushchenko's bullying. This left the physicist J.D. Bernal to take the lead, supported by lesser lights such as the botanist Alan Morton, via his *Soviet Genetics*,[73] and the plant-breeder James Fyfe, who assured the world that *Lysenko Is Right*.[74]

Bernal had no qualms about doing that sort of work. On the death of Stalin in 1953 he would write "Stalin as Scientist" for the *Modern Quarterly*, eulogizing Stalin as one of the very greatest scientists of the twentieth century. "The true greatness of Stalin as a leader was his wonderful combination of a deeply scientific approach to all problems with his capacity for feeling and expressing himself in simple and direct human terms. His grasp of theory never left him without clear direction, his humanity always prevented him from becoming doctrinaire."[75] Bernal found Stalin worth reading "many times over" as a "master of Marxism," expert at "learning with the people," though his very simplicity was sometimes deceptive. Critics who accused Stalin of a love of power were deluded, for they had overlooked his "concern for the advancement of oppressed people and nationalities." Stalin had made great contributions in his own right to linguistics, using his practical common sense. It went on and on like this. Bernal was awarded the Stalin Peace Prize for 1953. By then he and JBS were back on friendly terms—though they might have made up right away, for all we know.[76]

Haldane was expressly forbidden by the Party to continue with his tepid defense of Lysenko in public. With elections coming up in February 1950, the Party was worried that he might defect as a result of this rough treatment. Incentives seem to have been offered. MI5 monitored a phone call between Harry Pollitt and John Gollan, who said that he had "done the negotiations with Haldane's secretary." Pollitt replied, "beg him to do nothing until he has seen me, be fatal if anything came out during this General Election."[77] There is no further clue as to the arrangement made. The Communist Party went on to lose both their seats in the election, forfeiting their deposits in each contest they entered, and commanding a mere 0.3 percent of the vote. But in April Haldane was still itching.

Part of the trouble was that Haldane was short of money now that he wasn't being paid for his regular *Daily Worker* articles. His father had left him a legacy in 1936, in joint trust with his then-wife Charlotte. They had divorced on grounds of desertion in 1945; but Charlotte, a defector from the Party, was no longer cooperating. Haldane could not draw capital from the trust. The microphones were live in the King Street offices when inducements were again on offer, and souls were bared. Harry Pollitt, Johnny Campbell, and Haldane were grappling.[78]

> There also appeared to be a question of *Daily Worker* articles. HALDANE mentioned that the *Daily Worker* he knew considered him hopeless and out of date, and probably reactionary in his outlook, but he said these were his views and he was making them show to the contrary.
>
> At this point JOHNNY CAMPBELL's voice heard, although had had no indication of his presence before—regret that voice distorted and very difficult to follow—he appeared to be giving his opinion of HALDANE's articles.
>
> HALDANE wanted them to get someone else to do a leading article.
>
> HARRY pointed out that they could not as HALDANE's articles were unique; he then spoke of an article they required which would strike a counter-blow (think this may be something in connection with CREECH-JONES[79]). . . .
>
> HALDANE seemed not to want to carry on as things were.
>
> HARRY next appeared to read from a letter (presumably written by HALDANE) which read—"Dear HARRY—I am in fundamental disagreement with the political line of the Party; it's not the slightest use you trying to make me change my mind and I hereby send in my resignation. . . ." HARRY pointed out how difficult HALDANE made things by writing like this. He again said that if he could get hold of the Yankee he would, but he could not get hold of him.
>
> JOHNNY CAMPBELL heard again, but regret voice too distorted to get gist.

HALDANE argued that he would not be an efficient Party member if he stayed on; he wanted to express his opinions freely, and could see he would not be allowed to do this. He would not say that the Party angle was wrong, but he would say that he thought the tactics wrong, especially in the last few years. The Party was not getting anywhere because propaganda was not being put over in the right way, especially with intellectuals. He thought results showed this, and they did not seem to be much better than the bandits.

HARRY then asked what he had in mind, because it was a difficult point, and perhaps they could attend to what he considered a weakness.

HALDANE then referred to the considerable amount of American propaganda, which was being put over here at the present time and which was very effective; he did not think the Party treatment in regard to this correct. It was no good thinking everything in the garden was lovely when it wasn't.

HARRY said "what do you think I must feel like then"—and proceeded to state that he had received a letter last week which told him that he was the worst leader the Party had ever had, and the results were those after 20 years of his leadership. HARRY continued that there were many things they had "got to take on the chin", and there were many things they would like to protest against, but the thing to remember was that they were moving; it might not be moving here as much as they would like it but "by ___[80] it was moving a lot in other places".

HALDANE agreed with this.

HARRY reiterated that there were lots of things they would like to do but could not; there had got to be a lot more give and take. He pointed out that this was a wrong time for HALDANE to be deserting them, and it would be no good thinking that this would not come out, it would. . . .

HALDANE again said that he was at the stage where he wanted to sit back a bit.

HARRY pointed out that HALDANE could not have had a more reasonable person to deal than himself. He reminded HALDANE of his

> obligations, and said anything on their side they could do they would, and he would have another "shot at it".
>
> HALDANE said he would not make any promises and would leave the situation as it was at the moment; he said he saw HARRY's point, but would not make a definite decision now.
>
> HARRY seemed perturbed that once HALDANE had resigned from the Party, they would not be able to conceal the fact.
>
> CAMPBELL mentioned that he had had the offer of a car for HALDANE but HALDANE stated that at the present moment he had not [been able to?] get a garage to put it in.

Writing for the *Daily Worker* must have been lucrative. Normally such work would pay modestly, if at all, but the paper had substantial subsidies from Stalin, funneled through Eastern Europe and other parts of the Communist empire, where bales of *Daily Worker* copies were ordered. But what were the "obligations" that Harry reminded JBS of?

A week later the Party was still concerned that Haldane might let it down. Harry Pollitt had a visitor. "Various sums of money mentioned. Later HARRY stated that HALDANE was always changing his mind; then [VISITOR] stated that HALDANE had asked him to find him accommodation in Sydney? and that he had written to his brother in Sydney. [VISITOR] appeared to be worried about sums of money, HARRY told him not to worry."[81]

MI5 waited to see if Haldane would present a possibility for cooperation. A "reliable source" was in touch in July 1950.[82]

> HALDANE is clearly going through a period of great mental stress. He is having continuous and troublesome arguments with the *Daily Worker*; all, of course, due to LYSENKO. HALDANE complains that he is not allowed to put a balanced case. . . . He maintains that some mistakes and some good work has been done by Western biologists, but at the same time some of LYSENKO's ideas are interesting, as HALDANE points out, "At least he knows one end of a chromosome from another". HALDANE has also been asked to support western scientists who themselves support LYSENKO. This he finds

himself unable to do, as he cannot take the absolutist attitude. HALDANE has not stopped altogether his articles for the *Daily Worker*, but they will not be once a week, but more probably once every few weeks. He has left the Editorial Board of the *Daily Worker*.

This analysis was supplemented by a curious note conveying J. D. Bernal's opinion on Haldane's mental state. No source is mentioned. Was it directly from Bernal, or relayed by a third party?

> Professor BERNAL is of the opinion that J.B.S. HALDANE is at the moment an ill man. BERNAL states that at the best of times HALDANE was always a difficult person and a man possessed of a very bad temper. HALDANE is, he believes, a great deal older than his actual age owing to the effect which his past experiments have had on his constitution; he is, for example, apparently suffering from some incipient disease which is the result of one of his earlier experiments. HALDANE has now, according to BERNAL, got himself a second and impossible wife who is a theologian. She writes in a newspaper called "The Lamp." BERNAL thinks that HALDANE's private life[83] has always been his downfall.

Not all of his information was accurate—Helen Spurway might have been "impossible" but she was no theologian. Perhaps he had said she was bibulous?

Alison Macleod, the TV critic for the *Daily Worker* and a stalwart Party hanger-on, ultimately broke faith over Khrushchev's revelations about Stalin in his secret speech of 1956. She later admitted that "We all knew the Russians were talking nonsense about our favourite subjects.... Haldane seemed to share this attitude." She said that around this time "a Sunday newspaper got hold of the fact that Haldane had left the Party. He issued an evasive statement and went into hiding. The chief sub-editor, Allen Hutt, told me that the Communist leaders could not find Haldane. 'We should have been much firmer with him!' Hutt snapped."[84] But she was mistaken; Haldane had not left the Party.

At the close of 1950, MI5 decided that it was still too risky to ap-

J.B.S. HALDANE

A source who has known J.B.S. HALDANE for years, and who at one time was associated with him both scientifically and politically, is of the opinion that HALDANE is most unlikely to make a co-operator as far as we are concerned. He is still in a highly nervous and emotional state, and he is described as being typical of the type of intellectual Communist who, far from having his political ideas dictated by his brain, has always been the slave of his emotions.

Should any change in this situation take place, I think we shall hear about it, and we could perhaps then consider the matter.

12th December 1950

An MI5 note about Haldane's state of mind, December 12, 1950. National Archives, KV 2-1832.

proach Haldane. A longtime associate of JBS had informed them that he was "most unlikely to make a co-operator" in their sense, and that he was "still in a highly nervous and emotional state." Besides, he was "typical of the type of intellectual Communist who, far from having his political ideas dictated by his brain, has always been the slave of his emotions." They preferred to bide their time to see if he would become more approachable.[85]

The stalemate between Haldane and the Party petered out into glum silence. No formal resignation was ever revealed, and Haldane preferred to remain evasive when asked directly by the press about the relationship. What is more, in late 1952 he again approached the Party. "Telephone bugging of CPGB BURNS and MARGARET[86] suggests that Haldane was short of money, wanting to buy a house, and tried to write for *Daily Worker*, but had been turned down even 'after all he had done' for the party, because of his political unreliability on Lysenko."[87]

It may have been around this time that Alison Macleod ran into him again. "I came on Haldane by chance, in an underground train. We had met several times, but I reminded him who I was. I told him how much my little girl was enjoying his book: *My Friend Mr. Leakey*. Then I said: 'Your old friends wish that you would get in touch with them.' Haldane gave me a look of absolute horror. I realised for the first time what he had been through. I also knew that nothing I said would persuade him that I was not a tail, set on him by the Party. I could do nothing but let him get off at the next station, and see for himself that I was not following him. I never saw him again."[88] Macleod may not have known which "old friends" Haldane was thinking of.

It was five years or so before Haldane appeared once more in the *Daily Worker*, with his photograph on the front cover of the issue of March 1, 1956, offering his opinions on civil defense in a nuclear war.[89] Perhaps the Party had Haldane in a spot because of his relationship with Montagu's espionage ring, the X Group: mutually assured destruction. Money and other perquisites were repeatedly on offer in

1950, and may have been accepted at some stage. In the end, neither got what they wanted most, but each avoided their worst outcome.

Haldane did not alter his finessing of Lysenko by much over the years, but was now more discreet, sticking to academic forums. He held out hope that the man was good in parts, and that something resembling acquired characters would surface somewhere.

Declining a trip to the USSR in 1951 (again citing the pressure of work), Haldane asked those who were going to try to get him some newt specimens—newts were a recent interest of Helen Spurway. The request suggests that Lysenko, and Michurin's notion of "distant hybridization," were still of interest.

> I am very desirous to get living specimens of the newt *Triturus cristatus* from the Soviet Union. There are two sub-species, *Triturus cristatus cristatus* in Russia, and *Triturus cristatus karelinii* from the Caucasian region. We already have a number of specimens of *karelinii* which are descended from the stock which belonged to Kammerer,[90] who, you will remember, arrived at conclusions similar to those of Lyssenko,[91] about thirty years ago. But we do not know whether some of the results obtained by crossing them with animals from Western Europe are effects of 'distant hybridisation', to use Michurin's phrase, or [are due] to the fact that their ancestors have been domesticated for many generations. Either conclusion would be of interest. . . . Our Russian colleagues may be interested to learn that some of Kammerer's stocks still survive, and will certainly understand the interest of work on distant hybridisation . . .[92]

At a symposium at Cambridge in 1953, Haldane commented that a paper presented by Cyril Hinshelwood "was useful in presenting results comparable with those on which Lysenko's criticism of Mendelism is based, in circumstances which permitted a cooler discussion than is often possible."[93] Hinshelwood thought, along with several other microbiologists at that time, that bacteria could be "trained" by exposure to environmental agents, though he did not connect that to heredity. In reality he was selecting mutant strains, a fact that had already been experimentally demonstrated by then.[94]

In the following year, in a review of Cyril Darlington's popularization *The Facts of Life*, Haldane brought up Lysenko again. Darlington had covered Lysenko crisply and acidly, as in his BBC broadcast of 1948. Haldane resorted to Hinshelwood's bacteria, which he said had been "trained" to ferment sugars, and that this learned capacity "can sometimes be transmitted for thousands of generations." Unspecified work by Jacques Monod, Winge, and Spiegelmann supposedly showed the same thing (although there seems to be no support in the literature for this attribution).[95] Haldane thought that "training" and other "facts" about bacteria would have "made Lysenko's results on higher plants less implausible." Even Michurin's ideas had some support. Being too critical of Lysenko would only strengthen him. "Lysenko has, I believe, discovered a number of new facts. But he has also denied a number of well-established facts, and made quite unjustifiable claims." Nevertheless Lysenko had "stopped a lot of research work of great biological interest by workers who wished to work on lines different from his own."[96] This was the first public criticism, however mild, by Haldane of Lysenko's role in suppressing genetics in the USSR.

Later in 1954, addressing the conflict between mutation and selection, Haldane omitted Lysenko by name, and referred only to "statements of the modern Russian school," which unfortunately depended on data published but "not translated into languages which I understand," in which "they claim to have obtained adaptive inheritable changes in a number of plants. It is probable that genes which for any reason are induced to greater or lesser activity than usual might alter in consequence, such alterations being particularly likely to appear as mutations in the germ line where, as in higher plants, the germ cells are directly descended from the highly active cells of the growing point."[97]

At the end of the 1950s, Haldane was still misrepresenting and dissimulating, roping in bacteria again. "Some readers will suppose that recent discoveries in the Soviet Union have overthrown Mendelism. Exceptions to it have been claimed, probably sometimes correctly and sometimes incorrectly. However, many Soviet geneticists who make such claims fully realise they are dealing with exceptions. . . . I think it has yet to be shown that the results obtained from such processes as grafting are likely to have been of any importance in evolution. For example, Michurin's claim that grafting may facilitate hybridization of species has been verified. And hybridization has played a part in evolution. But grafting must be very rare in nature, particularly between animals, and between herbaceous plants. On the other hand, in bacteria a process very like Michurinism is normal." Here Haldane refers solemnly to "a recent paper by Gluschenko [sic]."[98]

The claim that Soviet geneticists "fully realise" that they are "dealing with exceptions" is artful. They might realize it without saying it; the trouble is that the Lysenko complex actually said the exact opposite. The nonspecific claim that "grafting may facilitate hybridization of species" is unverifiable. What does "facilitate" mean, and where did Lysenko claim that?

Later we will consider Haldane's idiosyncratic self-obituary, broadcast on BBC television after his death in 1964.[99] What is relevant here is the section on Lysenko, which has been much misrepresented. Here Haldane offered some real criticism, though it was carefully qualified and considerably mystified, principally by inventing *Lysenko the passive agent*.

Haldane started out by assuring his audience that "Lysenko is a very fine biologist and some of his ideas are right." But those "ideas" are "much more often right for bacteria, in my opinion, than they are for larger organisms such as animals and plants with which we are familiar." So far this is a repetition of the attention-transfer defense. As he had from the beginning, Haldane stipulated that *some* of Lysenko's ideas are "wrong and badly wrong" but this was quickly qualified—other people, including himself, can be proved wrong too!

Haldane added that "it was extremely unfortunate both for Soviet agriculture and Soviet biology that he was given the powers that he got under Stalin, and that he used to suppress a lot of what I believe, and what most geneticists believe, to be valuable work." That which is "unfortunate" happens only by chance. Was Lysenko's suppression of genetics a chance event? No, but this phrasing allows the force to be muffled and the direction of events to be flipped. Even more so when Haldane helpfully adds that "if I had been made dictator of British genetics or British physiology I should have been equally disastrous." It could have happened to anyone.

Forgetting his own passionate advocacy of *planned science* Soviet-style for thirty years, Haldane then suggested that Lysenko had been a victim of overambitious scope: "I do not think that any one man is big enough for the job of directing a branch of science." Tactically mistaken perhaps—"one gets the best results from science by giving people a good deal of rope, and letting them go on with work which looks as if it were not going to be very fruitful but which sometimes is." Commission converts to omission; Lysenko, always active with rope, is made passive again. Haldane ends this section with an echo of X Group days, as we are warned that scientific programs "suffer from the evil of secrecy, which I have no doubt may be necessary but must slow down progress very considerably."

A deep sense of weariness is unavoidable when working through Haldane's arguments touching Lysenko; they pant and perspire with bad faith. Haldane no more believed that Lysenko was a "fine biologist" than he lacked time to actually read his works. Likewise, the idea that Haldane resigned or even separated from the Party because of the suppression of science by Stalin must be scotched once and for all. Haldane retreated in silence from the Party *because his own freedom to finesse Lysenko was removed*. While he was alive, he never showed much concern for the suppression of science or scientists in the Soviet Union. Instead, he actively provided covering fire for the suppressors.

As it happened, that "fine biologist" Lysenko went on to survive the fall of Stalin and inflict still more damage on Soviet agriculture. He acquired new sponsors in the form of Khrushchev and, to a lesser extent, Brezhnev. Despite a drastic loss of influence, he was never entirely discredited in the USSR, even after his death in 1976. Reports still circulate from time to time that new adherents have emerged somewhere in the remnants of the old empire. The temptations of Lamarck need no autojektor.

Since our understanding of genetics is continually expanding, it is quite safe to hold out for discoveries of unexpected phenomena along the way. We now know that the epigenome assists the genome in expressing genes. In some simple organisms, germ cells can sometimes be passed on in "pre-expressed" form, though typically, but perhaps not always, they do not persist down through the germ line to subsequent generations. It is not clear yet what the adaptive significance of this phenomenon is, if there is any. Almost all organisms rigorously ensure that the "epigenetic marks" in expressed cells are reset when they are passed on, to ensure that the resulting cells are "totipotent" and capable of the full range of genetic expression in offspring. Mammals seem especially good at this erasure, but errors will undoubtedly happen in any such process.

It will take many years to sort out the swirl of fantastic claims that have already arisen, and continue to arise, in this field, greatly assisted by the dangerous allure of statistical significance tests and a general lack of incentives for preregistered replication of experiments. Measures like preregistration are essential for controlling otherwise-disguised multiple comparisons, which give experimenters vastly more bites at the cherry than their "statistical significance" calculations purport to account for. It is safe to predict a fall into disrepute, in specialist circles, of fanciful claims to have transmitted psychologically mediated behavior epigenetically. It is also safe to expect that in some other circles such claims will simply live on forever, since they supply a persistent demand.

Contrary to Haldane, these developments will never vindicate Lysenko, any more than the discovery of jet travel vindicated those who claimed to have flown from Moscow to Algeria on broomsticks after rubbing jam on their elbows, even though they sometimes serve sandwiches on those flights. Lysenko didn't believe in genes, let alone epigenes, which are part of the same system. Thus it is known that vernalization is expressed through epigenetic means, but it is also known that those epigenetic marks are *not* transmitted to future generations. All of which would have greatly mystified Lysenko since, discounting Haldane's flights of fancy, Lysenko really didn't know one end of a chromosome from another.

11. SOCIAL BIOLOGY

The connection between genetics and human society, specifically politics, appears early and often in Haldane's discursive writings about science. Two aspects of this must be kept separate: the idea that genes matter, and the idea that the human gene pool ought to be manipulated in desirable directions—eugenics. Haldane held differing versions of both these beliefs throughout his life. A belief in eugenics presupposes a belief that genes matter, but the reverse is not true. One might believe that genes matter without committing to a program that would deliberately alter their frequencies. That aversion might be based on religious or ideological grounds, fatalism, humility, indifference, or a worldview that sees mankind as irredeemably flawed, in the sense of "original sin." Or one might doubt that man has the required wisdom. Haldane had no such aversion himself.

At Oxford in 1912, eugenics was part of the mental atmosphere for the progressively minded. As a student, Haldane had argued at the Union in favor of eugenics, and his friends included characters like Harold Laski, who as a schoolboy had been enthusiastic enough about eugenics to invite himself to visit Francis Galton. It was at Oxford that the prototype for *Daedalus* was written, and the final product issued a decade later was built around eugenic ideas.[1] In that future utopia, as a "rather stupid" undergraduate reports in an essay "150 years hence," humans come to be conceived outside the body in labs via "ectogenesis," en masse. For these breeding batteries, donors are rigorously selected to meet desired traits. The quaint idea of restricted or encouraged marriages is discarded for a factory-like process, with quicker results. Here Haldane the God-slaying Daedalus-scientist was being at least half-serious; one cannot help noticing the connection between assisted reproduction, a theme he returned to often, and Haldane's own physical condition.

A few years later in *Possible Worlds* (1927), Haldane revealed a less alarming preference for voluntary "positive eugenics," and an aversion to punitive "negative eugenics." The less intelligent poor appeared to be outbreeding the more intelligent middle and upper classes. "All investigators are agreed that mental capacity is strongly hereditary, though, as with stature, environment plays a part in its determination." As evidence he cited twin studies, the fact that siblings are too dissimilar to support a large role for differences in upbringing, and the fact that the well-fed upper classes differed from the well-fed middle as much as the middle differed from the poorly fed lower. Therefore incentives ought to be offered so that the more intelligent would have more children, through family allowances and the provision of free education and other socialist measures to remove economic barriers. At the same time, the poor ought to be encouraged to use contraceptives. "Civilization stands in real danger from overproduction of 'undermen'. But if it perishes from this cause it will be because its governing class cared more for wealth than for justice." All this was conventional fare for a period in which the only serious intellectual opposition to eugenic measures was provided by religious conservatives like G. K. Chesterton.

Charlotte Haldane had also become interested in the field, translating a 1928 study of criminal twins conducted by Dr. Johannes Lange, a German professor. JBS supplied an enthusiastic foreword for Charlotte's edition of 1931. He found Lange's evidence compelling, interpreting the genetic explanation of crime as a scientific demystification of evil. It suggested that free will was an illusion, that "crime is destiny," except insofar as freedom means lack of *external* interference. He conceded that there were weaknesses in Lange's modest number of seventeen twin pairs, and that the best evidence could be gathered from single-egg twins raised apart from birth, of which Lange had none. "So far as I know only four such cases have been investigated." Hermann Muller and Horatio Newman had been unable to find any criminal cases to study.[2] Nevertheless, Haldane found Lange's evidence persuasive. For crime, genes must matter.

The following year, Haldane released his second collection of essays. As the title *The Inequality of Man* suggests, he had warmed to the above themes. Whereas *Possible Worlds* contains one piece on eugenics, almost every piece now mentions, in varying detail, both the notion of genetic inequality and eugenic measures for manipulating inequality. The title piece lays out the facts as Haldane understands them about the sources of human differences.[3] These are of two kinds: genetic differences (nature) and environmental differences (nurture), by then a well-established distinction. Two differences in outcomes are most important to his argument: intelligence and character. With regard to intelligence, he mentions the pioneering work that had recently been done by Barbara Burks on adoption and foster homes, which he considers persuasive evidence that intelligence differences depended strongly on genetic differences.[4] Muller and Newman's work on twins supports this, too. Haldane is also careful to partition genetic differences into an additive transmissible component and a non-transmissible component due to interactions unique to the particular combination of genes that an individual has, which is broken by segregation of the genes when sexual reproduction takes place.

The notion of general intelligence, as in Spearman's g factor, shows that the effects of intelligence differences are pervasive and not isolated to niches.[5] Moreover, education has little effect on g, which it depends on. "In the course of the next century, if psychologists . . . cooperate with geneticists, it should be possible by the time a child is about seven to arrive at a fair idea of its capacities, and children will be sorted out accordingly."[6]

The evidence for character, or personality, he finds less persuasive, and he is prepared to find differences in upbringing more influential. However, the state has to accept the diversity of human traits, a "natural phenomenon like the weather" which is "predictable to a certain extent" but "very difficult to control." Haldane's up-to-date command of the literature in laying out his conclusions about differential psychology underscores his impressively wide reading in allied fields at this time.

His stance here on eugenics remains essentially the same. Differential fertility by social class threatens to steadily decrease average intelligence but should only be addressed positively, for ethical reasons, by artificially raising the social level of the lower classes (simplistically equated with wealth) to the point where they have the same number of children as the upper classes. However, once conceived, this government-engineered "classless society" is not examined very closely by Haldane. For, if differences in ability inhere in the lower classes they will still be there when they are formerly lower class and have the same number of children. If a higher average is worthwhile, then the average would be raised even further by arranging for the formerly lower class to now have far *fewer* children than the formerly upper class. A genuinely classless society with random mating would end up with a lower average than a class-based society with different proportions. Haldane obliquely recognizes the force of this argument, which follows at once from his own work in mathematical population genetics, by contemplating that "In the remote future mankind may be divided into castes like Hindus or termites." He prefers equality of opportunity, but recognizes that state intervention would be required, since differences in ability would otherwise translate into differences in social achievement, though this reason is only implicit in his argument. Moreover, this "positive" eugenics, assuming it would work at all, involves its own form of "negative" coercion, through state-enforced engineering of the class system by wealth confiscation.

Haldane is also well aware of the conflict between his growing attraction to communism and his recognition of human inequality.

"The test of the devotion of the Union of Socialist Soviet Republics to science will, I think, come when the accumulation of the results of human genetics, demonstrating what I believe to be the fact of innate human inequality, becomes important." But he immediately qualifies this with the claim that only "sentimental and unscientific views," which are "often associated with Socialism," are incompatible with a belief in human inequality, whether that inequality is transmissible (additive) or not (interactive).[7] This statement was false when he wrote it, as the Hermann Muller saga soon showed.

Muller had moved to the USSR in 1933, after being converted to communism by his visiting students Levit and Agol. As we have seen, he arrived just as the war on genetics there was getting serious. One of his aims in coming was to persuade the Soviets to adopt his eugenic scheme, which he had written up as early as 1910 and had been refining ever since. This was eventually published as the book *Out of the Night* in 1935, and Muller had it translated into Russian so that he could send it to Stalin for approval. Haldane supplied an enthusiastic blurb for the Left Book Club edition of 1936.

> You may regard it as a revelation, or, quite as likely, throw it into the fire. But do not dismiss it as a mere phantasy. The author is one of the world's leading biologists, and his proposals, whether or not they are desirable, are entirely practicable. If they are adopted, the results will be as important as those of the industrial revolution.[8]

Most of the book is taken up by a futuristic prologue of four chapters, anticipating manifold benefits from the advance of technology and priming the reader for the coming marvels to flow from the application of biological knowledge to social problems. As the marvels include the abolition of home cooking in favor of prepared meals from canteens, it is hard to have complete confidence in his judgment. The core eugenic ideas appear in the remaining three chapters. Muller introduces the inexorable buildup of genetic load, caused by the accumulation of recessive mutations that are no longer weeded out, thanks to the cushioning effects of modern civilization. He anticipates that in a thousand years these will really start to tell on the human condition.

Negative eugenics is dismissed by Muller as hopelessly ineffective, given the large number of generations that it would take to select out recessives, if one waits for carriers to get a double dose and thus reveal themselves. This echoes similar arguments made by Haldane himself, who is referred to several times in the text. Muller does not address R. A. Fisher's calculation that negative measures would work a lot faster in the presence of assortative mating. Nor does he notice that budgeting a thousand years to accumulate a problem is not much different from reckoning a thousand years to accumulate a solution. In any event, his own eugenic scheme proceeds along entirely different lines. It is the eugenics of the harem.

Muller starts out by stipulating that two human traits of universal importance are under genetic influence and ought to be increased: high intelligence and group-oriented "comradeliness." Aside from the provision of abortion, birth control, and surrogate motherhood to generally induce more capable women to have children, his solution is a sperm bank for donors of proven ability. For example, "such men as Lenin, Newton, Leonardo, Pasteur, Beethoven, Omar Khayyám, Pushkin, Sun Yat Sen, Marx." Where, one might ask, is Genghis Khan? Muller at once recognizes this difficulty. The men that women prefer need not match his own whims. *They* might want "a maximum number of Billy Sundays, Valentinos, John L. Sullivans, Huey Longs, even Al Capones." That women would not see things his way is due, he is sure, to "general propaganda" and the press—not, one assumes, to the prospect of a million Lenins machine-gunning families by quota, excitedly encouraging famines, and endlessly hectoring their opponents for "bleating" and "sniveling." To get around this backwardness, he concedes that human nature would first have to be reengineered to undo retrograde preferences for Billy Sunday over Lenin, before the sperm banks could be allowed to dominate reproduction.

All this is not encouraging, from a purely practical point of view. Muller's eugenic scheme requires bootstrapping, and all we are left with for that is, apparently, abortion, birth control, and surrogate motherhood. Turning to R. A. Fisher again, the long-term eugenic consequence of abortion as birth control is the removal of those who practice it from the population, leaving behind exactly those with an instinctive aversion to doing so (Christian Scientists, Hasidim, etc.). Muller does not explain why those who are averse to abortion would be more or less intelligent or comradely.

It is also not obvious why 99 percent of men would passively accept being crowded out of reproductive success by the sultans of sperm, except at gunpoint. Rape would surely skyrocket. By contrast, under negative eugenics a small fraction of people known to have already developed serious symptoms would not reproduce. But as it happened, Muller had grossly misjudged his intended audience. Vavilov advised him to flee, which he did. Stalin had the man who translated *Out of the Night* into Russian shot.[9] Muller and Haldane could have predicted all this by simply reading the *Great Soviet Encyclopedia* of 1932.

Haldane's next foray into this area came in 1938, by which time he was a concealed Party member and an open Marxist. *Heredity and Politics* is his only book-length treatment of the issues raised by social biology: human differences, eugenics, and (his first full treatment of) race. All three topics are now represented as inconclusive, ending in something like *it depends*. The elementary principles of genetics are recapitulated. Human differences do exist, but he cannot say for sure whether differences of nature or nurture are responsible. Negative eugenics gets no support, apart from limited measures for clear-cut cases, such as discouraging cousin marriages. Eliminating recessives takes too long. Positive eugenics, too, is dubious, since he now considers that differential fertility is *not proven* to lower intelligence and remedial measures to counter it might be ineffectual. He jokes that Muslim attempts to monopolize women for sultans have not produced notably smarter people—"a Turk should generally beat an Armenian or a Jew in a business deal. This is notoriously not the case."[10] Therefore, Muller's sperm bank is now held to be harmless for those who care to use it, but is no savior.

Races in the sense of Negro versus white he thinks can be well defined, but differences in abilities, if they exist, may be due to any number of causes. If races within Europe exist, they are not the same thing as the previous sense of race, and not "pure" in any useful sense. Even to say that races may be different depends on what "different" means (wholly distinct, or on average, etc.). More study on all topics is said to be required. *Might, possibly, could, perhaps, probably not, it depends*—his qualifiers extend through 202 pages. Even Haldane recognizes the unsatisfactory nature of his conclusions and apologizes for not being able to do better. He makes no attempt to reconcile earlier statements with current ones. They would keep changing anyway.

In the following year, 1939, the Science Service of Washington, D.C., asked leading geneticists worldwide to answer the question "How could the world's population be improved most effectively genetically?" Twenty-three geneticists replied in an open letter published in *Nature*. After a lengthy oration advocating education, wealth distribution, birth control, equality of opportunity, and other socialist concerns, they finally addressed the question they were asked.

> The intrinsic (genetic) characteristics of any generation can be better than those of the preceding generation only as a result of some kind of selection, that is, by those persons of the preceding generation who had a better genetic equipment having produced more offspring, on the whole, than the rest, either through conscious choice, or as an automatic result of the way in which they lived. Under modern civilized conditions such selection is far less likely to be automatic than under primitive conditions, hence some kind of conscious guidance of selection is called for. . . . The most important genetic objectives, from a social point of view, are the improvement of those genetic characteristics which make (a) for health, (b) for the complex called intelligence, and (c) for those temperamental qualities which favour fellow-feeling and social

behaviour rather than those (to-day most esteemed by many) which make for personal 'success', as success is usually understood at present. A more widespread understanding of biological principles will bring with it the realization that much more than the prevention of genetic deterioration is to be sought for, and that the raising of the level of the average of the population nearly to that of the highest now existing in isolated individuals, in regard to physical well-being, intelligence and temperamental qualities, is an achievement that would—so far as purely genetic considerations are concerned—be physically possible within a comparatively small number of generations. Thus everyone might look upon 'genius', combined of course with stability, as his birthright. As the course of evolution shows, this would represent no final stage at all, but only an earnest of still further progress in the future.[11]

They took care to stress voluntary measures only and to dismiss Lamarckism. The statement was probably written by Hermann Muller, as it summarizes most of *Out of the Night*. He was among the twenty-three signatories, as were Cyril Darlington, F. A. E. Crew, S. C. Harland, Lancelot T. Hogben, Julian Huxley, T. H. Dobzhansky, G. Dahlberg, Joseph Needham, and Conrad H. Waddington. And so was J. B. S. Haldane. We must assume, then, that JBS had resolved some of his doubts since *Heredity and Politics*.

Fast-forward to 1947, and a different audience, the *Daily Worker*. Now definitions lose clarity again: "someone asked me 'Can you change human nature?' I don't know the answer, because I don't know what 'human nature' means."[12] However, Haldane concludes that there is "very strong" evidence that education counts for a lot more than inborn differences, which are nonetheless real. He considers questions like "is intelligence inherited" to be meaningless. "I don't know what determines differences in human intelligence. No doubt heredity and environment interact."[13] (If they interact, the question cannot be meaningless.)

Haldane had embraced the theme of interaction a few years earlier in a more technical forum. It is possible that a (genotype × environment) combination can have different properties when both are present than either factor has alone, or in different combinations.

> We are not justified in condemning a genotype absolutely unless we are sure that some other genotype exists which would excel it by all possible criteria in all possible environments. We can only be reasonably sure of this in the case of the grosser types of congenital mental and physical defect. A moderate degree of mental dullness may be a desideratum for certain types of monotonous but at present necessary work, even if in most or all existing nations there may turn out to be far too many people so qualified.[14]

It is an empirical question whether such interactions exist and are non-negligible. In a statistical regression context, one seldom, if ever, models an effect as the result of an interaction alone: the main (additive) effects are usually modeled too. All may contribute with different weights. Interaction may be negligible, or it may not; but it is highly unusual for a factor to matter in an interaction *only*, and not in its own right. For the example that Haldane addresses, intelligence, what needs to be argued is whether, as an empirical fact, interaction matters for the outcome. And if it does, to determine how much its components contribute in their own right. Merely raising the *possibility* of an interaction gets us nowhere—which is where Haldane leaves the question.

By now, Haldane had evolved from his early position, in which he advocated the importance of innate differences and a positive eugenics program, to one in which nothing is clear about innate differences, due to the possibility of interactions, and nothing is clear about eugenics, except that negative eugenics makes sense only in limited cases. Along the way, there are some unexplained reversions to more definite ideas. In his 1954 review of Cyril Darlington's *Facts of Life*, he finds new zones of uncertainty.[15] He thinks that Darlington "greatly exaggerates the importance of genetically determined differences between human beings," but may be proved right in the long run, because of a phenomenon later known as "Herrnstein's syllogism," but clearly stated here two decades earlier.[16] "In so far as nutritional, educational, and other facilities are equalized, and, if the school of Freud is right, in so far as psychological damage by infantile experiences is prevented or cured, human differences within a community will come more and more to depend on differences determined by genetical rather than environmental causes." To this he adds a qualification that runs directly counter to his own work on population genetics. "But even if they were so determined today, they would not be hereditary, in the ordinary sense of that word, as for example the differences between greyhounds and bulldogs are. This is because men and women are not bred as dogs have been; so two short parents may produce a tall child and conversely, even though children tend to resemble their parents in this respect." What is heredity if not an alteration in probability and a shift in a distribution, and where did Darlington imply that "determine" meant certainty?

Darlington cropped up again in this debate when Haldane reviewed the 1962 reissue of Francis Galton's *Hereditary Genius*, to which Darlington supplied a foreword.[17] Here Haldane claims that Galton had refuted his own arguments in favor of eugenics, since he had already conceded that the extremes of his ability distribution were less fecund than the mediocre. "Centripetal selection was discovered in sparrows by Bumpus in 1898, and in snails by Weldon in 1901. It is a component of 'genetic homeostasis,' which makes it much harder than Galton thought to change the characters of a population by selection." Since Galton was proposing to artificially apply *directional* selection, and as Haldane himself had shown, along with Sewall Wright and R. A. Fisher, that directional selection would inexorably shift the gene distribution, the argument is inscrutable. If Haldane means that centripetal selection is *unavoidable* in the current human population, he is wrong, given what we know about variation in a normally distributed population. As a purely hypothetical exercise, employ the Stalinist methods that Haldane admired so much, but in the reverse direction, and simply shoot the lower half of the population. The new breeding population now has a higher mean. Nor is this an accurate representation of Galton's settled position, which was that the lower classes of ability were slightly more fecund in recent times, but had been slightly less fecund through most of human history, where fecund means leaving survivors. Hence the evolution of human ability, which Galton wished to actively direct through eugenics (though he did not use that word in his *Hereditary Genius*).

The following year, Haldane contributed a piece to a CIBA foundation symposium, on the theme "Biological Possibilities for the Human Species in the Next Ten Thousand Years." He returned to human abilities. "Some successful people believe that everyone could do as well as themselves if they tried, others that rare innate gifts are needed. On what are probably quite inadequate grounds I consider that the truth is between these extremes." Vague and safe enough. But now eugenics was reanimated, at least for physiology.

> The recognition of human physiological diversity may have enormous consequences. As soon as its genetical basis is understood large-scale negative eugenics will become possible. There may be no need to forbid marriage; few people will wish to marry a spouse with whom they share a recessive gene for microcephaly, congenital deafness, or cystic disease of the pancreas, so that a quarter of their children are expected to develop this condition. I cannot predict the later steps which will make positive eugenics possible, since we know the genetic basis of few desirable characters.[18]

With regard to race, he was agnostic. "I do not believe in racial equality, though of course there is plenty of overlap; but I have no idea who surpasses whom in what." He then indulged in a raft of speculations about breeding varieties of men optimized to perform well as full-time space travelers—for example, space travelers might not need legs in zero gravity. But his position was to be refined again the following year.

In 1964, Haldane wrote the chapter "The Proper Social Application of the Knowledge of Human Genetics" for the Penguin collection *The Science of Science*.[19] Considering whether "Good and bad qualities are hereditary," he calculates that the probability of inheriting a specific configuration of genes responsible for a "good" quality has to be very low; therefore it won't happen often and could not be a basis for social policy. His argument here simply assumes that a character is *entirely epistatic*, depending on the interaction of the specific configuration of genes involved. He assumes that the character is entirely present or entirely absent. Characters that are not binary, and are inherited additively to a greater or lesser degree (though they may have an epistatic component), do not fall under this analysis, specifically intelligence. But Haldane knew all this well enough.

He goes on to consider whether "Higher social classes are congenitally superior." He now supposes this is true to a modest degree, and therefore that the upper classes should be encouraged to breed—a reversion to his original opinion from the 1930s. His suggestions for positive eugenics remain the same: socialist engineering of the class system by limiting the ability to bequeath wealth and eliminating access to expensive schools. The idea seems to be that those parents would then have to hedge their bets by having more children, rather than investing all their resources in just a few. He also suggests cultural changes to the sexual mores of the upper classes, based on his enthusiastic reading of the Kinsey Report. He is noncommittal about whether mean genetic endowment for intelligence is falling over time. He accepts that modern medicine is slowly building up genetic load by neutralizing natural selection, but hopes that some solution will be found in the future. Addressing whether "Some human races are congenitally superior to others in respect of socially valuable innate characters," he reiterates his agnostic stance from the previous year, but adds that he doubts if historical performance can be used as a guide to inherent capability.

After following Haldane's endless twists and turns, now coy and now forthcoming, on the core issues of social biology, it is worth pointing out that he conducted no original research of his own on any of its topics. His long list of several hundred technical papers covers many other subjects, but not those. He did no relevant experimental or observational work on the subject—unlike Hermann Muller, who had worked with separated twins—nor did he do any theoretical work about the specific techniques used by behavior geneticists, nor did he even reanalyze data gathered by others. Though he commented often on the broader questions, and was regularly asked for his opinion, which was always taken seriously, he did so from only a reading knowledge of the field.

An evaluation of Haldane's references to the literature over the years shows that his reading was well up to date in the early 1930s but badly out of date and inadequate, even idiosyncratic, in the 1960s. People have a widespread inclination to take the opinions of anyone who is considered a geneticist seriously when it comes to any topic affected by genetics. Authority, especially in the case of an eminent scientist like Haldane, is seldom questioned. But it should be questioned. Haldane's own fields lay mainly in mathematical population genetics, the biochemistry of genes, general physiology, the statistical issues raised by small samples, and, to a lesser extent, some aspects of animal behavior—an impressive list, but not directly relevant to the topics at hand. The facts about human social biology are contingent, and can only be informed by real data. But even leaving all that aside, Haldane was also often flippant and insincere in his arguments. This suggests that he did not take his own opinions on the topic at all seriously.

12. ANIMAL BEHAVIOR FROM LONDON TO INDIA

Once outside the enveloping folds of the Party structure, but still convinced of Communist ends and methods, Haldane subsided into morose discontent through the rest of the 1950s. Professional frustration was involved in this. He had not been offered the command of organizations that could be used to carry out research programs. He was anti-establishment, but the Party didn't want him either. Some vignettes from this period illustrate his frustration and unpredictable behavior.

John Maynard Smith had written to Haldane in October 1947 to ask for advice, leading with "Dear Comrade" and closing with "Yours fraternally."[1] He was already twenty-seven years old and living in Reading, employed as an aircraft technician working on structural design. Earlier he had studied engineering at Cambridge and there had joined the Communist Party, which he had worked for during the war and was still a member of. Maynard Smith had decided that science interested him more than aircraft, and wanted to meet Haldane to hear his thoughts on possible programs of study at UCL. "As an explanation of the 'Comrade', I am at present secretary of the Reading Branch of the Party—a fact for which you are partly responsible." Maynard Smith had long been an admirer of Haldane from a distance, after hearing about him as a schoolboy at Eton, where

JBS was then considered disreputable. A copy of Haldane's *Possible Worlds* (or perhaps *The Inequality of Man*) in the Eton library drew Maynard Smith in further.[2] Haldane responded positively with "Dear Comrade Smith," and plenty of advice. Comrade Smith ended up starting all over by doing another B.Sc. at UCL, this time in zoology, with the long-term intention of applying mathematical techniques to problems in evolutionary theory, topics germane to Haldane.

Maynard Smith and Haldane seem to have become close remarkably quickly, more than just comrades, though Maynard Smith later wondered whether he would have gone to UCL if he had known in advance of Haldane's explosive temperament.[3] He was "a very difficult person to work with. He was one of those people who liked the feeling of adrenaline circulating, I think. He enjoyed being either angry or afraid." Nevertheless they got on. "I knew him well, obviously, I worked with him, I loved him. I was, I think, close to him. But I never quite understood him, and why should one? Why should one expect to understand one's friends?" Maynard Smith admired Haldane's intellect, "I mean, he was staggeringly intelligent, I mean, there's no question about that," but noted that "he was not a naturalist, his skills were not from natural history, they were essentially from mathematics." Haldane's method when approaching any problem, from biochemistry to genetics, was to formulate a mathematical model that captured the essential elements but was simple enough to solve and make deductions from. In this Maynard Smith thought JBS uniquely skilled, but only as an applied mathematician—"I don't think he was a particularly good mathematician."[4]

When Haldane wrote out his will on December 12, 1951, he made Maynard Smith his executor and the recipient of the *Journal of Genetics* in the event of Helen Spurway's death—she was the recipient of all goods and effects otherwise. One of the witnesses was Peter Medawar. Haldane estimated that the *Journal* would provide 100 pounds of spare money per year.[5] This act shows remarkable trust in Maynard Smith after just a few years. Moreover, Haldane made no mention of his own family in this will. The estrangement that had set in before the war was long and deep; his mother, Louisa Kathleen, was still alive, as was his sister, Naomi Mitchison. He left his body to science, and if they had no use for it, forbade any headstone.

Maynard Smith later confessed to Richard Dawkins that he, too, had been under the sway of Lamarxism at this time. "I was still enough of a Marxist to feel that . . . that there was something very suspect about the classical Mendelian attitude in genetics. And indeed, I spent a substantial amount of time doing an experiment which was not, so to speak, intended to demonstrate the falsity of the inheritance of acquired characters, but was deliberately looking for a situation in which an acquired character would be inherited." Inevitably, it didn't work. "I did quite a bit of work on the acclimatisation of fruit flies to a change of temperatures. I showed that they can acclimatise during their lifetime. And then I bred from flies that I'd acclimatised, and compared them with flies that I'd not acclimatised. And it will not surprise you to learn that there was no difference, in other words, the acquired character was not inherited." But he did not publish this failure. In retrospect, Maynard Smith thought that the Lysenko episode helped to drive himself away from the Communist Party, a process that culminated in his response to the Hungary Invasion of 1956, which made the break solid. But, as we have seen, neither of these events had that effect on Haldane himself.

In 1949, Spurway, and later Haldane, too, became interested in the study of animal behavior and its pioneer, Konrad Lorenz. That July, Lorenz was in Cambridge attending a Society for Experimental Biology symposium, where Spurway appears to have met him. According to a story attributed to the ethologist Niko Tinbergen, who was at the conference, an affair developed between Lorenz and Spurway, though both were married. Lorenz had a wife back in Austria who had to turn from gynecology to farming in order to support their family. He had already established his reputation in ethology before the war, but had spent five years in the Russian Gulag, ultimately transferring through thirteen different prisons before his release in early 1949. His affair with Spurway is said to have lasted over a year, but had later repercussions. In the meantime, Haldane himself had joined Spurway in working on the evolutionary aspects of ethology.

At a conference on instinct held in Paris in June 1954, Desmond Morris ran into JBS and Spurway—"spiky, shrill voiced"—and also Konrad Lorenz, though he noticed that Lorenz seemed to be avoiding the Haldanes. When he raised this evasion with them at dinner, he was given a frank answer.

> 'Shall I tell him?' screeched Helen Spurway. Haldane paused for a while and then nodded. 'I have been fucked by Konrad Lorenz!', she shouted at the top of her voice, staring at me with a strange smile on her face.[6]

Morris later learned from Tinbergen that there was more to it. The Haldanes had attempted to persuade Lorenz to inseminate Spurway so that she could have a child—JBS was infertile and going on sixty-three. When Lorenz refused to cooperate, Haldane is said to have threatened to report him to some of his Russian friends, who would have been able to pack members of Lorenz's family still in East Germany off to the Gulag. Whether this was a grotesque joke or not—Haldane may well have had those kinds of contacts—it supposedly gave Lorenz a horror of the Haldanes. Since much of this information is at third-hand, it should be treated with caution. But Tinbergen himself retained a strongly negative view of Haldane, perhaps because of this incident.[7] There are two other clues. First, when Ernst Mayr visited the Haldanes in India a few years later, he found to his embarrassment (so he said) that Spurway became infatuated with him, with Haldane's encouragement.[8] Second, Peter Medawar reported confidences from Julian Huxley that Haldane was not "physically equipped to perform sexually."[9] This matches earlier stories from "Chatty's addled salon" days about alleged incapacities (see above). It may be that Haldane's wounds from the First World War were responsible.

Lorenz featured prominently in the contributions made by Spurway and Haldane to ethology. Spurway published a minutely considered review of Lorenz's popular book on animal behavior *King Solomon's Ring* that suggested, through its exaggerated interest, that hostility had begun to dominate affection.[10] Politics had intruded. The title of the review, "Behold My Child, the Nordic Dog!," immediately suggested National Socialism. Nothing in *King Solomon's Ring* is political, but Spurway made heavy work of some common manner-of-speaking phrases used by Lorenz to explain ideas to his lay audience. He had casually speculated that some more aggressive domestic breeds of dogs owed this to their higher "amount" of "wolf's blood." Spurway took a harpoon to this guppy, though her phrasing strongly suggests ventriloquism on the part of Haldane. What could Lorenz possibly mean by "quantity" of "blood," and how could he ignore the possibility of recent selection for aggression, as opposed to persistence of ancestral aggression? Spurway gratuitously speculated that the problems with Lorenz's thinking were linked to his homeland of Austria, where "discussion of evolution has been impossible, not only under the Nazis but under the Roman Catholic Church." Here she blithely ignores the effects of communist orthodoxy in circles more familiar to her.

At the Paris conference, Morris and his wife were treated to an exhibition. "Haldane, for some reason, was at his most outrageous and rebellious, and we provided the perfect audience for his antics." JBS demonstrated how to cross a busy road by extending his arms and barging through traffic, on the principle that he was signaling that he was bigger; somehow he survived. He alarmed the conference organizers by setting piles of paper on fire to illustrate a point during a lecture. When Karl von Frisch described his bee-dance language experiments, he wept openly and rose to mystify the German by introducing information-theoretical improvements from the floor. Vexed by some slight during a meal, he threw fish on the floor and stormed around the room. Touring the French countryside in a bus, the Haldanes disconcerted the other passengers by intimately caressing each other, producing a novel mating signal via the roiling flesh on JBS's bald head, as Morris saw it. At newly discovered Lascaux,

Spurway confounded their tour guides by shrilly insisting that their rhinoceros rock-art was *not* a rhinoceros after all.[11]

The biologist Peter Medawar—a future Nobel Prize winner—knew Haldane when they were at UCL together in the 1950s, developing a mixture of wonder and horror.

> In some respects—quickness of grasp, and the power to connect things in his mind in completely unexpected ways—he was the cleverest man I ever knew. He had something novel and theoretically illuminating to say on every scientific subject he chose to give his mind to: on the kinetics of enzyme action, on infectious disease as a factor in evolution, on the relationship between antigens and genes, and on the impairment of reasoning by prolonged exposure to high concentrations of carbon dioxide.

However, Haldane was not a good experimenter, nor was he fundamentally original.

> He was not himself the author of any great new biological conception, nor did his ideas arouse the misgivings and resentment so often stirred up by what is revolutionary or profoundly original. On the contrary, everything he said was at once recognized as fruitful and illuminating, something one would have been proud and delighted to have thought of oneself, even if later research should prove it to be mistaken.

Echoing Darlington, Medawar thought that Haldane's "great strength" was "to see connections, to put two and two together, to work out the deeper or remoter consequences of taking certain theoretical views." Medawar's horror was reserved for Haldane's political opinions.

> People were wont to ask how such a clever man could be so completely taken in by Communist propaganda, but Haldane was not clever in respect of any faculty that enters into political judgement. He was totally lacking in worldly sense, a sulky innocent,

a whole-hearted believer in Them—the agents of that hidden conspiracy against ordinary decent people, the authorities who withheld the grants he had never asked for and who broke the promises they had never made.

This trait was harmful to his career. "It might have made a difference if Haldane in his lifetime could have been made to realize the degree to which his work was obstructed by his own perversity. He was so ignorant of anything to do with administration that he did not even know how to call the authorities' attention to the contempt in which he held them."[12]

Another view is given by Peter Marler, an ethologist who had taken Haldane's lectures on genetics at UCL in the 1930s, but drank on more familiar terms with both JBS and Helen Spurway in the mid-1950s—at The Mill in Cambridge, and later at pubs in London. He developed an admiration for their tolerance of beer and vintage cider. "Spurway is well equipped vocally, and became even more loquacious as evenings wore on. We had to leave one pub hastily, in the face of verbal abuse, when she threatened to drown out the resonant voice of Winston Churchill coming through the radio on the bar."[13]

An incident involving Spurway demonstrates her frame of mind shortly before the Haldanes emigrated from England to India. On Guy Fawkes Night of 1956, she had been out drinking in Bloomsbury with an American research assistant at UCL, William Carey Clarke. (It may be significant that JBS was not there, given that it was his birthday.) Walking back after three and a half pints of bitter (Spurway's count), they were stopped by a police constable at Torrington Place. Seeing the constable's police dog, Spurway stamped on its tail, producing a yelp. When asked why, she replied, according to the constable, "For fun, that's what dogs' tails are for." Then she raised her hand to her nose, in a Hitler mustache, and goose-stepped off. After both Spurway and Clarke were arrested for drunk and disorderly conduct, she assaulted the constable (Clarke seems to have intervened here). Spurway proclaimed in court that "The British Government made the word police the obscenest word in the English language. I thought it would be constructive if the police were to realize they were hated and despised." She denied hurting the dog, as she had had rubber soles on, or assaulting the constable. Found guilty on all charges, she was fined 10 shillings for public drunkenness, £5 for assaulting the policeman, and 10 guineas in costs. Clarke was found guilty of technical obstruction and fined a guinea. Spurway promptly refused to pay, preferring two months imprisonment. "I hope to go to India and I will be very much happier with many of my friends if I too have been in a British gaol."[14] Clarke must have paid his own fine. After a few days in Holloway Prison, Spurway was freed when her fines were paid by an undisclosed person—rumored to be a Mr. Waller, possibly from a newspaper hoping to get an interview.[15]

Shortly after this, the American evolutionary ecologist John Tyler Bonner met Haldane and Spurway for dinner out. He recalled that they "both loved to shock" and "loved to argue." Spurway "reveled" in the details of the police dog incident.

> We ate at a small restaurant in Soho and soon were embroiled in some very spirited arguments. One was about some biological aspect of sex. Mostly they argued with each other, and both of them had very penetrating voices as they became more intense, resulting in all the people at the neighboring tables staring at us. Even though they both appeared to ignore the stir they were causing, I could not help feeling they not only were aware of it, but enjoyed seeing the shock waves travel across the room. By the time we left the restaurant, we were on to discussing the recent work in animal behavior and its evolutionary implications. It was a subject of concern to all three of us, and indeed it was the central theme of one of my lectures, in which I drew parallels between behavior and development. We all had more to say, so they said they would walk me to Portland Place, but we still had not finished, so I walked them back toward University College. The whole process repeated itself again before we were ready for

bed. On one of the laps we passed the BBC building, and on the ground floor a low window was open at the top, and one could hear a radio blaring away. Haldane was talking, and suddenly he veered across the broad sidewalk, stood on his toes, shoved his enormous head into the open window, and yelled with tremendous force, 'SHUT UP'. He went directly on to his next sentence as he cruised back alongside us without skipping a beat.[16]

The following year Haldane resigned his UCL professorship and announced that he was emigrating from the UK to India. Plans for his retirement had long been germinating. As early as 1950, discussions in the Party headquarters at King Street had mentioned retirement in 1957.[17] That year he turned sixty-five. Haldane extracted as much publicity as possible by dramatically claiming that he was forced to leave the country on moral grounds because of the British role in the Suez Crisis of 1956, which he described as "criminal." His thoughts on the Soviet invasion of Hungary were not preserved. In reality, he was attracted by the dramatically lower living costs in India, by the close connections between Nehru's government and the Soviet Union, and by the new opportunities afforded by the deprived Third World for publically claiming the moral high ground. Elsewhere he also let slip that "One reason why I have gone to India is to avoid chronic 'rheumatic' joint pains."[18] The fact that India was still in the Commonwealth may have played a role, too, keeping him at least partially in touch with the rest of the world. He made no attempt to reestablish himself in Eastern Europe or the Soviet Union itself, despite his past praise and advocacy of their planned scientific and educational establishments and superior living conditions.

For India this was a major coup, but Britain shrugged. The geopolitical significance of the Weldon Professor of Biometry at UCL was obscure to them, his conceit perhaps a little ridiculous. Intellectuals have never had much cachet in the English-speaking world. Governments pay no attention when they publish articles expressing their opinions about this or that. By contrast, in the Soviet Union intellectuals were taken very seriously indeed (some say this is still the case in France). When Solzhenitsyn published something, the attention of the Politburo was engaged, to his detriment. In their domains, words and images supplanted reality. Which is not to say that the powers there ever had any *liking* for intellectuals. As we have pointed out above, Lenin routinely dismissed them as "lackeys of capital who think they're the brains of the nation. In fact, they're not its brains, they're its shit."[19] Those who moved abroad to make trouble were hunted down. But not in Haldane's case—the "establishment" in Europe and America would shower him with awards in his absence.

Haldane's destination was the Indian Statistical Institute (ISI) in Calcutta (present-day Kolkata), run by Prasanta Chandra Mahalanobis (1893–1972), an influential statistician of Haldane's own generation who had originally trained as a physicist at the Cavendish Laboratory in Cambridge. The country was in the throes of its second "Five-Year Plan," which must have pleased him. The *Journal of Genetics* went with him, as did Spurway. A post awaited her, too.

JBS now professed love for India and things Indian, including Hindu philosophy. If this was so then he had changed his mind from earlier years. In the late 1930s, writing in his unfinished autobiography, he was at best ambivalent about India, to which he had been invalided during the Great War. Then he was hostile to Hinduism. "The Hindu caste system is the greatest glorification of snobbery that the world has ever known. If I am a Brahmin I not only enjoy privileges in this world and probably the next but I deserve to do so for having been good in past lives." "The ideal position in India would seem to be that of a wealthy Brahmin with an Oxford degree, and a high rank in the civil service." He was drawn more to Islam than Hinduism. "I could sympathize less with Hinduism amongst other reasons because I am a potential Muslim but not a potential Hindu. An act of faith admits one to Islam. Hindus are born, not made. And Islam is a religion of universal brotherhood, whilst Hinduism perpetuates a complex hierarchy of classes. Further I am sufficiently prudish to find the human sexual organs unsuitable as religious symbols." He was not impressed by the Hindu temples at Benares. However, he was able to "greatly respect the intellectual achievements of some Indians" and express hope that "communism, like Islam, will be a source of vitality to the Indian genius."

Now he donned Indian dress and progressively gave up meat, espoused nonviolence, and took up the idea of painless research on animals, which he generalized to "nonviolent biology." His reasons, on being asked, often invoked the feelings of his hosts, a notion that had not been operational when he lived in England. With Spurway he took up observation of animal behavior to a greater degree and claimed to derive spiritual enlightenment from the process. For the alienated, it is a short step from revolution and the liquidation of classes to simply embracing the *other*.

J. B. S. Haldane in India, early 1960s. Photograph by Esther M. Zimmer Lederberg, estherlederberg.com.

When Joshua Lederberg visited him in India in 1957, they passed an evening on Haldane's roof watching a lunar eclipse. Looking at the moon, Haldane speculated that, if the Russians were to land there first, they might send a signal, such as a red star. Lederberg wondered if a thermonuclear explosion would be visible from Earth. Some quick calculations suggested it might barely be visible. Lederberg deplored the possibility. "He didn't disagree with me, although I think he would have still been very pleased for the pro-Soviet demonstration that I think he half hoped would actually happen."[20]

Richard Lewontin recalled meeting JBS in Europe for the first time in the late 1950s. "I thought him a bit of a humbug. Dressed in his dhoti, smoking a twisted black Burmese cigar and addressing a group of admirers in a Dutch café, he seemed unreal. Nor was I alone in that opinion." They met again in America, when JBS was shopping at a department store for Helen Spurway. "Haldane pulled out an Indian rupee note, announcing to the sales clerk 'I'm a Hindu, you know'. As he left, she turned to me and said 'That's funny. I could have sworn he was one of those high-class Englishmen'."[21]

Within India, Haldane was feted and much in demand. In a letter to Joshua Lederberg dated March 7, 1959, he betrayed a certain weariness at his new role and the usual tetchiness. "If you wish to use a given name please call me John. Jack is not my name, and nobody who has taken the trouble to ask me uses it. My sister does, but she always knows what others want better than they do. . . . I am rather isolated here. The only alternative is to chuck research and be a grand old man."[22] He was skeptical that India was overpopulated. "I was fool enough to believe the people who were worried by the falling birth rate around 1930. I think India is grossly under-cultivated and under-irrigated. I am not sure that it is overpopulated, though things would be easier if fewer babies were born."[23] John Tyler Bonner also kept up a correspondence with Haldane after their dinner in Soho, continuing when Haldane moved to India. "He seemed to enjoy living in India, although he could be as difficult with his new Indian friends as he was with the people he left. I asked him once why he had left Britain, and he said, looking at me as though I did not exist, 'Because there are too many damned Americans here, especially damned American soldiers.'"[24] Haldane's baiting of Americans was an old indulgence, which Peter Medawar had experienced with embarrassment when hosting visitors to UCL. By now it was becoming a profession.

In January 1961, a much-publicized mix-up occurred over a banquet to be held in honor of some visiting students sponsored by the

U.S. Information Service. Haldane claimed that the students were deliberately prevented by the Americans from seeing him and were told to meet with an American official instead, all supposedly for political reasons. A public week-long hunger strike in protest followed, though he did permit himself the luxury of coffee.[25] Then latent tensions at the ISI between Haldane and Mahalanobis came to a head. During a visit to India by the Soviet apparatchik Alexei Kosygin (1904–1980),[26] arrangements were made to take him on a tour of the ISI. Mahalanobis and Haldane had differing ideas about the best way to present research in Haldane's lab, so Haldane immediately resigned, in high dudgeon. Spurway followed suit.

In the interim, Haldane had turned down the offer of a trip to the United States. Writing to Lederberg, he gave a disingenuous reason. "I applied to the local American consul. Before I could even apply for a passport I had to state all organizations which I had joined since my 16th birthday (1908) with dates of joining and leaving them. As I cannot remember all these organizations or dates and object to (a) lying (b) putting myself in reach of U.S. criminal law by making a false statement, this keeps me out."[27] In another letter to a colleague he mentioned the Association of Scientific Workers and the Genetical Society in this regard.[28] Naturally, some organizations from the long list were rather more interesting than others.

Haldane worked from his own home in Calcutta, expecting to form a Genetics and Biometry Research Unit working for the government Council of Scientific and Industrial Research (CSIR). He took some loyal colleagues at ISI with him, but was taken aback when he discovered that he was now, as a government employee, forbidden to take part in politics. His absence from Europe finally stimulated a shower of honors from that direction. After receiving a large monetary award, the Feltrinelli Prize, from the Lincean Academy in Italy, he toured Europe for a few months. He addressed conferences there and accepted an honorary doctorate from Oxford and a fellowship of New College. At Oxford, he visited his aging mother. He had just

J.B.S. Haldane at Royaumont, France, in 1958.
Photograph by Esther M. Zimmer Lederberg,
estherlederberg.com.

returned to India when he got the news that his mother had died, leaving him a large sum of money. He was always generous in using his money to support research students in India. Within a year he had resigned from the CSIR, which had failed to build facilities in time for him and tried to impose red-tape regulations and other nuisances.

So it was on to Orissa, southwest of Calcutta, where in June 1962 he was to start his own lab in the capital, Bhubaneswar. Unfortunately it was "lousy with Americans" and British businessmen. Nevertheless, he continued to conduct research there and proved influential among the emerging Indian geneticists who were drawn to him and who still remember him fondly. In 1963, he lectured at the Hague on Human Society and Genetics and then went on a lecture tour to America, ostentatiously refusing again to declare whether he was a communist (the requirement was kindly waived). At a conference in Florida he discovered that he had rectal cancer, but the prognosis appeared good. The following year, 1964, he had several operations in London at the UCL hospital; they were unsuccessful, but he was not told so at the time and left there feeling unduly optimistic. It seems that "emigration" from the UK to India had not affected his use of National Health Service benefits. While he was in London, he recorded his self-obituary for the BBC and wrote some flippant doggerel, "Cancer is a Funny Thing," which was published in the *New Statesman*.[29]

I wish I had the voice of Homer
To sing of rectal carcinoma
Which kills a lot more chaps, in fact,
Than were bumped off when Troy was sacked.

. . . .

I asked a doctor, now my friend,
To peer into my hinder end,
To prove or to disprove the rumour
That I had a malignant tumour.

After Haldane returned to India, the cancer recurred toward the end of the year. When told this by Spurway, he was furious to discover that he had not been told the truth in London; it would have altered his research plans. One of his last acts was to write a letter of apology to Kosygin, who had just succeeded Khrushchev, about the incident from 1961.[30]

Dear Prime Minister,

This letter will not reach you, but might reach one of your subordinates who can convey some of its contents. In 1961 you visited the I.S.I. I had made rather full arrangements for you to see the work my colls. were doing. Prof. M. arrived on the previous day, countermanded the arrangements without consulting me, and asked me to arrange my colleagues' research material according to a plan of his own. Two of the three were absent taking students round an agri. exhibition. My wife and I instantly resigned. However, during the visit I was in the Inst. as I should have been [as] an Academic of the Sov. Union. However, Prof. M. did not think it fit to inform you of my presence. I must therefore have appeared extremely impolite, if not worse.

The letter shows that he was still a member of the Soviet Academy in 1964. When it came to certain authorities, Haldane was remarkably subservient.

Haldane died of cancer on December 1, 1964, not long after his seventy-second birthday. The BBC broadcast his recorded self-obituary a few days later. Few, if any, scientists would get that treatment today. It did not last, though. His memory has been kept alive in subsequent years mainly through the influence of his former students and colleagues in India, who continue to promote his works and stature.[31]

13. A CERTAIN AMOUNT OF MURDER

Some communists lost their faith gradually, almost imperceptibly, eventually realizing that they were no longer in sympathy with the Party. But most of those who broke decisively with the Party, and homologously the Soviet Union, were prompted by a specific episode. The decisive event may have been the Civil War, with its famines and "blocking units" behind the front lines; the show trials of the twenties and thirties (Whittaker Chambers); the "liquidation" of the POUM and others in Spain (George Orwell); the Nazi-Soviet Pact (Victor Gollancz); the invasion of Finland; the crushing of the Hungary Revolution of 1956; the invasion of Prague in 1968; or something else from that long list. Or it may have been a personal entanglement with the serious men, such as a sojourn in the Gulag after the customary torture and defilement, or such treatment of a relative or comrade.[1] The last case is surprisingly rare, as most victims of that sort of thing simultaneously believed that, while they were of course innocent—if only Comrade Stalin could learn the truth!—the others were certainly guilty. The endurance of the faithful through this grim procession of events could be used as a test of the depth of their commitment, and it often was by agencies like MI5. Those who kept on believing even after 1968 exhibited an extreme form.

Disillusionment led to a new system of coordinates, after which events previously waived became part of the narrative of the disillusionment, inverting their meaning. So a stalwart communist could keep faith from the Revolution through the purges of the 1930s and then break only after the invasion of Hungary in 1956, yet go on to revile Stalin retrospectively for the monstrosities of the show trials in the 1930s, the famine in the Ukraine, and so on, even though those events had always been known but encased in mental permafrost.

Reasons for breaking with the faith were never in short supply. The Bolshevik revolution was notoriously a violent coup d'état by a ruthless minority, originally inserted by the German army, who prevailed by killing and exiling their opponents.[2] Their grip on power depended from the beginning on political terror, pioneered by Lenin and Trotsky, along with the use of famine as a political weapon. The concept of a "class enemy" entailed a dehumanization of large swathes of the population. The imposition of ideological conformity in intellectual life was in full swing by 1918, after which it only reached further levels of refinement and penetration into all facets of life. It has been estimated that on Lenin's watch at least 1.5 million people were murdered, quite apart from the casualties of the Civil War itself. Estimates vary for Stalin's tally: choose, say, 20 million.[3] Deliberate famines; relentless quotas of arbitrary offerings for forced confession and shooting; slave laborers deliberately worked to death, or deported and abandoned to die in the Arctic wastes; a peasantry broken and stripped of everything to export grain and stock the meat-grinder slave camps; the entire 20,000-strong officer corps of the Polish army, executed one after the other by pistol-shot, and dumped in mass graves; the Old Bolsheviks, minced up as steak tartare and seasoned with their own ideology; executioners recursively executing previous executioners; foreigners, like the American Finns and others who had come to find "universal plenty," arrested wholesale and poured down the memory hole of the Gulag; scientists shot, banished, and humiliated into nothingness for pursuing "bourgeois idealism"; entire nations shattered as "disloyal," along with Soviet prisoners of war returned from Germany; the Jews, "cosmopolitans" tangled up into fantastic imaginary plots to poison Stalin in concert with foreign powers. Whole libraries cannot comprehend it all.

Few of the Communist faithful and their fellow travelers could have known all these details, but all knew the broad outline. Reading through personal accounts of their loss of faith, one is struck by how seldom a lack of information per se plays a role. It is not as if those who lost faith were not aware of events—they were only too well aware, at various levels. Consider Haldane's protégé John Maynard Smith, who was, like his mentor, a Communist Party member for twenty years during the height of Stalin's reign. Remembering those days in conversation with Richard Dawkins, he had explanations on hand.

> Just as Freudians have a sort of built-in defence against any criticism, if you disagree with a Freudian he says, 'Oh, that's all because you saw something nasty in the woodshed when you were a child,' and dismisses your arguments as being irrational. The Marxists dismissed evidence against their beliefs on the grounds that this was all capitalist propaganda. I know this, to you, will sound absurd, and to me, it sounds absurd today, but we, my contemporaries, knew of the stories about the purges, about the excesses of Stalinism, not what is known now but we knew that such stories existed, but we dismissed them on the grounds that that's the kind of thing you'd expect the capitalist press to say. We didn't believe them. And we were wrong not to believe them, but we didn't. We had a coherent world view, which fitted together, which explained things, and . . . enabled us to ignore those facts which didn't fit.[4]

Note the term "excesses," of which more later. The trouble with Maynard Smith's explanation is that more than enough information could be obtained from the *official Soviet press*. That was the point of the show trials, which spared no details of the cockamamie conspiracies supposedly "confessed" to. Often the officially blessed details were not even internally consistent as to dates and places. No matter, they were believed, or at least embraced, anyway. The same is true of the Gulag. Lenin had established slave labor camps early on—for example, in the Solovetsky islands. In the early days, before security was perfected, people once in a while escaped from there to Finland and published their experiences in the West. To take only one case, in 1926 S. Malsagoff published *An Island Hell* describing the Solovetsky camps with many of the gruesome details, such as "torture by mosquito infestation while tied naked to a tree overnight," now associated popularly with later works like the *Gulag Archipelago*. This raised enough attention to persuade Stalin to rope in Maxim Gorky for the usual Potemkin tour of the area to provide a glowing testimonial to soothe the West.

Enough of these reports had reached London by 1931 to raise a wave of protests. Russian timber was being exported to Britain under a new trade agreement concluded by the Labour government of Ramsay MacDonald, but the timber was produced by the forced labor of the camp system. The *Times* ran a long report on the system in three parts, titled "Russian Conscripts." They made a point of using only *official* statements from Soviet publications as sources.[5] Official sources had not yet been cleaned up to hide the details. The *Times* report had all the essential elements of the Gulag system accurately worked out, and it was popular enough to be republished as a pamphlet that ran through several editions. Questions had already been raised in Parliament, and protest meetings were called to exclude Russian timber produced by slave labor. One of these was held at the Royal Albert Hall on March 6, 1931. The *Times* reported that the house was full. The speakers included Winston Churchill. "The conditions there, he said, were tantamount to slavery. That Government possessed despotic power, and used that power against their political opponents, and sent them in scores of thousands to those hideous places of punishment." Firsthand accounts were also on offer.

> An escaped Russian political prisoner, speaking through an interpreter, explained that he could not give his name, because after he was arrested, his wife—in order to avoid being exiled to the north—escaped with their child and was now living in the south-

east of Russia. 'If my name were known,' he declared, 'she might be traced by the Secret Police and made to suffer because I have spoken to you'. He was, he said, a small farmer in Russia, but a little over a year ago was arrested as a political prisoner as part of the general attack which the Soviet Government had been making on the better-class farmers. He gave a moving description of the treatment of political prisoners, and said that as a result of insufficient and bad food, lack of pure water and the excessive tasks imposed (30 trees to be felled per day), disease was widespread and the mortality was very high.

(Conditions would get a lot worse later.) Throughout the meeting, counterprotesters disrupted the speakers and had to be removed by force.[6] The Communist Party had quickly mobilized and soon brought out their own pamphlets to argue that these reports were all inventions. Through entities like Ivor Montagu's Film Society, they distributed films like *Solovki* (1928) showing model facilities for prisoner "reeducation." But this is just a brief taste of the information published. A select bibliography of early sources on the subject of the Gulag and related persecution in the USSR, by year of publication, is provided at the end of this book. Though only items published between 1918 and 1961 are included, it is a long list, consisting mainly of personal accounts from camp inmates and terror survivors. (Haldane was alive when every single one of these was published.)

What communist devotion really amounted to was a decision, conscious or unconscious, to accept a certain amount of murder. The final calibration of devotion was only *how much*? That is what Maynard Smith meant by "excesses." A certain amount of terror was *not* excessive. With Haldane's help, a quantity might be arrived at. For someone like Bertrand Russell, who could plainly see in the early 1920s that the Soviet Union was a police state, even though he never exerted himself all that much in an anti-Soviet way, it was rather less than, say, Arthur Koestler, who would only draw the line at the purges of the 1930s, and far less than, for example, E. P. Thompson, who could not stomach the events in Hungary of 1956, *but had been able to embrace all that came before.* Along these lines it might be possible to derive some strictly numerical measure of just how much strain produced a breaking point, rough but useful—a normalized quotient of communist resolve.

All this analysis, the traditional measure of a post-Stalinist communist, is moot for Haldane, who never expressed any disillusionment with Stalin, let alone for communism or any of its works. Khrushchev's revelations in 1956 disappointed him only in Khrushchev. Lysenko was always and forever a "great biologist." He accepted his own suppression with nothing more than glum silence, perhaps because of the delicate matter of the X Group. He scratched out his own photographs in the *Great Soviet Encyclopedia*. No public break ever followed. Of course, he took diligent care not to accept that holiday in the Soviet Union, or any subsequent offer, but the need to wriggle from the grasp of Stalin didn't seem to embitter him in any way. At a personal level, Haldane's abandonment of his former friend Vavilov allowed him to show a finer appreciation of logical niceties than his Bloomsbury acquaintance E. M. Forster. Whereas Forster hoped to betray his country rather than his friend, Haldane knew that he could just as well betray both.

Finally, one may evaluate the posthumous reputation of Haldane's politics in the light of what he actually said and wrote. A small sample of misleading descriptions from the literature follows. The authors are far off the mark. Either they trivialize Haldane's long and active service for the Communist Party, or they subsume it under a vaguer term when the concrete one is called for, or they completely misrepresent the reasons for his lapse from the Party itself. (Emphasis in bold has been supplied.)

> The English Communists soon recognized in him one of their greatest assets. He continued to have an editorial post on the *Daily Worker* **until his break with Communism on the basis of the dogmatic biology of Lysenko and the Czechoslovak trials.** —NORBERT WIENER[7]

Although he eventually left the communist party **when he learned of the magnitude of Stalin's crimes**, he probably continued to be a believer in the basic Marxist ideals.

—ERNST MAYR[8]

Haldane and Muller resemble twins. . . . Their careers were roughly alike . . . both were **socialists** (self-proclaimed **Bolsheviks**); both were eventually **disillusioned especially by the rise of Lysenkoism** and the state destruction of science as they knew it. —ELOF AXEL CARLSON[9]

Long before the suppression of mendelian genetics by Lysenko (with the support of Josef Stalin) in the Soviet Union, **Haldane cautioned that the close relationship which existed in the Soviet State between the state and science might prove disastrous** if scientific theories ran counter to the official doctrine. —K. R. DRONAMRAJU[10]

For many years he was a member of the Communist Party. . . . But **his scientific integrity was too much for him** when Lysenko with his bogus ideologised genetics was praised and honoured by Stalin and the Party in Russia (and consequently by the Communist Party in Britain): and he resigned from the Party. —JULIAN HUXLEY[11]

It is, however, clear that a break was in any event inevitable. The reason was the commitment of the U.S.S.R. and the British Communist Party to Lysenko's views on heredity. Haldane slipped out of his connexion quietly without the, at that time fashionable, admissions of past guilt: **but not without forthright condemnation of scientific error.** —N. W. PIRIE[12]

The depth of his attachment to communism as a general philosophy made it extremely hard for him to face the situation within his own scientific speciality when Lysenko and his followers attacked the whole system of 'western' genetics and its followers in Russia. **But Haldane's fundamental honesty triumphed and he brought himself publically to admit that communist orthodoxy was mistaken**, although it must have cost him a most severe effort. —*TIMES* OBITUARY[13]

[P]eople who are tired of reading how lofty thoughts can go with **silly opinions**, or of how a man may fight for freedom yet sometimes condone the work of its enemies, have a simple remedy: they need read no further. —PETER MEDAWAR[14]

Although Haldane fell away from the Communist Party, he insisted upon maintaining an open mind about whether some of T. D. Lysenko's ideas might be correct and suggested that support for the possibility might be found in aspects of recent research in biochemical genetics. Still, **he declared himself unconvinced by Lysenko's sweeping contention—that environmental modifications of organisms were genetically transmissible.** —DANIEL KEVLES[15]

J. B. S. Haldane was one of the most **socially-conscious** scientists of this century. Throughout his life he was acutely concerned about the **social implications** of genetical findings and theories. —P. P. MAJUMDER[16]

All of us participating in this Symposium must be aware that J. B. S. Haldane was a man of **strong political convictions**, and indeed, it was those convictions that brought him here to India and ultimately set the stage for this meeting. Some of us, and I include myself, have, when he found human fallibility too much to bear quietly, on occasion followed him to Hyde Park. But his **political interests were almost as broad as his scientific**, and it would be most presumptuous of me to pretend on an occasion like this to guess in what intellectual direction Haldane would have loosed his barbs and wit. —JAMES V. NEEL[17]

Based on my research, I have concluded that Haldane's nod to Stalin was **simply another aspect of the mid-1940s adulation** of that [sic] was typical of this high point in Stalin's personal glorification. This was, after all, both the time of the battle of Stalingrad and the year that Haldane joined the CPGB.
—EDWIN ROBERTS[18]

APPENDICES

APPENDIX 1

WHY I AM [A] COOPERATOR

Haldane wrote this incomplete set of reminiscences in the late 1930s, intending to bring it out in book form. The typescript has remained unpublished ever since.[1] The text has been annotated here to clarify references and other obscurities.

WHY I AM [A] COOPERATOR
by
J. B. S. HALDANE

Dedicated to Adolf Hitler and Benito Mussolini, who converted me to cooperation—, where softer arguments had failed.

PREFACE.
By the word communist I mean not merely one who sympathizes with the general aims of Communism, and occasionally supports it with his vote or money. I mean a member of the Communist Party, which is a section of the Communist International. Readers of this book will soon see why this distinction is necessary.

This book is not intended for proletarians—or shall we say for members of families with an income of less than £4 per week. Some of them may read it out of curiosity, but my arguments are not aimed at them. Frankly I think that those of them who are not members of or active sympathizers with the Communist Party are mugs, deluded by the vast weight of propaganda which is poured over them daily.

But it is not so obvious why a member of the bourgeoisie (or Social Class 1, as the Registrar General calls it, or the Upper and middle classes) should support a movement directed against the class of which he is a member. It will be my object to explain why he or she should do so.

Communism involves both theory and practice. And one joins the Communist Party both on intellectual and emotional grounds. The intellectual conviction by itself would rarely lead to practice, and one would be left in the Limbo of chronic sympathizers. The emotion alone might lead one to anarchism, window-breaking, or drink. But emotions are the least objective of things, and a true account of them must necessarily be autobiographical.

However to be autobiographical is not to be egotistic. To admit that one has emotions is not to own oneself their slave. I am no Augustine or Rousseau, and most of my emotional life is, so far as I can see, irrelevant to my politics, though a psychologist might think differently. Further I am well aware that in different people the drive towards Communism may come from profoundly different emotions. My own case is certainly not typical. It may be instructive.

CHAPTER I.
Autobiographical. Youth.

I was born in 1892 in a house in North Oxford with a rental of £60 per year. My father was demonstrator in Physiology at Oxford University. Later he became a Gas Referee, with a salary of over £1000 per year, and finally inherited capital from several sources.

I went to a very good day school at Oxford between the ages of 5 and 12. When 12 years old I won a scholarship at Eton. This college was originally founded by King Henry VI for poor scholars. My annual bills there were seldom much under £100 per year, while "oppidans", that is to say non-scholars, cost 4 or 5 times as much. Our pious founder's gifts to catholicism and the poor have been pinched by protestantism and the rich.

From an intellectual point of view the education available at Eton in 1905-1911 was good. It was possible to escape being educated at all, and any oppidan who made the attempt to learn was liable to bullying which in some cases left him a nervous wreck for life.

On the other hand a colleger could learn a good deal, and I did. When I left at 18 I could read Latin, Greek, French and German. I had won a mathematical scholarship at Oxford. I knew enough chemistry to take part in research work, enough biology to do unaided research, and I had a fair knowledge of history and contemporary politics. I knew about the Provisions of Oxford and the League of Schnalkalde. I knew the name of the Hungarian Prime Minister and the relations between the German Reichstag and Bundesrath. I have forgotten them all since. On the other hand I knew no economics. I had been forced to read parts of John Stuart Mill's Principles of Political Economy, and found them dull and unintelligible. And I knew practically no psychology or technology. Nevertheless I do not conceal my debt to T.C. Porter, M.D. Hill, and C.H.K. Marten,[2] who taught me chemistry, biology, and history, respectively, nor to Hollway-Calthrop,[3] the bursar, who treated me as a human being.

From other points of view my education was not so good. There was a very great deal of homosexuality, occasionally reaching the point of sodomy. In College, and in some of the houses, a fair number of the liaisons between two boys did no great harm to the character of either. But where there was much disparity of age the younger boy was not always a free agent. The Eton Society, or "Pop" included the most distinguished and popular athletes. The shapely youths who were alleged to assuage the desires of this august body, often in return for presents, were known as "Pop bitches". Some of them have since risen to positions of high distinction and trust.

I did not participate in these activities, in fact my sexual life was confined to a very occasional kiss on the rare occasions during holidays when I was left alone with an attractive girl. It was a long time before I could associate with women as equals. I am convinced that the segregation of the sexes of the English ruling class at puberty has a bad effect on our whole social life.

I was not brought up as a member of any religion, however I took part in the customary religious ritual of my school. On week days we had twenty minutes in chapel in the morning and ten of prayers in the evening. On Sundays the two chapel services occupied two hours. I developed a mild liking for the Anglican ritual and a complete immunity to religion. It is open to question whether boys should be so drastically immunized against what is, after all, one of the great motive powers in history. On the whole I think the answer is "yes", but I am by no means certain. As a member of the guard of honour on such occasions as the funeral of Edward VII and the Coronation of George VI, I was also immunized against ceremonies of this type. But this is a personal immunity. As long as they go on, I have no objection to other people watching them. After all they pay for them.

Our education as regards social duties was most interesting. It was assumed that we would be rich, and there was no suggestion that it was wrong to live on an unearned income. But a good member of the ruling class, we were told, would spend several evenings a week doing "social work" among the poor, for example organising boys' clubs in the London slums.

Exercise took the form of that contradiction in terms, compulsory games. It was however possible to take long solitary rows in a light skiff in summer, and on the whole athletics were less of a tyranny than in most English "public" schools. I was utterly bored by all games, and did extremely badly at them.

I was a rebel against the normal modes of behaviour, and probably a nuisance, not, I think, from any set purpose, but because I simply was not interested in events such as house-matches, which aroused the keenest emotions in my contemporaries. At any rate the senior boys in College did not like me. On one occasion I was caned by them every night for a week, at least once on the soles of my feet. English people who believe that the tortures which occur in German concentration camps are impossible in England are apt to forget that such events are by no means unknown among the ruling class.

The most valuable part of my normal education was carried out by my father in the holidays. He was occupied with physiological questions connected with mining and diving. I think I was four years old when he first took me underground. I was certainly very frightened. However by the time I was ten years old I was at home in a tin mine. At 12 I went down in a submarine and at 13 in a diving dress, which incidentally leaked badly. Somewhat later I went down a coalmine which had recently exploded. I have described the experience elsewhere.

After going through it I did not find the Western Front in 1915 particularly alarming. I was also mildly vivisected from time to time.

At 13 years I began to do calculations for my father, and at 17 I read my first paper to the Physiological Society. Ever since then I have been engaged in scientific research.

At the age of 18 I went as a scholar to New College, Oxford, where I was much happier than at Eton. In my first year I nominally studied mathematics, but actually did part of the final honours course in zoology. In my second and third year I took the course in Literae Humaniones [sic], or "Greats". This curious course, mainly concerned with ancient history and ancient philosophy, provided an opportunity for weighing evidence and writing essays. The subjects studied had little relation to modern life, so thought on them was free. The successful Greats Man, with his high capacity for abstraction, makes an excellent civil servant, prepared to report as unemotionally on the massacre of millions of African natives as on the constitution of the Channel Islands.

I had the great privilege of writing essays which were criticized by Mr. H.W.B. Joseph. He did his best to teach me not to use a given word (e.g. "idea", "feel", or "good") with two entirely different meanings in the same paragraph. This has often laid me open to the charge of pedantry. Pedantry, as I have remarked elsewhere, is the term which, in controversy, we apply to the accuracy of our opponents.

At the same time, with A. D. Sprunt (killed in 1915) I discovered the first case of linkage in vertebrates, a phenomenon which enables maps to be made showing the positions of genes in chromosomes. During the first year of the war my sister (now Mrs. Naomi Mitchison) looked after the mice and obtained the results which conclusively proved our points. Twenty years later I discovered the first cases of linkage in man.

I rowed in the college second boat, finding, rather to my surprise, that voluntary sport was enjoyable. I took part in innumerable college societies, but was not profoundly interested in politics. I was a liberal with leanings further left. I was considerably influenced by my contemporary Herron (killed in 1915) who was a syndicalist. He had a clear mind, and a real passion for justice which would have taken him a long way. Beyond joining the University Liberal Club and taking part in debates I was inactive politically.

But I was a member of the University Cooperative Society (this too died in the war) and even served behind its counter on occasion. My only serious political gesture was, I think, in May 1913. Oxford was then served by horse trams, which could readily be overtaken by a runner, but went definitely quicker than most people can walk. Neither the drivers nor the conductors earned so much as £1 per week. Wishing to remedy this state of affairs, they struck. Their places were taken by blacklegs. On the first three evenings of the strike trams were stopped and the horses taken out. The police made baton charges, and finally order was restored. I was unable to participate in these riots, I think because I was in training for a race. On the fourth evening the streets were quiet. I walked up and down Cornmarket Street chanting the Athanasian creed and the hymeneal psalm "Eructavit cor meum" in a loud but unmelodious voice. A large crowd collected. The police ineffectively pushed pious old ladies into the gutter. The trams failed to penetrate the crowd and their horses were detached and wandered off in an aimless manner.

The strike was successful, and as the trams could no longer yield a profit, they were replaced by motor omnibuses, which were capable both of higher speed and higher wages. I was subsequently martyred by the proctor to the extent of two guineas. This was, I suppose, the first case for over three centuries when a man was punished in Oxford for publicly professing the principles of the Church of England.

I was a member of the University Officers' training corps, and in 1914, joined the signallers, intending to learn radiotelegraphy. But I did not get very far. On the night of June 1914, I was sleeping on the heather at Hartford Bridge Flats, near Camberley, when, about 1 a.m. an Austrian commercial traveller named Sobotka arrived on a motor bicycle attempting to interest us in various new radio devices, and also announcing the murder of the Archduke Franz Ferdinand, and the probability of a war between Austria-Hungary and Serbia. We had been highly educated. Some of us had vivid imaginations. But we had not envisaged the possibility that the Angel of Death should arrive on a motor cycle, announcing not only the death of many of us signallers, but the death of the culture into which we had been born.

On August 4th, 1914, the announcement that I had been awarded a first class in Literae Humaniones [sic] was somewhat overshadowed by other events. I had intended to take a six weeks' holiday walking on the continent of Europe. Since that date I have never taken a holiday of more than three weeks. On August 12th I received a commission as a Second Lieutenant in the 3rd Battalion of the Black Watch. Given my opinions, I was right to join. I was not mistaken in fighting for the various causes which figured among the allied war aims. I was mistaken in thinking that these aims could be realised under capitalism. At the time of writing the various things for which we fought are being handed back to Germany on a plate. I went to France in January 1915. In February I became bombing officer to the 1st battalion of the Black Watch (42nd foot). I was a singularly inefficient platoon commander, because I could never remember the names of all the corporals, let alone all the privates, under my command. However I probably made up for this by my efficiency in patrolling between the lines at night.

In March I became the first trench mortar officer of the 1st

(Guards) brigade. April was one of the happiest months of my life. So long as any battalion of the brigade was in the line I remained within a mile of it, always in the sector between Neuve Chapelle and Festubert. I did not take my boots off for three weeks on end, and became fairly lousy. I believe that my battery of 13 small and very mobile muzzle-loading mortars was rather efficient. For example on one occasion a volley of eight bombs onto a suspected machine-gun position provoked a reply in which I counted over 100 shells.

I lived a curious life with Lance-Sergeant Evans of the 1st Coldstream Guards, and a dozen or so men. We had no particular discipline or regular hours, and sometimes I only got one meal in the day. We had a little workshop, where we put the fuzes into our bombs and did minor repairs. I am possibly the only man who ever made smoking compulsory in a bomb factory. In this case I did so on psychological grounds, as I thought it important that we should have absolute confidence in one another and in our weapons. We had no accidents and few casualties. From this truly enviable life I was suddenly recalled to Hazebrouck about May 1st, 1915. The Germans had attacked with chlorine north of Ypres. My father had been sent out to tackle the menace. I met him at Hazebrouck, and we started trying respirators of various kinds in a room in the college there in which chlorine was liberated. The concentration was not sufficient to cause fatal injury to the lungs in less than 2 minutes or so. But it made one cough very much sooner. About half a dozen of us went in, trying a different type of respirator; and another would take his place when he had inhaled enough gas to incapacitate him for a few hours, or in one case, for several days.

By May 8th we had a crude respirator which was at least of a certain value against chlorine. I was sent, so far as I can remember, to report as gas officer to the first division. Had I done so, I should probably have come safely through the war (and my chest would have been covered with medals). However on the morning of May 9th the Division had moved its headquarters forward. The whole Division was concentrated in a small area within a mile or two of the front line. The great spring offensive had begun. On my way up I encountered Micky Scotland, our platoon bard, who composed, among other things, a rude song about the sergeant of another platoon, which we used to sing on the march, and of which the refrain was

> "Oh, Oh, Oh, Oh Harrison
> Put me in the guardroom if you can."[4]

Micky had been hit in one foot, and was dancing with joy on the other. He had married on the day before mobilization, and was thoroughly delighted at the prospect, which had sometimes seemed to him rather remote, of seeing his wife again. Harrison was killed an hour or so later, so he never put Micky in the guardroom.

I discovered that the Black Watch was due to go over the top in the course of the afternoon. With more zeal than discretion I determined to join them, and if possible to command my old platoon. I began to run. But fortunately the moderate dose of chlorine which I had inhaled prevented me from expanding my lungs. The best that I could achieve was a moderate trot worthy of an old gentleman with chronic bronchitis. The bombardment began when I was well behind our reserve line. The enemy replied with a barrage of high explosive shells from his howitzers. The noise furnished (though I did not so verbalize it at the time) an excellent example of the transformation of quantity into quality. In the other battles in which I have taken part before or since, one heard the reports of individual guns, the bursts of individual shells, and their screams as they tore their way through the air.

Imagine the loudest bang you have ever heard, say a clap of thunder from a house struck in your immediate neighbourhood. Now imagine this prolonged indefinitely, a solid bang without intermission. And behind this, like the drone of a bagpipes behind the individual notes, a sound as of devil-driven tramcars taking a sharp corner.

Lesser bombardments had frightened me. This entirely novel sound intoxicated me. I ran forward through the monstrous black bursts of smoke and fountains of earth and bricks where the German shells were exploding....

I woke up, and began to scrape the earth off me. I noticed blood on my face and hands, and pains in various places. I realised that I had been hit. This struck me as funny, an automatic psychological defence reaction of considerable value. I ran on to a house and took stock. I was wounded in the right arm and left side, but my face was only scraped. I climbed up to the first floor and watched the battle through a loophole in a largely ruined and heavily sandbagged house. The enemy's front line trench seemed to boil like a pot under our bombardment. It was hard to believe that any Germans were left alive in it.

Then our barrage lifted and the Black Watch went forward. Sometimes a whole line would lie down suddenly. It was difficult to realize that they were not obeying some order, that in fact all of them had been wounded and most of them killed. I saw one man, and one only, turn back. As he neared our trench a shell burst very close to him, and he fell. So great has been the influence of the various religions on my mind, that I cannot help attributing what is probably an undue importance to a man's last act. "Suffer us not, in our last hour, through any pains of death, to fall away from thee" I found myself quoting from the Anglican burial service.

In spite of very heavy losses, a fraction, perhaps nearly half, of our men got into the enemy trench. I believed that I had witnessed a victory. But very few indeed of those who got across either returned or were captured. Half an hour later the shelling died down, and I walked along our reserve trench, full of Coldstreamers, to a road which I knew led back to casualty clearing stations. I was in a curious mental condition, and I seem to remember remarking to an uncomprehending guardsman "The real is the rational," a saying of Hegel's which appeared to me to be refuted by the existing circumstances.

All the ambulances were full, and I walked back for two or three miles. At Le Touret I met the Prince of Wales (afterwards Edward VIII) in a high-powered car with a staff officer acting as his bear-leader. He offered me a lift to Army headquarters, where I gave a very misleading account of the battle. The unfortunate prince was, I believe, sufficiently intelligent to realise the hopeless anomaly of his position, and I think he honestly tried to get killed. But the authorities reserved him for a less honourable fate.

Somehow I found myself in a hospital in Bethune. The wound in my side was trivial. A splinter had gone through my haversack, bumping me considerably, but only penetrating just under the skin, where it still is. The wound in my arm was deeper. It was probed and roughly disinfected with hydrogen peroxide, which was momentarily painful. Then I was put in a bed in a ward full of wounded officers, some in great pain, others delirious. An occasional shell-burst penetrated their groans, but I slept.

The next afternoon I was placed in a hospital train and the following morning I reached Versailles, where I was put in the Trianon Palace Hotel. The congestion of the medical services was such that I passed two nights after being wounded before I got my boots off. Ten days later my arm wound was healed and I was sent back on leave to England.

I reached my parents' home at Oxford, and at last removed my uniform, which was both bloody and lousy. My arm remained stiff, and it turned out that the X-ray examination at Versailles had been inadequate. A piece of shell had run right up my fore-arm and lodged above the elbow. I was operated on, and the wound turned out to be septic. I awoke in great pain, which continued for a week, while green pus oozed from my arm. It is still occasionally painful. The shell splinter remains in my arm, and the scar tissue on my chin is very liable to bleed when I shave.

I went back to duty with my special reserve battalion in July. I believe that I was a genuine case of shell-shock, as opposed to the war neurosis which was usually dignified by that name, for I jumped like a shot rabbit at any explosion. However I volunteered to organize a bombing school, and fatigued this reflex out of existence in about a week.

I commanded this bombing school in the north of Scotland for nine months, during which I instructed some hundreds of officers and N.C.O.s in the technique of killing other people with hand and rifle grenades. My methods were unorthodox. I took the view that almost all accidents arose, either from unavoidable causes, such as gross faults in the grenade, or from carelessness or panic, both of which have psychological roots.

I began by lecturing on the anatomy of hand grenades, and made each pupil attach a detonator to a fuze with his teeth. Should the detonator explode in the mouth, I explained that the mouth would be considerably enlarged, though the victim might be so unfortunate as to survive with rather little face. Pupils who did not show alacrity when confronted with this and similar tests were returned to duty, as unlikely to become efficient instructors.

The principal danger of academic bombing is as follows. The grenade has a time-fuze, taking five seconds to burn in the Willis Grenade, 3 ½ seconds in those now used in Spain. If, at the end of this period the bomb and its thrower are not on opposite sides of a bank of earth or a stout wall, he is unlikely to throw any more bombs. A pupil occasionally drops his bomb, or it may fall back into the trench. Once or twice I have known a man, after lighting his bomb, hold onto it paralyzed with terror, perhaps fascinated by the hissing of the fuze, as birds are alleged to be fascinated by serpents.

In such cases the instructor has somewhat under five seconds in which to save two or more lives. He must decide whether to pick up the grenade, to dodge round a corner, to strike the paralytic pupil on the chin and throw his bomb for him, or to throw a sandbag, the pupil, or himself, on the top of the bomb, in order to save other lives.

If his mind is clear he will do one of these things, if not, not. If a bomber is filled with inhibitions, and hedged round with precautions, he will make an error of judgement at the critical moment. During my command we had no accidents beyond an occasional scratch not involving hospitalization. Three weeks after I left an officer was killed during an act of quite unnecessary heroism, the alternative to cowardice in a man whose soul has not been adequately dealt with by the instructor. After some months I found this job exceedingly dull. Once a month I got leave to spend a week-end in the great and wicked city of Inverness, where on one occasion I got disgracefully drunk, and might have been court-martialled but for the kindness of a superior officer.

Twenty years later I discovered that I was left at this school and later given a particularly silly administrative job which lasted for six weeks or so, instead of being sent back to France, through the intervention of a general at the War Office who was a friend of my late uncle, Lord Haldane. If he is alive and reads this book he may repent of his favouritism.

In September 1916 I was sent out to Mesopotamia to join the 2nd Black Watch (73rd foot). The voyage round the Cape in an overcrowded transport took eight weeks. As ship's adjutant I had charge of over 2,000 men. I also made the acquaintance of various brothels in Dakar, Durban and Bombay, from which I had to see that my officer comrades returned in due time. The hostesses in Bombay were European women. English prostitutes were forbidden on grounds of national prestige. However English women were permitted to manage the brothels. The distinction throws an interesting light on the official mind.

We reached Basra in December 1916, and I went up the Tigris in a small steamer towing two barges with Indian troops. On the ship I read the only two serious books which I could buy in Basra, Robert Doyle's "The sceptical chymist" and Robert Browning's longer poems. I belong to the small class of human beings who have read "Sordello," but not to the far more select band who have understood it. The river was low, and we constantly ran aground. I made my first acquaintance with the remarkable objects with which a local bard (an artillery Officer, I think) described as follows in "The Mesopotamian Alphabet."

> ["]B is the biscuit that's made in Delhi.
> It breaks your teeth and bruises your belly,
> And grinds your intestines into a jelly,
> In the land of Mesopotamia."

I soon discovered that about one tin in three of these biscuits was infested with weevils, which mined their way into them and softened them considerably. Many of the men under my command did not like eating live weevils. However I had long since discovered that a soldier must get rid of many emotions appropriate to a civilized man, if he is not to be so acutely miserable as to lose much of his efficiency. I was generally able to get biscuits which had been partially pre-digested by the weevils, and liked them well enough. Nevertheless I broke two of my teeth.

I joined the 2nd Black Watch (73rd foot) then in reserve behind the trench system which ran from the river Tigris to Lake Suwaikiyeh. Colonel Wauchope (who later held Pontius Pilate's office of governor of Palestine) was in command.[5] He was a real liberal, and proved the futility of most of the "good order and military discipline" which infests the British army, and which is in the main a mere by-product of the class struggle.

Even in the 1st battalion there had been a quite unusual solidarity between officers and men. This was shown by the fact that N.C.O.s who were given commissions remained with the battalion, though transferred to a different company. Serjeant Wallace, who was in charge of the officers' mess, was definitely biassed [sic] in favour of existence. He therefore resisted the offer of a commission as long as possible. But when he became an officer neither he nor we who had recently eaten his excellent meals, felt appreciably awkward. For, whilst a sergeant, he had been treated as a human being. He accepted his promotion with a humorous resignation, and survived it for two months.

In the 73rd things went a good deal further. Perhaps half the officers had started as privates, and several others were of working class origin. 2nd Lieutenants invariably wielded a stick and shovel with their men. While orders were enforced, any officer giving an unnecessary order, was censured. The colonel took the men into his confidence. If an attack was ordered he would draw up half the battalion at a time, and point out to it not only the grounds for this particular order, but the reasons against it, and the ways in which the operation might be expected to fail. This was valuable for two reasons. To quote the alphabet again,

> "W stands for the wonder and pain
> With which we regard the infirm and insane
> Old Indian generals who run the campaign
> We're waging in Mesopotamia."

The "Indian" generals in question were of course British officers in the Indian army, and their incompetence had been largely responsible for the capture of Townshend's army in Kut-al-Amara.

The Indian troops presented a different problem. We were the only British battalion in a brigade containing three Indian infantry battalions and innumerable Indian mule and camel drivers. During a battle any British soldier might have to lead or rally a number of Indians, and for this purpose a knowledge of the tactical situation was necessary.

In the officers' mess we spoke the language which was native to the majority of us; Lowland Scots. This keeps a far better correspondence with the written language than does the dialect of South Britain. For example the letter R is universally pronounced. And many features of its grammar, particularly the use of prepositions as objects, as in "I want out," appear to be in the natural line of development of our language.

Our conversation was often fairly intelligent. We would discuss incidents in mediaeval Scots history or topics in elementary physics, with a vehemence which was encouraged by the complete absence of reference books. I did not subsequently come across as intelligent a group of officers until in March–April 1937 I was with the English-speaking battalion of the Spanish People's Army. There I remember one evening when the conversation in the battalion head-quarters dug-out passed from adultery to telegony. The commanding officer (a former mutineer in the navy) upheld the view that the foetal and maternal circulations anastomose in the human placenta, and a fascist attack unfortunately terminated my effort to convince him that he was wrong.

Our economics were of interest. There was very little that we could buy. I spent a quite inordinate sum on Libby's tinned asparagus. Had I not been on duty and near the front line, I might have got drunk. I had taken up auction bridge on joining the army, and used to make an average income of about £1 per week by it for 1/- a hundred. In Mesopotamia money was irrelevant. We might hope for a month's leave in India after a year in Mesopotamia, but this was a remote prospect, like the resurrection of the dead, and indeed the latter perhaps concerned us more closely. So bridge became a bad game. No-one worried to pay his debts or to ask for payment, and we frequently went down three tricks re-doubled.

I was appointed second-in-command of a company, and also took charge of the battalion snipers. For a while we held some miles of the right bank of the Tigris while the Turks held the left. We occupied the site of the Battle of Beit Bissa, and occasionally played a crude football with the sun-dried heads of the unidentified dead of the previous year, whether British, Indian or Turkish we did not know. We had become psychologically adapted to war, and wholly accustomed to the idea that next year someone would be playing football with our own heads. I think that my snipers and I accounted for a few Turks. We also toured the neighbourhood. The [sic] The Bhopal rifles on our left complained that they were being killed by Turkish snipers. I went along with a couple of men to deal with their problem. These wretched Indians were completely and utterly fed up. They hated the war, and did not wish to kill the Turks, their fellow Muslims. They hated life, and allowed the Turks to kill them. By a judicious display of helmets on sticks we drew the Turks' fire and located them. We returned their fire and got our gunmen to give then a few rounds of shrapnel. The Turks, who had no desire to die for their country, but could not resist the opportunity to kill the Indians who were no more dangerous than rabbits, went away to a safer place. The majority of

Indian battalions were not so demoralized, and in particular the 6th Jats struck me as firstrate soldiers.

On normal days we would crawl among the scrub by the river bank and exchange bullets with the Turks opposite. We claimed to have hit some of them. They did not hit any of us, but sometimes came fairly near it. In the evenings, especially when we were at rest behind the line, I used to practice marching on bearings with a prismatic compass. I finally managed to achieve errors of only half a degree or so, and it would have been my business to guide the brigade on a night march had I taken part in the subsequent advance.

Life would have been pleasant enough had I not suffered from constant tooth-ache. I once got down the river to the solitary dentist who ministered to some 20,000 men, but he merely had time to enlarge some cavities.

The trench warfare was entirely uneventful. We had markedly superior artillery, though not decisively so. To destroy the Turkish trenches we relied on trench mortars. But the main attack was made on a different part of the front. On the day when the British forces crossed the Tigris at Shumran, a day's march to our north, our battalion attacked almost abandoned trenches, and advanced a long way with very small loss.

Along with two or three other officers, and some N.C.O.s, including a serjeant who had frequently been recommended for the V.C. and was very angry, I was left behind to reconstitute the battalion should it be wiped out. Fate however decided quite firmly that the intentions of my guardian in the War Office should be frustrated that day.

About noon the adjutant (Captain Blair) and I were whiling away the time putting fuzes into Mills grenades. I heard a shout, which conveyed little to me, and jumped up just in time to see a bomb which he had been fuzing burst about five yards from me. He had dodged behind some boxes, but I had not understood his warning.

I was not hit by any of the numerous splinters, but about 6 p.m. one of the hangers of the aerodrome near our camp caught alight.

With several comrades I ran to deal with the fire. We got a number of lorries out of the burning hangar (a very large tent) and then went nearer to the fire to see what we could do. The Air Force had somewhat imprudently stored their bombs and their petrol side by side. The petrol flared up magnificently. But whenever a bomb went off the blast from it blew out the flames. However enough glowing sparks were left to ignite the petrol again.

I was running towards this fire inside the hangar when a bomb went off with a particularly violent report. I could not breathe, and supposed that my throat had been blown away, in which case I could expect another half minute or so of life. However I felt my throat, which was intact, and now attribute my choking to a reflex spasm of the glottis. I noticed that I was hit in the leg, and proceeded to run away as fast as I could, which was not very fast, though a number of other explosions increased my zeal.

I found that I had a fairly deep but almost painless flesh wound. Another officer was slightly wounded, but one of the men had a splinter in his stomach, from which I think he died. I was put onto an ambulance, and jolted over an execrable road to a field dressing station, holding the hand of the man with the stomach wound, who was in great pain, and trying to prevent him from being jolted too intolerably.

But I had one more trial in store. I was dumped in a small tent, where I think I collapsed into stupor. Anyway an officer who was brought in soon after said that when he arrived I woke up saying "That was a proper bang". My comrade was wounded in both legs and one arm. He tried to light his cigarette with the remaining hand, but only succeeded in lighting the tent above his head, which someone had thoughtfully soaked with oil. The tent burned above our heads in a steady and undemonstrative manner, and neither of us was in a position to get up and put it out. However we managed to shout, and someone else extinguished the fire.

My wound was almost painless. To be accurate, it was more painful than my toothache for about two minutes while being cleaned up for the first time. This was lucky, for although the medical services were vastly improved since the previous year, they left a good deal to be desired. We were, it is true, afforded the services of a chaplain at the Casualty Clearing Station. He enquired my religion, and when I answered "none" put me down as C. of E. It is not obvious to me that military discipline is improved when wounded men are insulted by attributing to them opinions to which some of them object strongly.

However on the hospital ship things were somewhat primitive. We lay on the deck in rows on stretchers, and when I rushed to relieve myself I had to crawl downstairs to a lower deck, which, when one's leg is wounded, is definitely uncomfortable. I noted, with interest, that the officer on my right was reading Lamb's "Infinitesimal Calculus," whilst I was engaged on Kelland and Tait's "Introduction to Quaternions," which incidentally I do not recommend as a preliminary course in vectorial analysis.

I passed a week or so in hospital at Amara, and went down to Basra in a much better ship. The wounded were mostly very cheerful. A man next to me had a colostomy (i.e. an opening into his intestine) from which the products of his digestion escaped every few hours. He appeared to be particularly interested in the fashion pages of the London papers. Another had lost his hand, and was smiling constantly. Perhaps he was glad to be definitely incapable of returning to the front. And he certainly did not know that his disability pension was later going to be reckoned as part of his income under the means test.

As my wound would not heal, I was shipped to India, where I passed another month in hospital in Poona before it finally skinned over. There I learned Urdu, an easy language which is picturesque both in its script and diction. Thus the following conversation would occur when my servant wakened me (for I soon obtained a Muslim servant with a beautiful beard).

"Peace, Presences."
"Peace, Lord Glorified Longer."
"Which clothes will the Presences deign to wear?"
"We shall wear our khaki shorts and stockings."

It is characteristic of British imperialism that boys of the ruling class get no opportunity of learning any of the Indian languages at any "public" school known to me, and that English women in India are often unable to speak one grammatically even after many years' residence. It is not obvious why they should expect to continue indefinitely to go on ruling a group of peoples which they make so little effort to understand. There were of course exceptions. The wife of my cousin, Gerald Keatinge, in the I.C.S., who did much to improve agriculture in the Deccan, accompanied her husband through the country and loved it. But there were not enough exceptions.

In June I went to Simla, the summer capital, where I stayed till December, first in a convalescent home and then in Army Headquarters. The white society there had altered very little since Kipling described it in "Plain Tales from the Hills." Noone but the Viceroy was allowed a motor car. Others walked, rode, or went in rickshaws drawn by men whose hearts are said to give out in a few years.

The key to social life in India was snobbery. The Hindu caste system is the greatest glorification of snobbery that the world has ever known. If I am a Brahmin I not only enjoy privileges in this world and probably the next but I deserve to do so for having been good in past lives. Besides the Hindu classification there is the official hierarchy, headed by the Viceroy, and based on military and civil rank. And there is the classification based on wealth. A rich British merchant is looked down upon as a "box-wallah" and is not eligible for certain clubs, although he is the historical raison d'être for British rule. But he will naturally be able to afford a higher scale of living than a colonel. Finally the educated Indians attach an exaggerated importance to university degrees.

The official hierarchy includes Indians. For some time the senior officer in Simla was General Sir Shemshere Jung Bahadur Rana,[6] the hereditary premier of Nepal, and all British officers saluted him as a superior. The ideal position in India would seem to be that of a wealthy Brahmin with an Oxford degree, and a high rank in the civil service.

Each hierarchy asserts its own claims, and much can be done by exploiting them. Thus a British Officer travelling first-class was joined by a very holy Hindu whose asceticism forbade him certain ablutions which are usual (for the average Indian is much cleaner than the average European). At the next halt the British officer bought a first-class ticket for a servant of his who happened to be an untouchable, and installed him beside the holy man, who departed in horror and disgust.

I had very little personal contact with Indians in Simla. Indeed I have never yet been intimate with an Indian, though I hope to be so in future, as I greatly respect the intellectual achievements of some Indians. I can only become intimate with anyone on a basis of equality. So long as I am a member of a "ruling race" such equality is impossible. When an Indian and I are both genuinely trying to be polite to one another he probably suspects that I am being servile. The Indians with whom I got on best were Indian army officers holding King's (as opposed to viceroy's) commissions, with some of whom I used to play chess.

At Army Headquarters I was set down to revise various guidebooks to areas where military operations were likely, for example the Kurram valley on the Afghan frontier where fighting took place in 1919. I also wrote a most illuminating history, with graphs, of the transport of men, animals, food and munitions during a recent campaign against the tribesmen in Waziristan. The communications were continually interrupted by floods which cut both roads and the railway.

I continued this epoch-making work when I came down to Delhi for the winter. But I was not overworked. I was able, for example, to watch meetings of the Legislative Council, a body containing a majority of government representatives and a minority of Indians elected on a very restricted franchise. Among these was Mr. Muhammad Ali Jinnah, who had developed the Oxford manner to perfection. His questions to ministers were a model for undergraduates wishing to raise a laugh in the Oxford Union by interrogating its officers. He had completely mastered the art of scoring verbal points, and was well aware that his words would have little or no effect on actual policy. He was thus educated in irresponsibility. So completely was the futility of the proceedings realised that I was sometimes the only occupant of the public gallery during a debate.

This divorce of theory from practice is perhaps congenial to the Indian temperament, as witness the numerous philosophical systems, several of which include a fascinating physics completely divorced from fact (I think of the Nigodas of the Jains and the ___ of the Samkhya). But if I was ever tempted to despise the practical powers of the Indians, the buildings of Delhi soon corrected my error.

A whole cycle of architectural development, beginning with cumbrous but impressive buildings such as Tughlak's tomb in the Early Pathan style, roughly corresponds to our Norman, lay before me, mostly in cities now abandoned. It culminated in the Jama Masjid, a mosque now in use. This building is (in my opinion) such a masterpiece of design that, until one realises that one can see its detail quite well from three miles away, one does not notice that it happens to be the largest mosque in the world.

The Taj Mahal, which I saw later, is perhaps a better building. It has a more universal appeal, but I detect in it a sentimentality which is foreign to the spirit of Islam. One can no more appreciate a mosque until one has read the Koran and seen Muslims worshipping, than one can understand a Christian cathedral if one has never read the Bible or attended some services. Nor, incidentally, are those great and noble temples of Mammon, the sky-scrapers of New York and Chicago which dwarf their cathedrals and would dwarf the great pyramid, intelligible unless one understands enough of the nature of capitalism to realise its greatness while observing its decline and expecting its fall.

None of these buildings however, nor the great mosques of Cairo and Kiarouan,[7] impressed me so much as the Muslim architecture of Northern India. I have never been to Greece, and may have to revise my opinion. But until I do so I shall continue to assert that India has produced the world's greatest architecture. It would be ludicrous to despise a people with such an achievement, and uncivilized not to deplore their present sterility in the creative arts. One can only hope that the new India which is growing up will be as creative as the old, and that communism, like Islam, will be a source of vitality to the Indian genius.

I could sympathize less with Hinduism amongst other reasons because I am a potential Muslim but not a potential Hindu. An act of faith admits one to Islam. Hindus are born, not made. And Islam is a religion of universal brotherhood, whilst Hinduism perpetuates a complex hierarchy of classes. Further I am sufficiently prudish to find the human sexual organs unsuitable as religious symbols.

The Hindu temples which I saw, particularly at Benares, did not greatly impress me, but I was deeply moved by the popular religion which finds its expression in Pilgrimages. I went on one such, the Magh Mela which occurs at Prayag, just outside Allahabad, a holy spot where the Ganges, the Jumna, and an invisible river called the Sarasvati, meet. It takes place every twelve years or so, in January, when Jupiter is in the constellation Pisces. I could not get leave to go on the biggest day, so when I went there were only about a million and a half people gathered on a piece of open ground covering perhaps two square miles.

The processions were beyond the dreams of Hollywood. Some thirty elephants carried solid golden idols (non-Christian images are called idols) in solid silver howdahs. A few gorgeously clothed holy fat men were carried in litters. Thousands of even holier stark naked men, their skin covered in ashes, and their hair matted with cows' hair, walked down to the holy rivers. A few policemen separated rival sects who might have quarreled.

The people were fantastically gentle and orderly. There was no pushing or horseplay. Waves of religious emotion crept over them like gusts of wind over ripe wheat. "Sri Ram, Sita Ram," they chanted. Everywhere I was welcomed. Noone suggested, as temple priests had often done, that I was out of place. On the contrary I was told that everyone in the world must attend this Mela once. So, if Hinduism is true, I shall almost certainly be born a low-caste Hindu in my next life, especially as I sprinkled some of the water of the Ganges on my forehead. Though of course I have eaten a lot of beef, a sin which is not easily atoned.

Apart from policemen on duty, I was the only Englishman, perhaps the only non-Hindu, at the Mela that day, or so I was told. True, there was a war on. But officers, at any rate, could readily get leave for such Purposes as a day's shooting. And there were thousands of unemployed white women. As a show, the Mela was certainly unique. A hundred and fifty thousand is reckoned a big crowd at Lourdes. And no-one will suggest that the British in India are such good Christians that they would not attend a heathen ceremony. As a matter of fact things are the other way round. The better missionaries are certainly interested in Hinduism, and some of the Anglo-Catholics are very sympathetic to it.

On the purely intellectual plane a case might be made out for British rule in India if the rulers were passionately interested in the country and its people. If the lead given by the great Jones, whose studies on Sanskrit in the late 18th century inaugurated modern philology, had been followed up on a vast scale, a professor like myself might be tempted to justify British Imperialism. But at the present moment we cannot spare the money for a thorough investigation of the extremely interesting prehistoric culture of the Indus valley. At least one site is being excavated by Americans.

In 1918 I naturally thought mainly in terms of war. And it was gradually borne in on me that, from the military point of view, India was a liability and not an asset to the British Empire. If India were an independent state or group of states, the rest of the Empire would be appreciably stronger. The Indian army is of little use for modern war, for several reasons. The soldiers do not understand the technique. In successive weeks I watched the Scots Guards and some Indian troops (Dogras I think) being bombed by trench mortars in the orchard at Neuve Chapelle. The Guards swore with great fluency, dodged round the traverses, and were rarely hit. The Indians stood and waited to be killed, which they were. They apparently thought that the bombs were devils, and could not be dodged.

After an Indian unit has had heavy casualties it takes a year or so before the new drafts can be incorporated in it and its fighting value restored. A good British unit can recover in a few weeks. The Indian army would be a good deal better were it not for official snobbery in favour of Punjabis and other "fighting races." Half the Indian army is recruited in the Punjab. On the basis of my own experience I should prefer Jats, Mahrattas, and Gurkhas as comrades in a battle to any Punjabis that I have ever met. And if I were recruiting an army, especially a revolutionary army, in India, I should remember that Clive won his victories with a force mainly composed of low-caste Hindus from Southern India, including untouchables. In 1918 the British rulers of India were still concentrating on Punjabis. Sir Michael O' Dwyer, the governor of the Punjab, was pressing the local authorities for more recruits, and thousands of men were bullied into joining each month. They were of no military value, and many of them deserted. The discontent produced by this recruiting led to the insurrection of 1919, and the massacre at Amritsar.

In order to hold India the Indian army and the British army of occupation have to learn tactical methods quite unsuited to modern war. They have, in fact to be policemen rather than soldiers. From 1914 to 1919 the Wessex territorial division was kept in India, and many other British soldiers, including the whole 13th division, were sent to Mesopotamia to clear up the mess into which the Indian army had got during a side-show which, even if Kut-al-Amara had been a victory instead of a defeat, would not have been decisive. For these reasons, as a British patriot, I desired to see India, if not completely independent, at least enjoying Dominion status, and therefore not requiring a British garrison. Its natural frontiers render it more easily defensible than any other continental area of the same size. In fact, owing to the huge transport requirements of modern armies, an invasion of India through mountain passes would be far harder today than in the past.

Besides, I do not like the present influence of India on British culture. In the past India gave us one great thing, the daily bath. Even today Indians are far cleaner than most Europeans. An Indian crowd is not smelly. And Indians justifiably regard us as not merely unclean, but dirty. For they invariably wash themselves after going to stool, and are disgusted that we do not. Indeed it is to our failure to wash that they attribute the prevalence in Britain of "that most distressing and almost universal complaint, the piles." But today India exports retired officers who are accustomed to command rather than to persuade, and emasculated forms of Hinduism such as theosophy. Of the economic side of the Indian connection I will write later.

I stayed in India for nearly a year and a half, spending some months at a bombing school at Mhow. Here the commanding officer, straight from the Western Front, illegally constituted himself Mess President, and insisted on a meat diet in the hot weather, instead of the vegetarian or nearly vegetarian diet of the country, which sensible Englishmen adopt. In consequence I went down with jaundice, and was again invalided to the Himalayas.

In July 1918 I was sent back to England to learn certain methods employed by the Intelligence Department. I came back by Aden, where for the first and last time in my life, I won a chess tournament, defeating a number of British and two Indian officers by playing a safe game. I sweat immoderately and could at that time stand heat very well. My opponents could not, and generally presented me with their queens about the twentieth move. I spent three weeks in Egypt and reached England in September 1918.

My employment in London falls under the Official Secrets Act. Fortunately it did not last long. Intelligence work is dirty work. Somerset Maugham's "Ashenden" gives a not unfair picture of its high spots, including murder, though of course it is mainly routine snooping into other peoples' affairs in the hope that something may turn up.

But before I could be sent back to India, as was originally intended, the war was clearly coming to an end. I was on leave on November 11th, and was demobilized in January, after being sent for a week to Ireland, where our reserve battalion was now quartered. Nobody shot me there, and I became a civilian without undue fuss.

I liked the war, or rather those brief periods of it when I was actually in the front line. I hated army life, behind the line, but my specialist occupation at least saved me from battalion parades, and teaching recruits to form fours. I believe that these sentiments are fairly genuine, for they are borne out by my dreams. In my pleasant war-dreams I am in the trenches, in my unpleasant ones I turn up late on parade, or cannot find my tie or Sam Browne belt. Since I went to Spain I have had dreams of neither of these classes.

Only a few of my comrades admitted to liking fighting. I remember 2/Lt. Garden remarking, I think on the day before the battle of Neuve Chapelle, that he did so, and giving as his reason "Ye've nothing to worry about unless whether ye'll live till next day's meat." The next day I saw him walking back, his usual grin slightly exaggerated by the mild cerebral concussion caused by a bullet wound in his scalp. But Garden's remark suggests that there may be something wrong with a civilization in which people take refuge in war from the worries of everyday life in peace.

My own liking for war goes somewhat deeper. I get a definitely enhanced sense of life when my life is in moderate danger. On the other hand I do not get thrills in the sense of autonomic excitation causing goose-flesh and erection of the hairs. I get these from motoring, when I have a "narrow shave." But during five weeks under fire in Madrid in 1936 and 1937 the only thrill I experienced was from reading Rimbaud's "Soleil et Chair." The explanation of these facts (for I think they are facts) I must leave to psycho-analysts.

I also find happiness in practicing the virtue of courage. I am not a particularly brave man. I have not got the requisite courage to dive head first into water, and I am frightened by flying in bumpy weather, and by several other things. However I was taught courage by my father, and am sufficiently self-conscious not to pretend that I am doing something else on the rare occasions when I am being brave. I believe that this attitude is far commoner in France than in Britain, where the best people do not even admit their virtues.

Finally I enjoy the comradeship of war. Men like war because it is the only socialized activity in which they have ever taken part. The soldier is working with comrades for a great cause (or so at least he believes). In peace time he is working for his own profit or someone else's. If I live to see an England in which socialism has made the occupation of a grocer as honourable as that of a soldier, I shall die happy.

I believe that many more people would admit their liking of war if they could distinguish between liking and approval. I think that war is a monstrous evil, and yet admit that I enjoy it. This is an internal contradiction in my mind. I have found a number of other things enjoyable which I believe to be wrong. But I believe, with Freud, that if we do not admit the existence of these contradictions we shall merely invent rationalizations to cover them up, including the loftiest moral reasons for involving our country in war.

I am also convinced that temperaments such as my own will be useful to humanity for a long time to come. We have still for example to explore our own insides and that of our planet. And some day we shall have to explore the moon and the other stars. At the present time there is plenty for us to do. Those of us who can sublimate our bellicose tendencies may be useful to our fellows. But those of us who think that we are in any way superior to people who hate war with their whole souls are a nuisance and a danger to mankind.

CHAPTER II
Autobiographical
Science.

Before the armistice I had already been offered a fellowship at New College, Oxford. Four fellows had been killed in the war, and my academic career justified the offer. Further, my father was himself a fellow, though at that time he drew no salary. This doubtless helped me.

I was not however to become a fellow till next October, but to live in college as a post-graduate student, helping to revive college and university life. For example I helped to resuscitate the Oxford Union, becoming Junior Librarian (Vice-president), the President being Hore-Belisha, now Secretary of State for War.[8] He was then rather to my left in politics. I also revived the New College Essay Society. Having to produce an essay at three days' notice I rewrote one I had originally written in 1914. It was later published under the title of "Daedalus."

While in Delhi I had done a little rather second-rate theoretical work on genetics, working on results obtained by Morgan[9] and his colleagues in New York. I now started on biology in real earnest. I began learning the technique of gas analysis under my father's direction, and finally became rather good at it. It is important to realise that the majority of scientists are highly skilled manual workers. Some of us are much more. For example, Aston,[10] the discoverer of isotopes, has a manual virtuosity worthy of a Van Eyck or a Stradivarius. We are, in fact hand workers as well as manipulators of verbal or mathematical symbols. The exceptions to the rule are mathematical physicists such as Sir James Jeans,[11] whose curious theory that God is a pure mathematician has recently been hailed as a scientific pronouncement.

I decided that I would teach physiology, and began learning that subject with about six weeks' start on my future pupils, though I already knew a small section of it, the physiology of breathing, rather well. My father put Pete Davies[12] (now a Professor in Adelaide) and me onto a really stiff problem involving blood gas analysis. We had to learn to take blood from a finger, defibrinate it by stirring so that it did not clot, and expose it to a known gas mixture. Then a measured amount had to be transferred into a special apparatus without contact with air; acid was added to it, and the quantity of carbon dioxide driven off was accurately measured.

I am not naturally skillful with my hands, and it was three months before we could get our duplicate determinations to agree, and our real work began. Since then van Slyke,[13] an American biochemist with a genius for designing apparatus, has made the technique a good deal easier, but I do not think more accurate. Our three months of failure taught us a lot. Some scientists have a moral lesson to teach the world, because we are up against Nature, and Nature may be defined as "That which does not accept excuses." Never before in life had I been in a situation where there was noone to whom I could give some plausible reason when I failed.

One of my principal objections to religion is that most religions teach that the Author of Nature accepts excuses. We are pardoned as the result of our own repentance, or Christ's blood, or the prayers of the Saints, or the holiness of the river in which we bathe. The Almighty is represented as equipped with the moral prejudices of a human judge, and with complete access to all evidence. So there must be some way of getting round him, or we should all be damned.

But nature does not behave in this manner. It punished us inexorably for the oddest things. For not boiling or sterilizing water one is condemned to death by cholera. For some offence or offences whose nature we have not yet discovered, death by cancer is the penalty. We are just beginning to realize such facts as these in the sphere of biology. We resolutely refuse to do so in the sphere of human relations. We believe that if only we had enough "men of good will" (a hateful phrase) we could operate our present economic system, or lack of system. We expect a young couple to succeed in the complicated psychological and physiological adventure of marriage when their knowledge of the relation of the sexes is based on dirty stories and sentimental books. In fact we refuse to be materialistic, and pride ourselves upon our refusal.

I also developed, at least to a small extent, the craftsman's conscience, that feeling of responsibility for the quality of my work which is one of the marks of a good manual worker. In 1935 and 1936 a number of pieces of machinery on British warships were intentionally damaged. A naval war with Italy was then possible. When the British Government ceased to oppose Mussolini's conquest of Abyssinia, in any practical way, the sabotage stopped too. A less subtle mind than that of Sir Samuel Hore might have suspected fascist agents. Sir Samuel at once saw the hidden hand of Moscow, and five skilled dockyard workers were dismissed. None of them were communists, but all had a record of active work for their class. In the debate in the House of Commons which followed, Mr. Gallagher stated that a good workman would not spoil his own work. This statement was as intelligible to the conservatives as an appeal for the preservation of scenery would have been to an audience of blind men. And it was, I think, an exaggeration. A good antifascist artisan might spoil his work to overthrow fascism, though it would go against the grain. He certainly would not do it to help fascism.

This technical conscience, this refusal to turn out second-rate work, will, as William Morris realised, be one of the pillars of morality in the workers' state. I do not remember that it was even mentioned as a possible moral principle at Eton, where morality meant not going to bed with a person to whom one was not married. I am, so far as I know, the only person who has ever got duplicate determinations of urea by a volumetric method to agree to within one part in a thousand. And I am a better communist because of it.

Davies and I also developed a certain degree of intellectual conscience, of which more later. As soon as we had learned our methods we set to work. Our business was as follows. In a normal man the lungs and kidneys cooperate to keep the blood at a certain degree of alkalinity. The lungs are driven faster if the blood gets too acid, so as to blow out carbon dioxide from it. If the blood gets too alkaline the breathing slows down and carbon dioxide accumulates, thus reducing the alkalinity. Similarly if the blood gets too acid the kidneys turn out a more acid urine, and produce more ammonia to neutralize some of the acid which must be cleared out of the blood. In some diseases, including lung and kidney diseases, the regulation breaks down.

It was our task to upset the alkalinity of our blood in various ways, and for a variety of reasons. First we had to verify the principles stated above, some, but not all of which, were then known to be true. Secondly we had to make them quantitative, to answer such questions as this "If I diminish the amount of bicarbonate (alkaline reserve) in my blood by 30% how much more will I breathe, and how much more acid will my kidneys secrete?" Thirdly we had to find out how far the symptoms in various diseases could be explained by changes in the alkalinity of the blood. For if so we might find out how to regulate it when the natural method was upset. And lastly we had to keep our eyes open in the hope that something would turn up. For the last two purposes man is the ideal experimental animal. Even a dog, let alone a rabbit, cannot tell you if he has a headache, or an upset of his sensations of smell, both of which I obtained as symptoms during these experiments.

All these expectations were fulfilled in greater or less degree. Our colleague Kennaway,[14] who later discovered the substance in lubricating oil which gives rise to mule-spinners' cancer, happened to confuse a sample of my urine with one which he was testing for other purposes. He found in it a substance which is usually present in the urine of diabetics. Every scientist makes mistakes. A good one profits by them. Kennaway followed up this clue, and we found that when the tissues are too alkaline neither sugars nor fats can be burned completely. Our work was incomplete because, although we could increase the alkali in our blood by eating sodium bicarbonate, we could not drink enough hydrochloric acid to diminish it appreciably. Or at least we would not, as it would have meant an unusually painful death.

Meanwhile I had a certain amount of laboratory instruction in physiology and organic chemistry. But it was scrappy; and I have never taken a scientific degree, nor have I passed an examination in science since I left school. I also took part in the social life of the university, and of various circles connected with it, including that of the Morrells at Garsington, rather cattily described by Aldous Huxley in "Crome Yellow." Aldous Huxley, whom I have known since he was twelve years old, is a most instructive person.

He was two years junior to me in college at Eton. At the age of sixteen he took up biology. He was a born observer, and would have been as good a biologist as his brother Julian had opportunity permitted. <u>Dis aliter visum</u>.[15] After one term's work he developed a staphylococcus infection of both eyes. The pupils went white like those of an old dog, and he was almost wholly blind for five years. During this time he was left very much to himself, as his mother was dead. No wonder smells mean more to him than to the rest of us. His sight slowly recovered, and about 1915 he could read with a magnifying glass, and came up to Oxford, where he took a brilliant first class in English Literature. But he was, and is, embittered. It took not only superb natural gifts, but a staphylococcus, to make the man who most perfectly voices the spiritual muddle of the English middle class intellectuals.

In 1919 Barbusse was writing[16]

> "Nous sommes ceux cui n'ont jamais eu de lumière,
> Que l'ombre universelle a repris chaque soir,
> Ceux dont le sang vivant, le sang profond, est noir,
> Ceux dont le reve obscur salit tout, ce qu'il touche,
> Et nos yeux sont aussi ténébreux que nos bouches.
> Vides et noirs, nos yeux sont, aveugles, nos yeux
> Sont éteints: il leur faut le grand secours des cieux."

> "You and I are of those who have never had light,
> Whom the shadows have covered anew every night,
> Whose deep living blood is stagnant and black,
> And whose dim dreams defile all they touch in their track.

Black and empty, our eyes too are blinded. Our eyes
Are as dark as our mouths are. We need the great help of
the skies."

But Barbusse found the way out. Huxley has not. He can only repeat "Oh! Comme j'ai souffert, comme j'ai dû souffrir!" And he will go on persuading his compatriots that their unhappiness comes from within, that "The heart is deceitful above all things and desperately wicked."[17]

As if Aldous Huxley was so much wickeder than the rest of us as to deserve blindness, or his generation so much wickeder than that which came before them as to deserve fascism. Fascism, like blindness, is a symptom of conditions for which we are not individually responsible. The responsibility is a collective one.

If I am writing so freely about a friend it is only because he has taken similar liberties with me. Mr. Scogan, in "Crome Yellow" voices a curious mixture of my own opinions and those of Bertrand Russell. I am also, I think, one of the principal sources of Shearwater, the physiologist in that extremely amusing book "Antic Hay." And just because some of my characteristics are noticeable in him, I must point out that the portrait contains other elements. Shearwater is given a charming, but neither faithful nor intelligent, wife called Rosie. I was not married when the book was written. My present wife avenged Rosie by reviewing "Brave New World" in "Nature" under the title "Dr. Arnold and Mr. Huxley," the suggestion being that the souls of Aldous ancestors Dr. Arnold (headmaster of Rugby) and Professor T. H. Huxley, fought like Jekyll and Hyde for the possession of his pen. Shearwater is further endowed with a hopeless passion for a not really inaccessible lady. This is also contrary to fact.

And he will go on persuading his compatriots that their unhappiness comes from within, that "Th[18]

In October 1919 I became a fellow of New College and have held academic posts involving teaching ever since. Our Warden was the justly celebrated Spooner, who, on Armistice Day "opened the windows and flung out all his hags," and was once discovered wandering about Greenwich enquiring the whereabouts of the Dull Man. It turned out that he had an appointment to meet a friend at the Green Man at Dulwich. Such, at any rate, were the stories whose manufacture was one of the chief occupations of the undergraduates of the college. The nearest thing to a Spoonerism that I ever heard him make was the statement, in the course of a sermon, that "Now we see through a Dark, glassly." Freud attributes this habit to a deep-seated inner conflict due to causes which cannot be decently disclosed. A more materialistic explanation is that Spooner was an albino, and had very bad vision. He overcame this handicap, and was an efficient and beloved head of the college. But the strain may have told on those parts of his brain concerned in speech.

I lectured on genetics, and later in physiology, committing the usual fault of callow lecturers in being far too abstruse. Simplicity in lecturing is an achievement. I also had to give individual instruction to those medical students who were studying physiology, including some from other colleges. I was rather good at this. In 1922 three New College men out of 60 candidates sat for the Honours examination, in physiology, and secured three of the six first classes awarded.

Among my pupils were Brain and Strauss,[19] now well-known neurologists, and Janet Vaughan.[20] Janet gave me an appalling insight into female education in England. She came to me with an essay on the nervous regulation of the heart beat, or some such subject. She had read innumerable authors, mostly German, of whom I had never heard, and stated the various theories which they held. But she would not give her own opinions. Apparently at her educational establishments, girls were taught to know their place, and not to think for themselves. I pointed out that when she became a doctor she would have to form opinions on rather little evidence, and act on them; and I systematically bullied her into expressing her opinions, which I than [sic] criticized. After a few similar experiences with women pupils I am absolutely sceptical as to innate differences of ability between the sexes.

They may exist for all I know, but I know no more about them than about the back of the moon. Given the existing educational and social system, my experience of women colleagues is clear enough. I would always sooner choose a woman for accurate routine work, and a man for work requiring initiative and action based on incomplete evidence. Thus I prefer to go to a woman dentist and a man doctor. It was, I think, Langevin who said that men were better scientists than women because they are lazier. The women will plow through oceans of routine, the men look for a short cut. And science consists largely of short cuts.

The great women scientists have mostly done arduous routine work which admitted of no short cuts. Thus Mme. Curie cleaned up radium out of six million times its weight of dirt. She had the mind of a first-rate chemist and the soul of a first-rate charwoman (which is a pretty good kind of soul). I met her in the year before her death, presiding over a conference in Madrid. She was aged and immensely tired. She probably knew that she was dying. And she worked for eight hours or so per day. Her life had included not only the isolation of radium and polonium, but a very tempestuous love affair. Like those other very few people who have lived maximally, I have little doubt that she was quite ready to die when her time came. If such qualities as hers are the peculiar property of her sex, women are not merely useful in science. They are indispensable.

In 1920 I returned to the problem of making myself more acid. The problem was how to smuggle an ounce or so of HCl into my blood in such a form that it would not dissolve my teeth, gullet, and stomach on the way. I argued as follows. "If you give a dog a little hydrochloric acid[21] he excretes part, but not all, of it in the urine as ammonium chloride. Now incomplete reactions are generally reversible. I am a set of chemical reactions some of which are reversible. I will drink a solution of ammonium chloride, and some of it will be turned into hydrochloric acid."

I was probably the first person to drink something because he thought he was a system of chemical reactions. This is a piece of materialistic thinking. Nevertheless I am not perhaps more materialistic than the average man, merely more consistently so. If you say, "I am in this room, and weigh twelve stone" you are not branded as a crass materialist. Your idealistic friends do not at once retort that you are an immortal spirit outside space and do not weigh anything. Some kinds of materialistic thinking are regarded as common sense. A materialist is a man who wants to see a little more common sense of this kind in general use.

Putting my theory into practice I drank 5 grams of ammonium chloride in 100 cc. of water. But the effect was not to make me more acid, as I had hoped. It merely made me violently sick. Later on I diluted the stuff still further, and got the predicted effects, and a good many others for which I had not bargained, though none of them did me any permanent harm.

These experiments, and others of the same kind, have often been regarded as showing unusual courage on my part. I do not think that this is the case. I had the firm conviction that I (or my body, if you prefer that phrase) obeyed the laws of chemistry like any other piece of matter. I acted on this conviction. If I had been wrong I should have been killed. If an engineer designs a motor car to go at 120 miles per hour without shaking to pieces no one regards him as a hero if he goes for a passenger in the car on its trial run. They would regard him as a coward if he refused to do so. My behaviour was analogous. But it appeared unusually brave because we are accustomed to think in an accurate and materialistic manner about motor-cars but not about ourselves. This is not to say that we should regard ourselves as machines. We are to some extent machines. But we are also alive. However because we are alive it does not follow that we are immaterial, or that the matter of which we are composed ceases to behave as matter.

Now, curiously enough, at this time I did not think that I was a materialist. I thought that I was an idealist of the Hegelian kind (which I still think is the best kind). After studying the philosophy of Marx and Engels I examined my own conduct, and changed my mind. I should be willing to define a man's philosophy as the set of theories on which he is willing to stake his life. He will often say that he believes in some other set of theories. This may not matter a great deal, but it may lead to confused action, and it may deceive other people. So it is desirable to adjust your theory to your practice, as well as <u>vice versa</u>.

Thus in Spain today I have met comrades who said that they were Catholics or Anarchists. They also said that they were anti-fascists, and acted as such. I did not dream of arguing them out of their principles. For one thing we were too busy. Indeed political commissars have orders to do their best to prevent such arguments in the army. Communists hope that Catholic and Anarchist comrades will come to realise that the Communist party line is the only one that can lead to victory, and that in fact without knowing it, they have been acting as Communists. Then they will examine Communist theory, and conclude that it has some bad holes in it, but is a pretty good guide to action. A little later they will perhaps find that the holes are not such very big holes after all. And then they will ask to join the party.

Besides my biochemical work on myself I did some genetical work on poultry and rabbits, and also produced some genetical theory involving rather elementary mathematics. I was the first person in Britain, perhaps in Europe, to support Morgan's theory that the gene, the unit of heredity, is a definite thing at a particular place in a particular chromosome.

This theory, by the way, was not invented by Morgan in 1910, but probably by Correns in 1902.[22] But Correns never followed it up or used it to predict observable phenomena or to suggest experiments. And very likely it was not re-invented by Morgan individually, but appeared as a social rather than an individual product in conversations with his junior colleagues Sturtevant, Bridges, and Muller.[23] At any rate that is how theories have often appeared during my own work with colleagues.

The obvious alternative possibilities were that genes were immaterial, or that they were properties of gametes (spermatozoa or eggs) as a whole, like the characteristic note of a stretched string, which is not located in any particular part of it. These theories were certainly incorrect. And I was justified in backing Morgan's theory in 1919. Nevertheless it is wrong, at least in part. Muller and Prokovieva[24] have since shown that some genes, at any rate, are not material objects, but rearrangements of a very small part of a chromosome. That is an example of the dialectical way in which science grows. I was historically right to back Morgan's theory. Muller and Prokovieva's emendation of it was only possible after the original theory had proved a satisfactory guide to practice for twenty-five years, and for almost all practical purposes, Morgan's theory (like Newton's theory of gravitation or Dalton's theory of the indivisible atom) can be used with complete confidence. But each carried within it the seed of its own negation. Needless to say I did not think of such themes while counting my chickens or considering how to mate my rabbits.

In 1921 I spent three months as a biochemist in the Edinburgh Royal Infirmary. I was acting as temporary assistant to Professor Meakins while Davies, whose job I took, was in Australia. By singular good fortune a strike had delayed the completion of the laboratory, and I had a good deal of spare time, during which I was allowed to go round two wards with the medical students, and thus learn some medicine as very few laymen can learn it.

The sociology of medicine is most interesting. The poor get fairly good service from their panel doctors, though certainly not such good service as the rich. On the other hand they are not encouraged to think themselves ill when there is nothing wrong with them, nor given large amounts of expensive and useless medicine. As outpatients in hospitals they are kept waiting for hours on hard benches, and do not generally get good attention. Often too they are attended by students who are only learning their jobs. And they may have to wait a long time before admission to a ward when seriously ill. Once in a ward of a good hospital they get better treatment than money can buy. But they are often turned out before cure is complete, and rarely spend long enough in convalescence.

The middle classes get better medical attendance, but when ill they are generally sent to nursing homes most of which are far worse than the average hospital. Further the horrors of solitary confinement during most of the day are added to those of disease. The most expensive physicians and surgeons generally hold a hospital post, and therefore serve the poor as well as the very rich. The same man may remove the King's appendix and the dustman's. A member of the middle classes cannot afford him. But he may obtain the services of a very good doctor who has not yet made his or her name.

The very rich are treated at home or in good but extremely expensive clinics. They can afford expensive doctors, but are liable to be exploited if their ailments are imaginary.

I do my best to get the advantages of all these classes. When there is room, as at Oxford, I go into an ordinary hospital ward for an operation. For diagnosis, for example examination of my heart if it drops beats, I also gate-crash a hospital, where the electro-cardiograph and other instruments are up-to-date. As a communist I want to see the poor getting as quick service as the rich, and the middle classes as efficient as the poor. I get them myself, and others will get them when we have an adequate hospital system free for all. Free medical attention, by the way, is communism rather than socialism.

I took little part in politics before 1933. In 1918 or 1919 I made my last appearance as a liberal sitting behind Mr. Asquith on a platform at Oxford. And in June 1919 I acted as chucker-out at a meeting addressed by George Lansbury and Austin Harrison to protest against the origin [sic] version of the Treaty of Versailles, which was even more unworkable, and an even more flagrant breach of Wilson's promises to Germany, than the final form of the treaty. The interrupters, who were the sort of people who now hail Hitler, threw tomatoes. I had my tactical scheme prepared. I approached one of the smaller ones from behind, placed a finger in each nostril, and dragged him backwards, hooked and struggling like a salmon, and too agitated to hit me in a vital spot. The rest followed, but before they rescued him they were half way to the door.

I am no boxer or wrestler. My tactics are to grapple with a man and use my weight to bump him against a wall or floor. I remember being down once or twice, and there was some rather half-hearted fighting with chairs before we cleared the Corn Exchange. On my way home the interrupters counter-attacked. I took refuge in a jeweller's doorway between two plate-glass windows, determined to break them if attacked. My opponents did not assault this position, and on the approach of a policeman, who scented danger to Property, they dispersed.

But most of us were far too busy for politics. I have never known a group of university students who worked as we ex-soldiers worked. There will perhaps be a similar group after the Revolution. It had taken a combination of enthusiasm and discipline to change us from civilians to soldiers. Now we had to reverse the process, which required a great moral effort. It took me six months to break myself of a craving for regular exercise, which I do not need, since, in the early forties, I can still climb or swim a mile untrained.

Some of us never made the psychological change back, and remain with a permanent war fixation. Many of those who did not make it would have fitted very well into a socialist society. In our age of inhuman competition they have drifted into drink, crime, and fascism. I found, first in research into my own physiology, and later in communist party work, the moral equivalent of war.

In 1922 Professor Hopkins (now Sir Frederick) asked me to come to Cambridge as Reader in biochemistry, that is to say second-in-command of the department of biochemistry, of which he had just been created head. I accepted, and worked at Cambridge for the next ten years. Although I did some research and some teaching, my most important duty was the supervision of research work carried on by others on a great variety of subjects, including the chemistry of men, animals, higher plants and bacteria.

My salary was £600 per year, and Trinity College gave me a bedroom with a sitting room, board and amenities such as newspapers at an extremely cheap rate. This was a great advance on New College, which gave me £150 per year, rising to £250, free rooms, and a subsidy towards my evening meal. Besides this I had a pension of £40 per year for 25% disability till 1921. Moreover prices were falling. I was far better off than most of my colleagues of the same age. And I have never known poverty or unemployment.

On the whole the economic position of scientific workers has worsened in the last thirty years. They have become cheaper because there are more of them, and they are unorganized. The highest paid positions in universities are chairs at £1000 - £1500 per year. A few medical chairs are worth £2000. But these are the plums. Fellows of the Royal Society may earn £400-£600 per year. And salaries have not risen appreciably since the war. In fact they have fallen. About 1921 the government added 10% to the salaries of university teachers. But this, and 5% of the former income, goes into an insurance and pension scheme. I find it hard to believe that during the twenty years which have to run before my pension becomes due there will not be considerable inflation. If so I shall lose my money, which I should not do if I had bought a house or a grand piano with it. Meanwhile however the insurance companies can use it to finance rearmament, British films, and other important objects. In earlier years some of them lent a good deal to Germany.

If however such considerations occurred to the university staffs, they were forgotten in the joy which was felt that professors would in future retire at 65. It is not easy to become a professor, but once one has reached this exalted rank the need for serious work is over. It was then understood that a professor could be dismissed for anything but inefficiency. I shall tell later the story of how I broke this precedent. Professors elected under the old statutes have no retiring age. Some of them still hold their fiefs at over 80. The chair of pharmacology at Cambridge was held by one such, Professor Bradbury. The work was done by his Reader, Dr. Dixon, some 20 years his junior, one of the acknowledged leaders of his science.[25] Dixon hung on waiting for Bradbury's death. But he was a worker, and had strained his heart. So he only survived Bradbury by about two years.

For the younger scientific workers one of the central phenomena of life was and is unemployment. They are a characteristic example of over-production, or rather under-consumption. During the latter part of the 19th century there was a scarcity of scientifically trained men and women. They were wanted to solve industrial and other practical problems. Some of them made a good thing out of their scarcity. My father though badly paid in his early years, later earned £1500 per year on a part-time job.

This scarcity was remedied by the endowment of numerous university laboratories, technical colleges, and so on, and of not very generous scholarships to keep them filled. The demand for teachers, who are, so to say, production goods in this industry, created a temporary boom. This was accentuated by the number of young scientists killed in the war. I got in on the boom. If I had been born five years later I should not perhaps have been so fortunate.

From 1923 onwards I assisted in turning out an annual dozen or so of highly trained biochemists onto the market. Those who secured first classes in Part II of the Natural Sciences Tripos (a post-graduate examination) could be sure of a grant from the Department of Scientific and Industrial Research amounting to about £150 per year, if they had not private means. They then received two or three years of training in research, which led to a Ph.D. degree. They were then put on the market, the Rolls-Royces of the biochemist manufactury.

And what a market! I only watched the marketing of the high-grade stuff. Articles with London [or] Manchester degrees do not command such high prices. A few got college fellowships or demonstratorships in our laboratory. They are assured of secure if not lucrative academic posts. Others drifted through various scholarships, studentships, and grants into and out of industry and medical research.

It was not as if they were not needed. They were, and are. In medicine we are up against thousands of problems which need biochemical knowledge for their solution. The surgeons still haven't got the ideal antiseptic which kills bacteria and is quite harmless to human cells. If you think such humanitarian considerations old fashioned, the national defence of Britain is largely a biochemical problem. Plants make wood, straw, and fibre out of sugar. They also make sugar into starch. If we could reverse the first process and imitate the second, Britain could feed itself, at any rate for several years. However, such an invention is not a likely source of profit to anyone, so noone is working on it, and if a lot of us starve to death in the next war, that will be just too bad.

The struggle for jobs is conducted with the aid of testimonials, so I had to write several of these every week. At first I hoped that the slump would be temporary, and that we should get back to the situation of early 1920, when almost everyone was at work. I interested myself in economics, and might have become an inflationist, had it not been for the consideration that my salary arose from Local Loans yielding a fixed annual sum in sterling. About this time I was concerning myself with the growth of populations, and with the question why some populations wobble, and others remain fairly stable. Thus the rabbits of Canada increase in numbers until every eleven years, an epidemic wipes all of them but about one in a thousand out. The foxes which eat them naturally oscillate in the same way. The rabbit plague causes a famine among them in which some starve to death, and some are caught in traps. On the other hand the numbers of beavers do not fluctuate to any great extent.

Now the equations for the growth of a population with birth-rates and death-rates determined in a certain way obviously resemble those for the production of a commodity. Lotka[26] had shown that, under certain specified conditions, birth-rates and death-rates will settle down again after a shock. For example if a lot of young people are killed in a war the death-rate will be lowered thirty years or so later, when they would otherwise be dying of old age. But oscillations due to such a cause are damped out, like waves in treacle.

I asked myself, under what conditions, in a highly simplified and theoretical economic system, oscillations would die out, and when, on the other hand, they would increase, like the wobbles in a shaft which is whirling beyond its limit of safety. Actually I assumed that the mathematical economists had answered the question, which is after all fairly fundamental. And I wanted to use their answer for my work on populations. I found that they had not. On the contrary, no mathematical economist had ever employed an integral equation, such as occurs in this problem. I found that under certain conditions small oscillations would increase indefinitely. I offered to vote conservative in the 1924 election if anyone could prove that I was wrong.

But the Cambridge economists were not interested, and I found no journal to publish my results. Ten years later Tinbergen,[27] an economist at _____ introduced integral equations to investigate the determination of the period of trade cycles. My own investigation, on stability, was complementary to his. Integral equations now became respectable in economics, and I published my paper in 1934 in the "Review of Economic Studies." It admittedly deals with a grossly simplified system, like the mythical men who exchange boots for corn. But it does show that in a system which is stable according to the principles of text-book economics, fluctuations may become larger and larger. For the benefit of mathematicians I may say that the condition for instability is that the real parts of at least one pair of complex roots of a certain transcendental function should be positive. Later on I discovered that Marx had managed to say in words a great deal more than I had said in symbols. I do not of course claim to have proved mathematically that capitalism is unstable. I have only shown that it may be rather less stable than had been supposed. But I can perhaps claim the distinction of being the only person in whose conversion to Marxism functions of a complex variable played any important part.

If I had confined myself strictly to science and mathematics, except when playing golf, bridge, or other amusements suitable to my station in life, I should now, I suppose, be ripening into respectability, well-known in my profession, and looking forward to a knighthood a little before my funeral. But I have never played golf, and have a unique record at lawn tennis, at which I have never been beaten, having played exactly one game and won it. I probably dislike ball games because I am very bad at them. My weight of 245 pounds is by no means adapted to most forms of athletics; while a couple of wounds handicap me to a slight extent, and furnish an admirable excuse for not playing games, which I dislike. I am also a conscientious objector when it comes to shooting birds and rabbits. After all I have been shot myself. If I had been a sport addict I might never have taken either to popular writing, or to politics.

In 1923 the Century Magazine published "Daedalus" and I embarked on a career of popular writing. The immediate effect was an unexpected piece of luck. A female reporter from the Daily Express, having read the article, came down to interview me and to ask for information on biological questions, for a book which she was herself writing. To my astonishment, the resulting paragraph in the Daily Express not only kept to the facts, but, as had been stipulated, did not mention my name. For this and other reasons I fell in love with the reporter, and my love was reciprocated.

She was already married, but not successfully. Her husband's conduct would have enabled her to secure a divorce under Scottish, but not English, law. It had, at any rate, terminated the marriage as a practical proposition. However, in order to dissolve a marriage in England, adultery is necessary. I informed the Vice-Chancellor of Cambridge University that I proposed to commit this act, to which he replied "Oh."

A few months later the stage was set for the necessary formality. My future wife and I arrived, at a certain hotel. She did not like its appearance, and insisted on going to another. It was, however, essential that evidence should be furnished. We could not run the risk of losing the detective. But my wife has not been a reporter for nothing. With eagle eye she marked down a rather nondescript young man in the hotel lounge. He was, as she surmised, the sleuth. I asked him to accompany us to the other hotel, which he did, carrying one of our suitcases. The next morning he appeared in our bedroom with the morning papers. Save for one moment when I had feared that we might lose sight of the detective, everything had passed off without a hitch. Such is the sad reality concerning the profession of Sherlock Holmes and Inspector French.

In the autumn of 1925 a Decree Nisi was pronounced, my name appearing as co-respondent. I then received a letter from the vice-chancellor suggesting that I should resign my post. And here I must pause to explain the peculiar conditions under which professors and readers held their posts at Cambridge. They could be deprived of them if a body called the Sex Viri (meaning Six Men, not Sex Weary) found them guilty of gross or habitual immorality. But so far every professor who had found himself in my position had resigned. I might have done so had the Vice-Chancellor recommended it a year earlier. But I felt his present suggestion might not be quite spontaneous. I did not resign.

I was then called before the Sex Viri, who consisted of heads of Colleges and Professors of law. The Reverend Dr. Pearce, who afterwards became a bishop, appeared to be their leading spirit. They produced a cutting from the Times, describing the case. I did not plead that the Times is not infallible, though as I have read my own obituary notice in it, I know that it is not. I brought some documentary evidence that I had not broken up a home, because there was not a home to break up. I offered to call witnesses and suggested that I might be legally represented. These suggestions were not accepted.

It was once more urged that I should resign. And all the best people thought that that was the very least that I could do. Unlike King Edward VIII in somewhat similar circumstances I dug my heels in, although it was pointed out to me that after such a scandal I could never hope for a good position elsewhere. I was formally deprived of my office and fell back on my second line of defence. I had the right to appeal from the Sex Viri to a tribunal of five judges appointed by the Council of the Cambridge Senate. I appealed. During the first term of 1926 I continued to live in College, for which I must thank my colleagues at Trinity, and to deliver my lectures, for which I must thank Professor Hopkins. Meanwhile Mr. J. M. Keynes[28] took up my cause in the Council of the Senate, and insisted that the tribunal should be a little less Anglican in its make-up than the Sex Viri. As however its president was Mr. Justice Avory[29] (who had the reputation of being a hanging judge) and included the Provost of Eton, an eminent authority on the Fathers of the Church, it was not a body of revolutionaries.

When I came before it I was very ably represented by Mr. Stuart Bevan K.C.[30] Further, Professor Hopkins and my father gave evidence on my behalf. The case was heard in the law courts, but the public was not admitted. A majority of the court decided in my favour, and restored me to my position at Cambridge. I do not know hwo [sic] they voted. I am not sure whether Mr. Justice Avory was more horrified by my conduct or by the fact that the Sex Viri, like the jury in Alice in Wonderland, had delivered their verdict before hearing the evidence.

The court may or may not have been influenced by the knowledge that a verdict in the opposite direction would not have ended the case. I intended to stand for reelection, and my Union, the National Union of Scientific Workers, after investigating the case, had decided to advise its members and others not to apply for the position. As my chief wanted me back, this would have created a situation of some interest.

I was able to carry my fight through for three reasons. I had some private means from a legacy, much of which went in damages and costs. I had an alternative livelihood, for I had discovered that I could write for the daily press. And I had a Union. Soon after this the conservative government passed the Trade Unions Act of 1927. Our union contained a number of government employees, and in order to include both them and such persons as myself, it was forced to transform itself into the Association of Scientific Workers, thus giving up the rather limited powers of interference in "trade disputes" such as my own case, which it previously possessed.

I finally obtained the blessing of the State on my union with my dear wife and comrade in the middle of the General Strike of 1926, and we lived at Cambridge for the next six years. Unfortunately we have no children of our own. However her son Ronald lived with us, and we soon had a couple of students who were doing research without visible means of support. One of them stayed on for some time after getting a job, and helped to keep us young.

By 1932 I knew some biology. I had worked under two great biologists, my father and Hopkins, who were in a way complementary to one another. My father was not a materialist though in his later years he was not an idealist either. He saw through the fallacies of the mechanistic theory of life, and was at his best when investigating a function such as breathing. Hopkins is a materialist, in the laboratory at any rate, but I find his point of view rather too mechanistic for my liking. His greatest work has been on chemical lines, as when by investigating what was needed to supplement a diet of known composition in order to keep rats alive, he proved the need of small quantities of the substances which are now called vitamins. It was obvious that in practice both my father and Hopkins were somehow right, although they disagreed on many fundamental points. It was not until I read "Feuerbach" and "Anti-Dühring" that I saw how their contradictory views could be reconciled.[31]

From 1927 to 1937 I held a part-time advisory position at the John Innes Horticultural Institution near London. On the whole my time there was wasted, as I had little actual control over the work done. However I was able to help some colleagues, particularly Miss de Winton, in really fine work, and I had the good fortune to discuss Darlington's remarkable discoveries concerning the chromosomes with him, and may have assisted some of his embryo thoughts into the world. I also started Miss Scott-Moncrieff on her brilliant research on the chemical side of plant genetics.[32]

But my main contributions to genetics were theoretical. With R. A. Fisher, Sewall Wright, and a few others, I have built up a rather complicated mathematical theory of population with special reference to evolution. Many of our colleagues think that we have gone too far ahead of the facts. This may be so, but at least our theory has made us look for facts which were not previously suspected, and find some at least of those for which we were looking. It has also made us investigate the exact meaning of a number of words in common currency. Thus Darwin wrote about the survival of the fittest, but never defined fitness exactly. Fisher and I have had to do rather complicated calculations about natural selection. So we had to define fitness fairly rigorously.

Some of my calculations led to surprising results. Thus it appears that, as a result of the survival of the fittest, a population may become less fit, just as the effect of gravity on a spinning top is to make it stand up, instead of falling down. I came across so many paradoxes of this sort that I was rather reluctantly compelled to adopt the dialectical terminology which Engels, and to a less extent Marx, took over from Hegel and applied where Hegel had failed to apply it. As soon as I did so, my ideas about evolution began to arrange themselves.

Meanwhile I carried on with biochemistry. I demonstrated the presence in insects, rats, and green plants, of a peculiar respiratory ferment discovered by Warburg in yeast, and with Cook and Mapson I later studied it in bacteria.[33] I also wrote a book on enzymes (or ferments) which had the merit of being so much shorter than any of the vast German works covering the same ground that it was translated into German and Russian. Unfortunately I backed the wrong horse in at least one case where I had to choose between two competing theories. But my main work was the supervision of other people's research.

In 1933 I came to London as Professor of Genetics at University College London. It was only a half-time job, and I might have had considerable difficulty in starting a department. But I had an immediate stroke of luck. Herr Hitler supplied me with two first-rate junior colleagues, Dr's. Grüneberg and Philip, who were of Jewish origin.[34] I had some difficulty in finding them salaries, but when this was done through the generosity of the Rockefeller Foundation, we got a real school of Genetics started in London. Interestingly enough, neither is a communist, and I have never tried to convert them. I should be unlikely to succeed where Hitler has failed, even if I thought it correct to influence my subordinates.

Among my main discoveries in London was the rate of mutation for a human gene. It is often believed that human abnormalities which are handed down to descendants are always due to heredity, that is to say derived from an ancestor. This is not so. If it were, eugenics would be a much easier task than it actually is. But the severer abnormalities, such as haemophilia (the bleeding disease found in many of Queen Victoria's descendants) would soon be wiped out by natural selection if they did not constantly arise afresh by a process called mutation, whose rate I was able to measure. I also discovered the type of human inheritance called partial sex linkage, and made the first, very rough, map of the positions of genes on a human chromosome.

In 1937 I was appointed to the full-time post of Professor of Biometry in London University. My salary is £1000 per year, less £50 paid into the superannuation fund. I am particularly interested in the genetics of wild animal populations, to which my junior colleagues Gordon, Street, and Spurway are making important contributions.[35] At present I am largely engaged in developing really powerful statistical methods for the study of such populations. Some of these methods are also applicable to man.

I have had a remarkable opportunity of studying the administration of science under our present social and economic system. It is a wonderful muddle. Most of the research work in "pure" science is done at universities by people whose principal duty is to lecture. The system of teaching by lectures has been out-of-date since the invention of printing, though it was of course necessary in the middle ages. Most of the time which students spend in lectures would be better spent in reading text-books, and discussing them in small groups with a teacher of whom they can ask questions. This system is actually followed in the scientific honours courses in certain Scottish and English provincial universities. The only indispensable lectures are those in which the lecturer critically discusses work which has not yet been dealt with in textbooks. By far the best lecture course that I have ever attended was Professor E. S. Goodrich's advanced course on vertebrate anatomy at Oxford.[36] But that is mainly because Goodrich is a superb blackboard artist. However, professors are paid to lecture, and often have no other specific duties. In a properly organized educational system the writing of first-rate textbooks would be subsidized. They would be cheaper, and more constantly brought up-to-date. In England the mediaeval method of teaching is well paid by the State and the universities, while textbooks are left to private enterprise. I lecture to audiences often of only a dozen students. I am not at the time of writing, provided with a secretary. I should like to be paid to write textbooks, and to lecture mainly to workers, many of whom enjoy a lecture a week, rather than students, who are bored by two or three per day.

I believe that a large proportion of the money spent on research is wasted. Much of the industrial and military research is secret, and therefore the same problem is tackled by workers employed by different firms and governments. A good deal of the so-called pure research is useless for other reasons. These reasons will become clearer if I describe some features of a really admirable body, the Medical Research Council. This council administers a sum of about £___,000 per year, which is spent partly on the National Institute for Medical Research, and partly on subsidies to individual workers or groups of workers.

The first secretary, Sir Walter Morley Fletcher, had the reputation of being an autocrat, probably because he stood up to the bigwigs of medicine and science. On the other hand he treated junior workers like myself with the greatest consideration and courtesy. But courtesy is not a warrant of efficiency. He took the revolutionary course of appointing committees to advise him which were composed, not of Old Gentlemen, but of the men and women concerned in in [sic] the research in question, some of them being under forty. In fact he introduced democratic methods. These committees keep a jealous and efficient control over the money allotted to them, and remarkably little of it is wasted, because their members are technically competent to criticize all suggested expenditure. The Royal Society also spends most of its income on research which is carefully watched by its committees.

This efficiency is not, however, found in all such bodies. I have seen men receiving grants from the Development Commission engaged in activities very different from those for which they were paid. I have noticed the utter absence of planning of genetical research in this country, which would presumably fall into the province of the Agricultural Research Council.

One reason for this inefficiency is not far to seek. In June 1937 I read that Professor H.E. Armstrong had been appointed Chairman of the Lawes Agricultural Trust, which administers the Research Station at Rothamsted.[37] Professor Armstrong is a venerable figure with a wide knowledge of chemistry, but he is over 80 years old, and I was not surprised to read in "Nature" a few weeks later that owing to ill health he had been unable to preside. So far from being an isolated case this is typical of a whole group of institutions. In one of these the chairman, a distinguished scientist now 80 years of age, receives £600 per annum for attending four meetings per year and signin [sic] some papers in between, whilst research workers of ten years standing and with international reputations under his care receive £300 for a year's work.

Finally there is a certain amount of real corruption, mainly among administrative workers. I do not propose to give examples for two very simple reasons. I cannot afford it, and if I were rich enough to risk a libel action, no publisher would be willing to share my risk. A jury might consider that Mr. A. was fully justified in using his position as an employee of a scientific institution to establish relations between it and a firm from which he also received emoluments. Or they might think that Mr. B's little perquisites were entirely legitimate. Our standards of honesty in such matters are at present altering rather rapidly for the worse. Any reader is therefore entirely at liberty to say that my statements are a baseless expression of political prejudice. And I certainly think that corruption is exceptional in scientific institutions, whereas it is extremely common in many other branches of our national life.

It may however be pointed out that both inefficiency and corruption are far less likely to occur in laboratories or other scientific organizations which enjoy a measure of self-government, and where the junior workers have a say in the direction, than in those which are controlled either autocratically or by a committee appointed from outside, and with little first-hand acquaintance with the work in progress.

I see little prospect of any very great improvement in British scientific research under capitalism. For efficiency we need a combination of planning and democracy. And these are incompatible under capitalism. Scientific research is certainly more planned (though not necessarily better planned) in Germany than in England. But it is carried out on the Führer-Prinzip (leader principle) and the leaders are not chosen primarily for their scientific attainments. And the plan being imposed from above, inevitably tends to neglect fundamental research, that is to say any research into basic principles which will only yield results after a long period, in favour of research which will yield immediate results. Originality in subordinates is also discouraged.

In England we have far too sharp a division between pure and applied science. The universities are mainly concerned with the former, and very properly discourage work in their laboratories on behalf of individual firms. The majority of firms do not encourage research which may yield no profit for many years, and then perhaps to their competitors. In the Soviet Union the same worker is often engaged in "pure" (i.e. long-term) and immediately applicable research, to the great benefit of both. My father's work, which led simultaneously to the saving of thousands of miners' lives and the discovery of how breathing is regulated is an example of what might be and what will be when the internal contradictions of our existing society have been abolished.

This was possible because the practical application of his work was to life saving, and not directly to production. The problems arising in industry are just as scientifically interesting as those of hygiene, but as long as industry is conducted for private profit there will either be a gap between university research and factory research or the university laboratories will be run for the benefit of capitalist groups. In a socialist commonwealth science takes its natural place as a combination of theory and practice for the common good.[38]

APPENDIX 2
HALDANE ON THE NAZI-SOVIET PACT

The manuscript for this letter to the New Statesman *is in the Haldane Papers, University College London.*[1] *The strikethroughs in the text indicate Haldane's hand-written erasures.*

IS THERE A RUSSIAN ENIGMA?
by J. B. S. Haldane

A number of writers to the New Statesman appear to find the foreign policy of the Soviet Union difficult to explain or to reconcile with their theories ~~on the~~ regarding that state. Now I can quite understand people hating the Soviet Union. If I were rich enough or believed a quarter of the anti-Soviet propaganda which I read I probably should hate it myself. But I cannot understand how an intelligent person can find its policy in any way inconsistent. On the contrary it appears to be almost fantastically consistent, certainly far more so than those of the British, French, or German governments. Of course the Soviet policy changes, and changes dramatically. But these changes ~~are~~ occur in response to changed circumstances.

The Soviet policy is based on an objection to two things, capitalism and wars, which the rulers of the Union, and the vast majority of its people, believe to be intimately connected. They want to see other countries adopt socialism ~~and since~~. Further they believe that attempts to bring socialism about by constitutional means would be countered by fascist revolution in most countries ~~they~~ where parliamentary government exists, ~~and~~ while in fascist countries ~~it could only come~~ socialism could only be established by revolution from the left. Hence they hope to see revolutions in other countries.

Lenin naturally hoped that the revolutions of 1919 in central Europe would lead to socialism there, as in Russia. When this hope failed two policies were open, to devote their ~~efforts to~~ main effort to stirring up revolutions abroad, or to building socialism at home. The former was the policy of Trotsky, the latter is that of Stalin. Stalin realises that genuine revolutions must come from within, and that in most circumstances ~~the~~ an attempt to impose ~~it from without~~ socialism from without would be regarded as imperialism. For this reason Outer Mongolia has not been abruptly socialized. It is a democratic republic with a strong tendency towards socialism, but capitalism has not been fully abolished there. The government of the Soviet Union could ~~do so~~ abolish it tomorrow if ~~they~~ it wished, but ~~they~~ it prefers to wait for the Mongolians to do so of their own free will. A peaceful and constitutional transition to socialism is far more probable in Outer Mongolia than in Great Britain.

Those who admit the validity of the preceding paragraph may yet argue as follows. The Soviet Government hopes for ~~revolution in~~ socialist revolutions in the countries defeated in a war. Hence they have deliberately maneuvered France and Britain ~~into a~~ and Poland into a war with Germany. Their policy has been devilishly clever, but if it is a peace policy for themselves, it is a war policy for others. The idea of a peace front including Russia, like that of a popular front including communists, was merely a cunning trap designed by Stalin to lead up to such a war as this. If this argument is true Stalin is a warmonger like Hitler, only a much cleverer one. We must therefore examine it carefully.

Let us begin by comparing the positions of Stalin and Roosevelt. Roosevelt has to deal with two problems. On the one hand he has a vast army of unemployed and underemployed. On the other he must placate American capitalists who would like to see profits as high ~~and taxes as low~~ as in the golden days of Coolidge. Now as a matter of fact both these aims will be at least partially achieved as the result of the present war. If the arms embargo is repealed ~~vast~~ large numbers of Americans will be employed in making munitions. The rise in value of "war babies", as

~~securities in the~~ shares in the armament and allied industries are called, shows that the American capitalists look forward to a boom. In addition American exporters will capture neutral markets from their British, French and German competitors, and even if no American ship sails to Europe, they will be freed from competition in their own hemisphere.

When it suits them, the Nazis will probably say that Roosevelt is an unscrupulous warmonger, who egged on Britain and France against Germany for the above reasons. This is of course absolutely false. Not only is Roosevelt a fundamentally decent man, but he realises two things. His country may yet be involved in the war; and the war boom is from its very nature a temporary boom followed by a peace slump, or else by lending to the war-struck nations on a scale which can only lead to ~~a slump bankruptcy~~ their bankruptcy and further slumps later on.

~~But the Soviet Union has no~~ Similarly, war is not an unusual event for the rulers of a capitalist nation. It solves the problem of unemployment, and even if profits are restricted on the whole, certain groups of capitalists, often in very close touch with the government, amass immense fortunes.

But the Soviet Union has no unemployment, and very little foreign trade. It can export certain raw materials, such as oil and manganese ore. It needs machinery, and other raw materials such as rubber. A war between other nations cannot cure its unemployment, for it has none to cure, and is just as likely to raise the price of its imports as of its exports. And a war in which it is itself engaged, or even a mobilization, as at present, simply means a ~~loss of labour~~ diversion of labour from production. Thus the Soviet Union is the only nation which ~~has~~ can make no economic gains from war.

Politically, Stalin might be pleased enough ~~if~~ that Britain and France should fight Germany if he could be quite certain that the war would be localized. But this is very far from sure. On the contrary, the Soviet Union is more likely to be drawn in than the United States. And, if we take the lowest possible view of Soviet foreign policy, and suppose that Moscow aspires to be the capital of ~~a world~~ an empire covering the whole earth (which is nonsense, if only because the Russians have a sense of humour), it would still pay the Soviets to keep out of war for a very simple reason. Their production per man and their population are both increasing faster than those of any other country. They can afford to wait.

For this reason I believe that the Soviet Government was ~~genui~~ sincere in its support, first of the League, and even after Britain and France broke the covenant, of a peace alliance, provided it was an alliance of a really solid and practical character. As ~~this was in~~ the British Government would not form such an alliance, they signed a pact with Germany which automatically localized the war. For once ~~Japan was~~ the British navy did not have to face Japan, it could be concentrated in the Mediterranean, and instead of Italy cutting British communications in the event of a war, Britain could cut Italian communications. So Italy and Spain stayed out, and ~~the threat to~~ the British and French empires are only threatened at their centres.

Given this situation, the Soviet policy with regard to Poland followed inevitably. If the Red Army's westward march had been delayed for a week, I, for one, should have ~~lost my confidence in the essential sanity of~~ confessed myself unable to understand the Soviet policy, and should almost have begun to believe that the Soviet-German ~~pact~~ agreement was an alliance instead of a non-aggression pact. The ~~Soviet~~ Red Army has occupied the Byelo-Russian and Ukrainian-speaking ~~districts~~ areas of Poland, and ~~these will be incorporated with the~~ their inhabitants will be incorporated ~~with the~~ into the existing Byelo-Russian and Ukrainian republics without ~~dif~~ much difficulty. If there is a plebiscite in these regions after the war, I do not doubt that an overwhelming majority will vote for the Soviet Union, provided the plebiscite is as free as that in the Saar ~~district~~ region.

But the occupation of Polish-speaking districts creates a different problem. If Stalin really proposed to partition Poland along the line of Narew, Vistula, and San, then he ~~is~~ would be a successor of Catherine the Great, and a century and a half behind the times. But he is well ~~of~~ aware of the fact that the Poles have an intense national sentiment, even if they have not always respected that of other peoples. And as a specialist in the problem of nationalities within the Soviet Union, he will doubtless pay full attention to this sentiment. The treatment of the Polish-speaking part of Poland which they are occupying will be the most searching test of Soviet policy.

The Soviets will start with one advantage. ~~The~~ Many of the Poles in the occupied districts will hate the Russians. But they will certainly hate the Germans worse. And although it is ~~doubtless~~ possible to explain the inactivity of the French and British armies on military grounds, the Pole will find it harder to comprehend the doubtless cogent reasons why there was no aerial retaliation for the bombing of Warsaw, although there certainly would have been for that of Paris or London. (Even the Pope, who ~~by his advice to~~ a nuncio [sic] at Warsaw in 1920 condoned Polish imperialism and is said to have supported it, may lose adherents now that this imperialism has borne its fruits.[2]) So the Poles may find that their Russian invaders shine, at least by comparison.

~~And in any case the lot of the Poles in the~~ The contrast will soon be deepened. In the German occupied area there will be starvation, and the men will be drafted to work in Germany ~~factories.~~ In the Russian-occupied area this will not be so. ~~And thus the new Poland will almost certainly start with~~ If ~~landlords are abolished~~ the land is given to the peasants, as seems likely, the Russians will soon have millions of friends. And thus ~~when~~ if a Poland shorn of its eastern and western provinces is reconstituted, it will have a ~~strong~~ distinct Russophil bias, and be ~~a considerable menace to the Third~~ ill-disposed to the Third Reich, though not to a peaceful Germany. In fact if the Soviet policy in Poland is correct the Polish nation should become an element of peace and stability in Europe, whereas since 1919 it has been the opposite. But if this ~~is to be done~~ end is to be achieved, it is not idle to pretend that the task of the Russians in Poland will be easy or simple.

~~But it is very simple compared with Hitler's task~~. But their position in Poland is at least intelligible from their own point of view. Whereas a study of Hitler's policy convinces me that he has not read the important passage in "Alice in Wonderland" dealing with the Owl and the Panther, or if so has not been able to complete the final line.[3]

I believe ~~however~~ that the main reason why people of the left in Britain cannot comprehend Soviet Policy is this. "Why" they ask "does Moscow draw no distinction between the Western democracies and Nazi Germany! ~~After all we are vastly~~ We have an active labour movement. The German trade unions are abolished. The Communist party is not illegal in Britain or France, but communists are beheaded in Germany. The present ~~Rus~~ Soviet policy is only explicable if Stalin has a sympathy for Hitler's methods." The answer is simple. "What do you mean by <u>we</u>? Do you mean Britain or the British Empire, France or the French Empire? I would sooner be a Jew in Berlin than a Kaffir in Johannesburg or a negro in French Equatorial Africa. If the Czechs are treated as an inferior race, do Indians or Annamites enjoy complete equality?"

Until the British and French Empires become Commonwealths, ~~yo~~ they can only expect Soviet friendship ~~as a~~ if ~~they~~ their foreign policy is a hundred percent peace policy. With their existing record they ~~cannot expect~~ could ~~reasonably~~ only expect Soviet help ~~in a genuine peace front~~ on Soviet [terms] in the past. But they cannot hope that, to aid them today, the Soviet Union will antagonize the ~~German people~~ people of Germany, ~~who~~ which may, when Hitler falls, become at least as democratic as England and much more so than the British Empire. The British and French people may prepare to fight their battles ~~to~~ under the leadership of the men of Munich, with the firm resolve to preserve their rule ~~of~~ over the coloured peoples of their own empires. But if so they can hardly complain that the Soviet Union remains neutral in the struggle, and occupies itself in stemming Hitler's advance and abolishing feudalism in Eastern Europe.

It may be that the ~~Russians~~ Soviet troops are at present occupying some parts of Poland where the majority of the people speak Polish, ~~feel themselves~~ think of themselves as Poles, and would like ~~to be unite in a~~ once more to be members of a Polish Republic. If so these areas constitute only a small part of the occupied territory. And if they exist they will very probably be given back to a free Poland, as ~~Vilniu has been~~

Vilnius has been given back to Lithuania. Certainly the Soviet-occupied areas include large numbers of ~~Poles, and many of~~ Polish speakers, in regions where they form a minority. The treatment of this minority will be a searching test of Soviet policy.

Another group of critics complain of Soviet imperialism in the Baltic states. We must remember that in 1918-1920 large numbers of people in these states wished to ~~remain~~ belong to the Soviet Union on the same terms as the Georgians, Ukrainians, and other nations. And the fact that ~~the most important man in the Soviet Union~~ Stalin is a Georgian is a guarantee that this desire would not mean forcible Russification. Nevertheless the Soviet Union does not propose to incorporate the Baltic states. Before we join a crusade for their freedom it would be well to find out what their inhabitants think about it.

I was recently dining with a leading Lett in London. He is not a communist, and appears to be largely concerned in trying to reestablish the export of dairy produce from Latvia to Britain. He told me that ~~he had~~ before the war, he had asked the Soviet Minister in Raunas whether ~~he had~~ the Soviets would not use the port of Ventspils (Windau) which is too big for Latvian needs. "No," said the diplomat "it would make you too rich". ~~In the~~ He welcomed the pact between his country and the Soviet Union because he hoped that it would give employment in Ventspils, whose export trade has gone, and enable his country to export dairy produce through Murmansk. I ~~expect the~~ believe that many other Letts share his views, and now that the League of Nations is impotent, welcome Soviet protection.

APPENDIX 3
SELF-OBITUARY

*Recorded at University College London on
February 20, 1964, and broadcast by the BBC on December 1, 1964.
This transcript was published in* The Listener.[1]

It is now February, 1964, and this is supposed to be my own obituary, so I hope it won't be shown, say, until 1975, when I shall be eighty-two years old, which is perhaps old enough. However, I have just been operated on for cancer, and if the operation has not been successful, you will be seeing and hearing me a lot sooner.

I am going to begin with a boast. I believe that I am one of the most influential people living today, though I haven't got a scrap of power. Let me explain. In 1932 I was the first person to estimate the rate of mutation of a human gene; and my estimate was not far out. A great many more have been found to mutate at about the same rate since.

Please don't think I have done nothing but mathematical theory. I have done some animal breeding, some plant breeding, and at least worked out a few human pedigrees of various abnormalities. But beside that I have done plenty of other work. A lot of it has been physiological work on myself and my friends. I am not going to talk about the scientific details of that work, I will just give you some examples of its practical application.

In 1942, Dr. E. M. Case and I were the first people who spent forty-eight hours shut up in a miniature submarine with our own air supply. We were confident as to what would happen, because we had tested our apparatus out for six or eight hours; but the Admiralty, very properly, were a little sceptical; they wanted to be quite sure that at the end of forty-eight hours a crew of two would have some oxygen left, which we had. But we were rather uncomfortable because of two things. First of all the miniature submarine was just a little bit cramped, only one of us could stick his legs out to go to sleep at a time, the other one had to sit up on a bench; and although the carbon dioxide was absorbed, the water condensed on the sides of the submarine and ran down our necks, and we were rather stiff and cold by the end of the show. In addition, we did a lot of work on oxygen poisoning. That seems a funny thing—you do not think of oxygen as a poison, but it is a very severe poison when you breathe it at high pressures. What happens is that you get convulsions rather like epileptic fits. My wife is the only person who has ever had five such convulsions. I have had only three, but that was quite enough.

But I have been very much of a dabbler, as is obvious. I may say I have done some almost pure mathematics, and I have even ventured to push my nose into astronomy. But I am not ashamed of being a dabbler. It sometimes comes in very useful. Let me give you just one example. In 1933 there were a number of refugees from the Nazis. Some were Jews, some were liberals and socialists and communists, and so on, and I did my best to find the scientific ones jobs. Among the people who came in was a man called Chain. We talked, for an hour or two about the work he had been doing, and I said: 'I don't think I can help you much, but there is a man called Florey at Oxford who is certainly interested in this kind of stuff, and I would advise you to have an interview with him'. Chain did, and, as perhaps you know, Chain and Florey shared the Nobel Prize for the isolation and preparation of penicillin. As is always the case, there were innumerable people concerned. I do not claim credit for the discovery of penicillin, but even if I have half of one per cent of the credit I must have saved a good many thousand lives. So it is worthwhile, perhaps, being a dabbler and knowing a bit about what one's colleagues are doing in various branches of science.

My main work has been on genetics, and I am often asked what

I think about Lysenko. In my opinion, Lysenko is a very fine biologist and some of his ideas are right. Curiously enough, they are much more often right for bacteria, in my opinion, than they are for larger organisms such as animals and plants with which we are familiar. But again, in my opinion, some of Lysenko's ideas are wrong and badly wrong, as, of course, some of mine or any other biologist may well prove to be. And I think it was extremely unfortunate both for Soviet agriculture and Soviet biology that he was given the powers that he got under Stalin, and that he used to suppress a lot of what I believe, and what most geneticists believe, to be valuable work, much of which has been started up again but with a considerable lag.

I was for some time a member of the Communist Party and I am still a Marxist, but it does not seem to me that Lysenko's biological ideas follow from Marxism at all, rather the other way around. That again is only a personal opinion. But I am quite sure that if I had been made dictator of British genetics or British physiology I should have been equally disastrous, except that I do know one gets the best results from science by giving people a good deal of rope, and letting them go on with work which looks as if it were not going to be very fruitful but which sometimes is. And I do not think Lysenko quite realized that. But I do not think that any one man is big enough for the job of directing a branch of science, and that is one of my criticisms of these enormous programmes in nuclear physics, cosmical research, and so on, which in addition, of course, suffer from the evil of secrecy, which I have no doubt may be necessary but must slow down progress very considerably.

I am sometimes asked to whom I owe most in my scientific career, and I have no reasonable doubt about that. I owe most to my father, the late J. S. Haldane. He was, like me, only more so, a dabbler. He was, in my opinion, a great physiologist, though he certainly made mistakes, as we all do. I had been my father's bottle-washer during the holidays for twelve years, and he always discussed his work, and I found no difficulty in starting scientific research with the science I had learned at school. I never took a degree in science at the university. I took my degree in other subjects—mathematics, and a curious Greek-Latin hotchpotch called Greats, or *Literae Humaniores*, which we studied at Oxford before the last war, and, as I say, I did not find any difficulty in starting scientific research, and curiously little difficulty in teaching it.

I went to India in 1957 partly because I had fallen in love with the country when I was there after being wounded in Iraq, or Mesopotamia as we called it then, in 1917. That was the first opportunity I had to visit it, or rather to settle down there, because I had made several earlier visits, after independence. It was no good my going there while I could not associate with Indian colleagues on a footing of equality as I now can. Now I am told that some of my scientific colleagues thought I was committing scientific suicide as I only had three more years to run as a professor at University College. I can only say that I regard my work in India as extremely fruitful and useful. There are enormous difficulties—you cannot get apparatus and it is very difficult to get things done, harder than here even. But there were tremendous opportunities for outdoor work, on plants, on animals, on men, and there are magnificent young men available, every bit as good as I had at Cambridge—and I was there for ten years—and probably, on an average, rather better than my post-graduate students in London during the twenty-five years or so that I was a professor in London.

If I am not forgotten completely a hundred years hence, I shouldn't wonder if I should be remembered for something which I have not mentioned today. It might be something like, let us say, a letter to *The Observatory* entitled, 'Is space-time simply connected?' I am not going to try to explain to you what that means. It is a rather abstract geometrical idea. It might be the clue to new approaches to cosmology, though I should think it is more than twenty to one that it will not be: it might be—but, still more likely, it will be something which I have completely forgotten now. Some little remark I made in some paper which perhaps someone will dig out and say: 'Oh, but that explains what I found out last year'. Or perhaps some historian will find out and say: 'Haldane's remarkable anticipation of Chew Wong', or something like that. We do not know. But to take an example, the estimation of human mutation rates was, so to speak, a footnote to what then seemed to me more important.

But I don't really very much care what people think about me, especially a hundred years hence. I should not like them to be too critical of me as long as my widow and a few friends survive me. But the greatest compliment made to me today, I believe, is when people refer to something which I discovered—for example, that eating ammonium chloride causes acid poisoning in men—as a fact the whole world knows—to quote good old Aunt Jobiska in Lear's poem about the Pobble—without mentioning me at all. To have got into the tradition of science in that way is to me more pleasing than to be specially mentioned. But what matters, in my opinion, is what I have done, good or evil, and not what people think of me.

APPENDIX 4
VENONA INTERCEPTS

Sections that were redacted on the first release of the decoded intercepts are **highlighted like this**. *The unredacted versions are held in the GCHQ records, National Archives, HW15/43.*

0812. July 25, 1940

USSR   Reference: 3/PPDT/T11[1]

THE HON. IVOR MONTAGU (COVERNAME INTELLIGENTSIA) AND THE X GROUP (1940)

From: LONDON
To: MOSCOW
No: 812

                                                                                                            25th July 40

To DIRECTOR.

    I have met representatives of the X GROUP [GRUPPA IKS][i]. This is [a] **IVOR MONTAGU [MONTEGYu]**[ii] **(brother of Lord MONTAGU [iii]), the well-known local communist, journalist and lecturer. He has [1 group unidentified] contacts through his influential relatives.** He reported that he had been detailed to organise work with me, but that he had not yet obtained a single contact. I came to an agreement with him about the work and pointed out the importance of speed. He (INTELLIGENTSIA [INTELLIGENTsIYa] [iv]) reported the following:

    1. HITLER's speech [v] will not make a great impression here. The press has taken it unfavourably. He considers that the SAUSAGE-DEALERS [KOLBASNIKI][vi] were given their answer earlier by CHURCHILL since the SAUSAGE-DEALERS' conditions passed to the British through KhOSKhOR [vii] and WILSON [viii] proved

unacceptable. HALIFAX's speech [ix] was not only an additional answer to the SAUSAGE-DEALERS but [also][b] the outline of a political programme. The slogan "For Christianity" is similar to one of the conditions in ROOSEVELT's declaration on the freedom of religion, will also impress PÉTAIN, FRANCO and the Pope and has an anti-fascist edge [to it][b]. INTELLIGENTSIA considers that there is an anti-SAUSAGE-DEALER mood in the army and the knowledge [of this][b] is so strong that the Conservative Party is afraid of risking talking about peace. They might start talks if Britain were to suffer some serious defeat or the SAUSAGE-DEALERS start effective bombing. CHURCHILL continues to stay the main figure of the war in the SAUSAGE-DEALERS' eyes, but he intends [c] but is supporting CHAMBERLAIN's group so as to give the anti-SAUSAGE-DEALER mood no chance of developing into a movement of the left.

2. Generals IRONSIDE [x] and GORT [xi] were removed from the leadership because at a parliamentary [C% secret session] on the question of the cause of the defeat in FLANDERS the Minister of Supply proved with documents that the Expeditionary Force [ĖKSPEDITsIYa] was supplied 100 percent in accordance with the instructions of these generals, who are personally responsible for the [B% development] of the mechanized armies.

No. 200   BARCH [xii]

Notes:

[a] The translation of this could possibly be "[In] this is", i.e. in this group is; the plural "representatives" in the first sentence is certainly correctly recovered and seems odd otherwise.

[b] Inserted by translator.

[c] Part of the text has obviously been omitted here.

Comments:

[i]   X GROUP: Not identified; see also MOSCOW's No. 450 of 7th September 1940, and LONDON's Nos. 895 of 16th August 1940,

987 of 6th September 1940, 1071 of 26th September 1940, 1099 of 2nd October 1940 and 1188 of 18th October 1940.

[ii] **IVOR MONTAGU: Hon. Ivor Goldsmid Samuel MONTAGU.** See also LONDON's Nos. 895 of 16th August 1940, 987 of 6th September 1940, 1099 of 2nd October 1940, 1149 of 11th October 1940 and 1165 of 15th October 1940.

[iii] **Lord MONTAGU: Stuart Albert Samuel MONTAGU, 3rd Baron SWAYTHLING.**

[iv] INTELLIGENTSIA: **i.e. Ivor MONTAGU, see comment** [ii].

[v] HITLER's speech: to the REICHSTAG on 19th July 1940.

[vi] SAUSAGE-DEALERS: the Germans.

[vii] KhOSKhOR: Not identified.

[viii] WILSON: Sir Horace WILSON, Chief Industrial Advisor to CHAMBERLAIN. He acted as a special envoy between the latter and HITLER.

[ix] HALIFAX's speech: the Earl of HALIFAX's speech of 22nd July 1940 on the radio.

[x] IRONSIDE: General IRONSIDE, Chief of the Imperial General Staff.

[xi] GORT: General Lord GORT, commander of the British Expeditionary Force.

[xii] BARCh: Possibly Simon Davidovich KREMER, whose official post was Secretary to the Soviet Military Attaché in LONDON. He was appointed in 1937 and is thought to have left sometime in 1946. The covername BARCh occurs as a LONDON addressee and signatory between 3rd March 1940 and 10th October 1940, after which it is superseded by the covername BRION.

0895. August 16, 1940

USSR   Reference: 3/PPDT/T1^2

    1. INTELLIGENTSIA, **PROFESSOR HALDANE** AND THE X GROUP

    2. MARTHA, MARY, DICK AND MARK

    3. RECRUITMENT OF HEIN (COVERNAME BAUER)

                       (1940)

From: LONDON
To: MOSCOW:
No: 895   16th Aug. 40

To DIRECTOR.

1. INTELLIGENTSIA [INTELLIGENTsIYa][i] has not yet found the people [a] in the military [C% finance department][VOENNYJ FINANSOVYJ OTDEL][b]. He has been given the address of one officer but he has not found him yet. He has promised to deliver[c] documentary material [MATERIAL] from **Professor HALDANE**[ii] who is working on an **Admiralty [MORSKOE MINISTERSTVO] assignment concerned with submarines and their operation**. I have taken the opportunity of pointing out to the X GROUP [GRUPPA IKS][iii] that we need a man of a different calibre and one who is bolder than INTELLIGENTSIA.

2. MARTHA [MARTA][iv] has handed over material on experience [gained from][d] using artillery in France. The material is in French [and][d] we are looking for a translator. In view of MARY's [MĒRI][v] departure I request permission to put DICK [DIK][vi] or MARK[vii] in touch with MARTHA.

3. A Lieutenant HEIN [viii] of the Czechoslovakian Army has presented himself at the METRO[ix]. He [1 group unidentified] that in October 1939, when he was in an internment camp near KAMENETs-PODOL'SKI[x], he was recruited by somebody for special work [SPETsRABOTA][xi]. His covername is BAUER [BAUĒR] and he has been instructed to [B% look for][KOLYaRSKIJ] [xii]. BAUER is now in the Czechoslovakian camp near MALPAS [xiii]. The total number of Czechoslovakian troops in Britain is about 4,000 other ranks and up to 700 officers. He himself is on the reserve for the time being. In France there was only one Czechoslovakian division. He has confirmed that up to 600 Czechs have been removed from the camp and sent somewhere or other because of their left-wing views. Czecheslovakia [sic] has no weapons. Send instructions at once.

Notes:

[a] Or "not yet found any people". The first interpretation seems more likely, however.

[b] Possibly means "War Office Finance Department".

[c] The form of the Russian verb indicates repeated action.

[d] Inserted by translator.

Comments:

[i] INTELLIGENTSIA: **Hon. Ivor Goldsmid Samuel MONTAGU. See** also LONDON's Nos. 812 of 25th July 1940, 987 of 6th September 1940, 1099 of 2nd October 1940, 1149 of 11th October 1940 and 1165 of 15th October 1940.

[ii] **HALDANE: Professor J.B.S. HALDANE, then Professor of Biometry, University College, LONDON. See also LONDON's No. 987 of 6th September 1940.**

[iii] X GROUP. Not identified. See also MOSCOW's No. 450 of 7th September 1940, and LONDON's Nos. 812 of 25th July 1940, 987 of 6th September 1940, 1071 of 26th September 1940, 1099 of 2nd October 1940 and 1188 of 18th October 1940.

[iv] MARTHA: Alta Martha LECOUTRE (with aliases), secretary of André LABARTHE and wife of Stanislas SZYMONCZYK (with aliases); was mistress of Pierre COT, French Minister for Air 1933-1937. See also LONDON's Nos. 776 of 17th July 1940 (3/NBF/T1472), 807 of 24th July 1940 and 987 of 6th September 1940.

[v] MARY: unidentified covername. See also MOSCOW's No. 482 of 21st September 1940, and LONDON's Nos. 755 of 11th July 1940 (3/NBF/T1455), 807 of 24th July 1940 and 875 of 13th August 1940.

[vi] DICK: Covername of unidentified LONDON signatory and addressee between 6th July 1940 and 11th October 1940.

[vii] MARK: Unidentified covername. See also LONDON's Nos. 776 of 17th July 1940 (3/NBF/T1472), 798 of 22nd July 1940,

895 of 16th August 1940, 998 of 11th September 1940 and 1107 of 3rd October 1940.

[viii] HEIN: Not further identified. See also LONDON's Nos. 998 of 11th September 1940 and 1107 of 3rd October 1940, and MOSCOW's Nos. 450 of 7th September 1940 and 469 of 16th September 1940.

[ix] METRO: the Soviet Embassy.

[x] KAMENETs-PODOL'SK: 48°35'N 26°35'E.

[xi] Special work: possibly synonymous with SPETsRAZVEDKA (literally "special intelligence"), which is used to mean illegal operations.

[xii] KOLYaRSKIJ: No Russian surname of this form can be found. It is probably a transliteration of the relatively uncommon Czech surname KOLÁŘSKÝ (cp. PRAGUE Telephone Directory, 1964).

[xiii] MALPAS: approximately 15 miles SSE of CHESTER.

[xiv] BARCh: Possibly Simon Davidovich KREMER, whose official post was Secretary to the Soviet Military Attaché in LONDON. He was appointed in 1937 and is thought to have left sometime in 1946. The covername BARCh appears as a LONDON addressee and signatory between 3rd March 1940 and 10th October 1940, after which it is superseded by the covername BRION.

0987. September 6, 1940

USSR   Reference 3/PPDT/T10³

    1. JÉRÔME AND MARTHA

    2. INTELLIGENTSIA, PROFESSOR HALDANE, NOBILITY AND THE X GROUP

    3. MUSE

                        (1940)

From: LONDON
To: MOSCOW
No: 987

                                                 6th Sept. 40

To DIRECTOR.

1. JÉRÔME [ZhEROM][i] stated that he has been feeling bad [a] recently. He knows that someone has gossiped about him to the British to the effect that he is a left-winger and was in Spain. I advised him not to embark on anything, but to improve his work for the British and the General [ii]. I understood that although he has a post he will nevertheless for the time being be active only in the plan of establishing a factory. He hopes that the British will bring him into [B% organising] the supply of the Allied troops in the French colonies. He handed to us data and drawings of the HISPANO-SUIZA type 404 cannon. The material was taken from [b] in the French Ministry for Air and is dated 1939. State urgently whether this material is of interest to you. From a conversation with MARTHA [MARTA][iii] I understood that JÉRÔME considers that we are not paying him much (£50 for both). It would appear that high prices and the need to maintain a flat are involving them in a lot of expense. I told her that we should help but [the amount of][c] intelligence must be increased.

2. INTELLIGENTSIA [INTELLIGENTsIYa][iv] has handed over a copy of Professor HALDANE's [v] report to the Admiralty on his experience [d] relating to the length of time a man can stay underwater. However he does not deny the main point that for a month he has not been in touch with the British Army colonel [vi] picked out [VYDELENNYJ] for work with us although the latter does come to LONDON. I have told the X GROUP [GRUPPA IKS][vii] via NOBILITY [ZNAT'][viii] to give us someone else because of this [e]. INTELLIGENTSIA lives in the provinces [ix] and it is difficult to contact him.

320   APPENDIX 4

3. Yesterday MUSE [MUZA][x] picked up a 30-group telegram from you [xi]. Only individual parts of the telegram were understood.

No. 259.   BARCh [xii]

Notes:

   [a] The Russian phrase here is normally used in a physical sense in regard to health. The next sentence, however, makes it clear that the conspiratorial meaning of "health" in the sense of "security" is what is in the writer's mind.

   [b] There is obviously an omission at this point in the text.

   [c] Inserted by translator.

   [d] **Or "experiment".**

   [e] Or possibly "to do this".

Comments:

   [i] JÉRÔME: André LABARTHE, Director-General of French Armament and Scientific Research at General DE GAULLE's Headquarters until 12 September 1940. See also LONDON's Nos. 741 of 8th July 1940 (3/NBF/T1773), 776 of 17th July 1940 (3/NBF/T1472), 791 of 20th July 1940, 798 of 22nd July 1940, 807 of 24th July 1940, 865 of 10th August 1940, 1056 of 23rd September 1940 and 2151 of 6th August 1941 (3NBF/T1477).

   [ii] The General: i.e. General DE GAULLE.

   [iii] MARTHA: Alta Martha LECOUTRE (with aliases), secretary of André LABARTHE and wife of Stanislas SZYMONCZYK (with aliases); was mistress of Pierre COT, French Minister for Air 1933-1937. See also LONDON's Nos. 776 of 17th July 1940 (3/NBF/T1472), 807 of 24th July 1940 and 895 of 16th August 1940.

   [iv] INTELLIGENTSIA **Hon. Ivor Goldsmid Samuel MONTAGU.** See also LONDON's Nos. 812 of 25th July 1940, 895 of 16th August 1940, 1099 of 2nd October 1940, 1149 of 11th October 1940 and 1165 of 15th October 1940.

[v] **HALDANE: Professor J.B.S. HALDANE, then Professor of Biometry, University College, LONDON. See also LONDON's No. 812 of 25th July 1940.**[4]

[vi] British Army colonel: see also LONDON's No. 1188 of 18th October 1940 for a British Army colonel (covername RESERVIST) connected with the X GROUP (see comment [vii] below).

[vii] X GROUP: Not identified; see also MOSCOW's No. 450 of 7th September 1940, and LONDON's Nos. 812 of 25th July 1940, 895 of 16th August 1940, 1071 of 26th September 1940, 1099 of 2nd October 1940 and 1188 of 18th October 1940.

[viii] NOBILITY: Unidentified covername. See also LONDON's Nos. 1024 of 17th September 1940, 1170 of 16th October 1940 and 1232 of 30th October 1940 (3/NBF/T1742, first sentence, where NOBILITY occurs as "[1 group unidentified]".).

[ix] Provinces: **At this time Ivor MONTAGU was living in WATFORD.**

[x] MUSE: Unidentified covername. See also MOSCOW's Nos. 5839 (internal) of 18th September 1940 and 479 of 20th September 1940; and LONDON'S Nos. 767 of 15th July 1940, 816 of 26th July 1940, 865 of 10th August 1940, 876 of 13th August 1940, 949 of 27th August 1940, 1071 of 26th September 1940, 1107 of 3rd October 1940, 1321 of 25th November 1940 (3/NBF/T1618) and 2035 of 31st July 1941 (3/NBF/T1619).

[xi] 30-group telegram: this is probably MOSCOW's internal serial No. 5264 of 30th August 1940, referring to LOUISA [LUIZA], which is thought to be an error for MUSE [MUZA].

[xii] BARCh: Possibly Simon Davidovich KREMER, whose official post was Secretary to the Soviet Military Attaché in LONDON. He was appointed in 1937 and is thought to have left sometime in 1946. The covername BARCh occurs as a LONDON addressee and signatory between 3rd March 1940 and 10th October 1940, after which it is superseded by the covername BRION.

1099. October 2, 1940

   USSR   Reference: 3/PPDT/T15[5]

   1. BRITISH GOVERNMENT VIEWS

   2. BREAKING OF A SOVIET CODE REPORTED BY INTELLIGENTSIA AND THE X GROUP

   3. GERMAN BOMBING

                           (1940)

From: LONDON
To: MOSCOW
No: 1099

                                              2nd Oct. 40
To DIRECTOR.

1. ATTLEE [i] and GREENWOOD [ii], two members of the War Cabinet, have been to the METRO [iii] and stated to the MASTER [KhOZYaIN][iv] that the danger of invasion was not past but the possibility was growing less every day. The Government intended to follow SASHA's [v] example in supporting China financially in the hope that the latter would use the money to buy armaments in the U.S.S.R. An intensification of German and Italian activity in the Middle East was expected.

2. INTELLIGENTSIA [INTELLIGENTsIYa][vi] has reported that the X GROUP [GRUPPA IKS][vii] has reported to him that a girl working in a government establishment noticed in one document that the British had broken [RASKRYLI] some Soviet code or other and apparently she noticed in a/the document [a] the following [words:][b] "Soviet Embassy in Germany". I stated that this was a matter of exceptional importance and he should put to the group the question[6] of developing this report [further][b].

3. The SAUSAGE-DEALERS' [KOLBASNIKI][viii] night bombing is having a bad effect throughout the country. Trains are running slowly and late. Loading and unloading are slowed down because of the black-out. Some lines [1 group unidentified] do not work for one or two days after a raid. Goods have been sent by canal. The trains are overflowing with refugees from LONDON and other bombed centres.

4. [1 group unidentified][c] the SAUSAGE-DEALERS have now mostly been bombing LONDON and LIVERPOOL. In the period 28th-29th September some of the

hangars on HESTON Aerodrome, the PHILIPS Map Works on the corner of WESTERN AVENUE and VICTORIA ROAD, the BOWDEN factory located at WILLESDEN JUNCTION, which makes components for aircraft and ships, and an aero-engine factory in SOUTHAMPTON were heavily damaged. The BRUNSWICK DOCK in LIVERPOOL was destroyed. Following a raid 7,000 workers at the SIEMENS electrical instruments factory at WOOLWICH [ix] are idle. The SAUSAGE-DEALERS are continuing to bomb the railway stations at junctions in LONDON.

5. Latterly the SAUSAGE-DEALERS' heavy bombs have shown that none of the government types of above-ground shelter is any good. At night the Underground is overflowing with people, notwithstanding the instructions of the authorities. An epidemic is possible because of the start of the cold weather and the unhygienic conditions. The Government's plan is to reduce the population of LONDON to five million. Use of the Underground and other places can provide quarters for 500,000 people. It is difficult to understand the Government's statement [4 groups garbled][d].

No. 290[7]

Notes:

    [a] It is not clear from the text whether this is the same document as the one first mentioned, or a different one.

    [b] Inserted by translator.

    [c] This group probably has the general meaning of "according to reports".

    [d] The signature of the message is probably among these groups.

Comments:

    [i] ATTLEE: Clement ATTLEE, then Lord Privy Seal.

    [ii] GREENWOOD: Arthur GREENWOOD, then Minister without Portfolio.

    [iii] METRO: The Soviet Embassy.

    [iv] MASTER: The Ambassador.

    [v] SAShA: the United States. For an explanation of this coverword see LONDON's No. 998 of 11th September 1940 (3/PPDT/T12).

324   APPENDIX 4

[vi] INTELLIGENTSIA: Hon. Ivor Goldsmid Samuel MONTAGU. See also LONDON'S Nos. 812 of 25th July 1940, 895 of 16th August 1940, 987 of 6th September 1940, 1149 of 11th October 1940 and 1165 of 15th October 1940.

[vii] X GROUP: Not identified. See also MOSCOW's No. 450 of 7th September 1940, and LONDON's Nos. 812 of 25th July 1940, 895 of 16th August 1940, 987 of 6th September 1940, 1071 of 26th September 1940 and 1188 of 18th October 1940.

[viii] SAUSAGE-DEALERS: the Germans.

1149. October 11, 1940

USSR   Reference: 3/PPDT/T14[8]

    1. GERMAN BOMBING OF LONDON

    2. INFORMATION FROM THE FRIENDS

    3. INTELLIGENTSIA

(1940)

From: LONDON
To: MOSCOW
No: 1149

11th Oct. 40

To DIRECTOR.

1. In recent nights the SAUSAGE-DEALERS [KOLBASNIKI][i] have been stepping up the bombing of LONDON, whilst the gunfire has fallen off slightly. The SOUTH HAMPSTEAD railway junction has been damaged [and][a] traffic [1 group unidentified]. Admiral EVANS [ii], who is in charge of the shelters in LONDON, stated to the MASTER [KhOZYaIN][iii] of the METRO [iv] that a one-ton bomb dropped there had made a crater 21 metres deep. A 250-kg bomb dropped last night in the HATTON GARDEN area destroyed a chemical works and badly damaged our trade delegation [building][a]. Information on a map captured from the SAUSAGE-DEALERS reveals that LONDON is divided up into small squares and each has a numbered target in a definite [C% category]. The SAUSAGE-DEALERS are over LONDON throughout the night, one aircraft at a time on each flight path.

2. According to the FRIENDS' [DRUZ'Ya][v] information the SHORT aircraft factory in ROCHESTER has been completely destroyed and 5,000 workers are being transferred to SWINDON, where new production of [1 group garbled] and transport aircraft will be developed on the basis of the PHILLIPS aircraft factory where trainer aircraft were produced previously.

3. INTELLIGENTSIA [INTELLIGENTsIYa][vi] confirms that the British really do render delayed-action bombs safe by freezing [b] the bombs' exploder mechanism.

No. 306   BRION [vii]

Notes:

    [a] Inserted by translator.

    [b] The Russian term may be taken literally or metaphorically.

Comments:

    [i] SAUSAGE-DEALERS: the Germans.

    [ii] EVANS: Admiral Sir EDWARD R.G.R. EVANS, appointed LONDON Regional Commissioner for Civil Defence in 1939.

    [iii] MASTER: The Ambassador.

    [iv] METRO: The Soviet Embassy.

    [v] FRIENDS: members of the Communist Party.

    [vi] INTELLIGENTSIA: **Hon. Ivor Goldsmid Samuel MONTAGU.** See also LONDON's Nos. 812 of 25th July 1940, 895 of 16th August 1940, 987 of 6th September 1940, 1099 of 2nd October 1940 and 1165 of 15th October 1940.

    [vii] BRION: Col. I.A. SKLYaROV, Soviet Military and Air Attaché in LONDON 1940-1946. This is the earliest occurrence of the covername BRION as a signature (or address) in a LONDON message.

326   APPENDIX 4

1165. October 15, 1940

   USSR   Reference: 3/PPDT/T19

                   1. GERMAN BOMBING

                   2. INTELLIGENTSIA

                             (1940)

From: LONDON
To: MOSCOW
No: 1165

                                                              15th Oct. 40
To DIRECTOR.

[1.][a] In the area that we can see we observed the following method of
attack by the SAUSAGE-DEALERS [KOLBASNIKI][i]: first of all they dropped six
flash bombs [OSVETITEL'NAYa BOMBA][b]; after they had fallen a considerable
part of KENSINGTON was [B% dotted] for a time with a large number of small
fires [SVETOVYE TOChKI][c] which were extinguished with sand. As they fell
the flash bombs were fired at and machine-gunned with tracer bullets. One
was shot down. The first aircraft after the alert flew lower than usual,
obviously so as to be able to aim as it dropped its flash bombs. A few
minutes after [the start of][a] the raid the SAUSAGE-DEALERS' flash bombs
had established a seat of fire in several places, and afterwards throughout
the night they dropped bombs in various parts of the city while flying at a
height of about 4,000-6,000 metres. Explosions of special bombs that produce
a strong light for 5-6 seconds were observed for the first time. From
observations of the gunfire it can be concluded that the SAUSAGE-DEALERS
come into LONDON from the east, south-east and south. No more than three
series of explosions [at a time][a] were observed. Despite the moonlit,
cloudless night British fighter aircraft did not take part in the defence.

2. INTELLIGENTSIA [INTELLIGENTsIYa][ii] has reported [the following][a]
as a result of a conversation with an officer of the Air Ministry:

   The shortage of trained night navigators is confirmed by the fact
that the SAUSAGE-DEALERS have not used strong forces of aircraft in night
air-raids on Britain. He stated that the British pilots who fly by night
over Germany have an extra four months' training in addition to the usual

six months' training. Apparently when there has been no [anti-aircraft][a] fire the SAUSAGE-DEALERS have been bombing from a height of up to 5,000 metres and at a speed of 290kph, and from 6,000-7,000 metres at a speed of 400 kph when there has been no firing. He considers that the SAUSAGE-DEALERS proceed towards the target along a radio beam.

No. 309   BRION [iii]

Notes:

[a] Inserted by translator.

[b] Presumably "flares" [SVETYaShchAYa BOMBA] are actually meant.

[c] Literally "points of light".

Comments:

  [i]   SAUSAGE-DEALERS: the Germans.

  [ii] INTELLIGENTSIA: Hon. Ivor Goldsmid Samuel MONTAGU. See also LONDON's Nos. 812 of 25th July 1940, 895 of 16th August 1940, 987 of 6th September 1940, 1099 of 2nd October 1940 and 1149 of 11th October 1940.

  [iii] BRION: Col. I.A. SKLYaROV, Soviet Military and Air Attaché in LONDON 1940-1946. BRION's first appearance as a signatory (or addressee) is in LONDON'S No. 1149 of 11th October 1940.

1188. October 18, 1940

USSR   Reference 3/PPDT/T21

RESERVIST AND THE X GROUP (1940)

From: LONDON
To: MOSCOW
No: 1188

                                                    18th Oct. 40

To DIRECTOR.

1. BARCh [i] had a meeting today with one of [the members of][a] the X GROUP [GRUPPA IKS][ii]. This was an artillery colonel [iii] who had been

in the British Expeditionary Force [ĒKSPEDITsIONNAYa ARMIYa], but who at present is out of active service and doing a job at the Ministry of Supply because he was seriously wounded. He thinks that he will be returning to service in the regular army in two months' time. RESERVIST [REZERVIST][iv] has agreed to work with us. He has promised to bring to the next meeting in a fortnight's time his notes on the campaign in France and information on the way the British Army is organised.

2. RESERVIST reports that the process of reorganizing the army continues. The organisational structure of the Home Forces [ARMIYa METROPOLII] command [ShTAB OKRUGA], army corps, division. There are individual garrison districts which are directly subordinate to the C.-in-C. Home Forces, e.g. the DOVER Garrison which includes two infantry brigades and one tank division. The composition of an infantry battalion is: two infantry companies, one machine-gun company consisting of two heavy machine-gun platoons and one anti-tank defence platoon. A brigade consists of three battalions. Each brigade deployed in a war zone has its own independent defence sectors and divisional reinforcements are placed under the command of brigades.

3. Tanks are the main item of the work of the Ministry of Supply. [The following][a] are in mass production: 14-ton cruiser tanks and 25-ton ARMSTRONG-WOLSELEY MK-8's armed with one 37 mm gun and two BROWNING machine-guns (it has been suggested that a machine-gun be tried on the turret for anti-aircraft fire). There were five tank divisions in the British Expeditionary Force in France; he does not know how many tank divisions there are now. 300 light tanks are expected to arrive from SAShA [v]. A battalion of light tanks consists of three companies with 30 tanks per company. A battalion of 25-ton tanks is 40 strong. The different kind of organisation can be explained by the shortage of officers.

4. In France the British had 25 and 36 mm anti-tank guns. They proved to be too light against the German tanks and were replaced by 75 mm French guns which were successful in action. The arming of units with 49 mm anti-tank guns is now beginning. Anti-tank units are now being brought together into a brigade for training.

5. Artillery batteries are armed with 18 and 25-pounder guns. The LONDON air defence is now using new 6-inch guns as well as the 4.5-inch A.A. guns. They are experimenting with new 4.9-inch A.A. guns.

6. He considers that the War Office [VOENNOE MINISTERSTVO] is not making the slightest use of the experience of the French and the coastal defence is based on a network of blockhouses that are weak in design with no allowance made for the manoeuvrability or strong artillery and tank equipment of the SAUSAGE-DEALERS [KOLBASNIKI][vi].

No. 312   BRION[vii]

Notes:

   [a] Inserted by translator.

Comments:

   [i] BARCh: Possibly Simon Davidovich KREMER, whose official post was Secretary to the Soviet Military Attaché in LONDON. He was appointed in 1937 and is thought to have left sometime in 1946. The covername BARCh occurs as a LONDON addressee and signatory between 3rd March 1940 and 10th October 1940, after which it is superseded by the covername BRION.

   [ii] X GROUP: Not identified. See also MOSCOW's No. 450 of 7th September 1940, and LONDON's Nos. 812 of 25th July 1940, 895 of 16th August 1940, 987 of 6th September 1940, 1071 of 26th September 1940 and 1099 of 2nd October 1940.

   [iii] Artillery colonel: see also LONDON's No. 987 of 6th September 1940 for mention of a British Army colonel.

   [iv] RESERVIST: i.e. the artillery colonel; not further identified. For a similar casual introduction of a cover-name see LONDON'S No. 812 of 25th July 1940 (INTELLIGENTSIA).

   [v] SAShA: the United States. For an explanation of this coverword see LONDON's No. 998 of 11th September 1940 (3/PPDT/T12).

   [vi] SAUSAGE-DEALERS: the Germans.

   [vii] BRION: Col. I.A. SKLYaROV, Soviet Military and Air Attaché in LONDON 1940-1946. BRION's first appearance as a signatory (or addressee) is in LONDON's No. 1149 of 11th October.

APPENDIX 5

IN SUPPORT OF LYSENKO

The date of this document is circa December 15, 1948. It is unsigned, but accompanying material suggests the authors included Angus Bateman, who may have drafted it. The text was marked up by Haldane, whose critical annotations appear here as notes prefaced by [JBS]. His handwriting is illegible at times. The original is in the Haldane Papers, University College London.[1]

A DISCUSSION STATEMENT

1. Bourgeois Science and Soviet Science.

2. The Science of Biology.

3. Michurin Biology and Mendelian Genetics.

4. The Success of Michurin Biology.

5. A Note on Dialectical Materialism and Heredity.

1. BOURGEOIS SCIENCE AND SOVIET SCIENCE

In the early days of capitalism science flourished and achieved many triumphs. Even from the first, however, bourgeois social relations imposed limitations upon the advance of scientific practice and theory.

When capitalism began to decay much of its science, from being a weapon in the fight to master nature, became a weapon in the conduct of imperialist war, fascist racialism, the fight against the working class and the U.S.S.R. Unable to control, understand, or predict its own violent crisis, capitalism infected scientific theory with a pessimism which tended to regard things as unknowable, beyond human control, and thus subject to chance. Scientists, increasingly separated from the working people, from practice, had neither opportunity nor incentive to wrest ever more from nature for a society that dreaded conditions of over-production as disastrous. In these conditions Biology, the science of living organisms, became limited, pessimistic, and relatively sterile.[2]

In the Soviet Union the victory of the revolution permitted (meant) the liberation of science from academic trends, meant its full deployment in the struggle with nature, and meant the participation of all the population in the work of science. Armed with the theory of Marxism-Leninism Soviet science had all the means for unlimited advance. With the achievement and consolidation of Socialism, Soviet science is ridding itself of the last limitations imposed by reactionary bourgeois theory, and is now free to make advances of a kind impossible elsewhere in the world.

2. THE SCIENCE OF BIOLOGY

One of the greatest achievements of capitalist science was the theory of Darwin which gave a partially materialist explanation of the way in which animals and plants have evolved. Darwin showed that if changes occurred in living organisms which could be inherited by their offspring, and if a favourable selection of the fittest individuals took place, then constant improvement of the species would follow. He pointed out that the practice of selecting the best individuals for breeding had resulted in the continued improvement of domestic animals. He collected much evidence to show that in the wild state a continued selection of the fittest took place, and suggested that the evolution of organisms could thus be explained.

After Darwin's death his theory was distorted by capitalist science. Weismann and the Mendelian geneticists produced the theory that the nature of every organism was controlled by the chromosomes contained within every cell. Mendelians believe that the chromosomes bear particles, the genes, that are self-reproducing and unaffected by changes in the mode of life of the organism. In this way Mendelians seek to account for the failure of the effects of gross mutilations of the organism to be reproduced in its offspring.³ For them the genes in the germ cells are not affected by changes in the function of the organism brought about by changes in its mode of life. In face of the fact that the nature of organisms has changed in the course of evolution, but unable to relate those changes directly with changes in their mode of life, the Mendelists are forced to the position that genes do change, but only rarely and by chance. Mendelists investigating the way in which gene mutations occur have used certain physical and chemical agents (X-rays, heat, mustard gas) to change the gene substance. Those agents being unrelated to the normal mode of life of the organisms subjected to them, cause only random changes in the genes, the great majority of which are deleterious to the organism.⁴ By this separation of the changes in the cell from changes in the mode of life of the organism the Mendelists have been self-condemned to failure to understand heredity and the nature of evolution.

The theory that only random changes in the genes can provide evolutionary novelty gives rise to a pessimistic view of the possibilities of directing evolution in a way valuable to man. Mendelists have to admit that practical breeders have enormously improved hundreds of species of animals and plants, but they deny that these results are connected with a systematic improvement of the conditions of life combined with selection.⁵ For them the improvement is the result of the sorting out, by selection, of a stock of favourable genes existing in the ancestors of the organisms before their domestication.⁶ Apart from these already existing genes, Mendelism also considers the possibility that in the course of selection, but independent of it, new genes of a favourable nature arise by mutation. Once the original stock of genes has been sorted out only these rare and chance mutations give hope of further progress, which must therefore be very slow.

Imperialist exploiters use the Mendelian theory as showing that colonial peoples have an inherent inferiority in consequence of their possession of a poor complement of genes.⁷ On the other hand, to the working people of capitalist countries Mendelism declares that their exploitation with its consequent malnourishment, poverty, and ill-health has no effect on their children, and may therefore be the better endured.⁸ Thus is Mendelism used as a weapon of Fascist Imperialist exploitation. The protests of honest Mendelians to whom this use is abhorrent are powerless to prevent it.

Capitalist biology further more states that in nature every member of a species is constantly engaged in a struggle against members of its own species for the limited supply of food.⁹ This theory, owing its origin to the unnatural relations of capitalist society, seeks to defend the barbarities of capitalism and, at the same time, to divide the workers in their struggle by making each the enemy of his fellows.

3. MICHURIN BIOLOGY

The definition of heredity given by Lysenko as "the property of a living body to require definite conditions for its life, and to respond in a definite way to various conditions" is accurate and complete.¹⁰ With this conception it is possible to define the ways in which heredity may be changed.¹¹ For Soviet biology those conditions may be explained in the following way.

Changes in the heredity of an organism are made by influencing the germ cells it will later shed. Those may be influenced by fundamental changes in the function of the organism caused by its adaptation to its environment. Not all such changes are inherited . . . only those which affect processes involved in the formation and life of the germ cells. The influence of the environment is thus for Soviet biology creative, not as for the Mendelians destructive.¹² Soviet biology accepts the Darwinian conception of selection as positive and creative, a conception which owes something to Lamarck but is free from the crude Lamarckian proposition that all acquired characters are inherited.

The conception of creative selection, of the important contribution to the heredity of an organism made by its relations with its environment, is full of hope for man's control of evolutionary processes. Furthermore, unlike Mendelian theory, it declares that the possibilities of mastering the heredity of organisms are limited by knowledge of the laws of their development and function. It thus demands ever greater study of every aspect of living function and promises that such study will be rewarded by the possibility of evolutionary advance. The pitiful philosophy of the unknowability of organic variation is replaced by a great incentive to biological study.

The methods used by Soviet biologists in applying Michurin concepts are

(1) Study of an organism's requirements in order to offer it those which best suit its potentialities.
(2) Change of the organism's environment as a means of inducing heritable variation.
(3) Study of the new requirements of organisms thus subject to change, to discover their new potentialities.
(4) Selection of organisms best able to realise the potentialities of new environments.
(5) Physiological treatment of the organism (graft hybridisation, partial vernalisation, distant hybridisation etc.) designed to increase its variability and thus form new material for the method of creative selection.

These methods have been productive of positive results unthinkable in terms of Mendelian theory.[13]

One tenet of Soviet biology which also has important theoretical consequences and a positive practical outcome, is its rejection of the Malthusian doctrine of competition within the species. Malthusians hold that limited food supplies are an actual limitation on the increase in the population of a species. From this the doctrine that competition within a species generally prevails is derived. Engels pointed out the absurdity of this conception and modern biology confirms his view. Long before the available food supply of a species is exhausted its numbers have been drastically reduced by other means.[14] Competition within the species is thus exceptional and of little evolutionary significance. On the other hand co-operation between members of a species is an important and general fact. The application of the understanding of this fact to agronomy has led to improved methods of sowing and higher yields in the U.S.S.R.

Soviet biology teaches the oppressed that in the achievement of better conditions by the overthrow of imperialism lies the possibility of unlimited improvement of their minds and bodies and those of their children.

4. THE PRACTICAL SUCCESS OF MICHURIN BIOLOGY

Mendelian genetics has to its credit a number of practical successes. This does not prevent some biologists claiming for Mendelism results which are in no way to be attributed to it. Thus selection practices that are a continuation of methods employed before Mendel was born are sometimes claimed to owe their success to the Mendelian explanation of them. For the rest Mendelian genetics has provided a partial explanation of many existing facts, but has proved of small value in controlling and moulding the heredity of organisms.[15]

The successes of Michurin in the field of horticulture have been repeated and extended in all spheres of agriculture by Lysenko and other Soviet biologists. In particular the past fifteen years has seen the adoption by the collective farms of many new and improved varieties of wheat. Michurinists have been able to claim successes also in the culture of tomatoes, fruit, potatoes, and in the rearing of sheep, pigs, and cows.

Before the increased yields over millions of acres obtained by Soviet biologists the complaints of Mendelians that these results lack "statistical validity" must inevitably fall down. It is significant that one objection raised by Mendelians to Michurin biology is precisely that the practical successes of Michurinists are impossible. This objection is a tribute to the achievements of Soviet biology.

5. DIALECTICAL MATERIALISM AND HEREDITY

Marxist theory correctly rejects a dominating role for chance in any process. While recognising that in the development of science chance must always _appear_ to govern aspects of a process, it demands the further investigation of apparently chance governed phenomena to disclose their real, inevitable, dialectical development. Mendelian theory, by accepting the chance origin of variation, has <u>ignored the very possibility</u> of a real investigation of material interactions within the organism as the basis for understanding heredity.[16] It is thus self-condemned to relative sterility.

Soviet biology, based on Leninist theory and advanced practice,[17] rightly finds it unthinkable that material changes within the organism should be incapable of being impressed upon any part of its living substance. In studying heredity it recognises not only the separation of successive generations and their relative independence, but also their continuity and absolute connection. Thus dialectical materialism seeks the understanding of both the relative stability and absolute change in the history of a species not in terms of chance mutations in part of the organism alone, but in terms of the material expression of the contradiction between conservatism and variability in all aspects of the function of the organism. One-sided, aspectal, approaches to natural processes can never be true, and in their falsity lies the basis for philosophical idealism, the ideology of reaction.

NOTES

INTRODUCTION
1. MI5 Personal File, J.B.S. Haldane, National Archives, KV 2-1832/264b. The text in the MI5 notes has been reformatted throughout for clarity.
2. Clark (1968).
3. Montagu (1969).
4. Haldane (1939b). See Appendix 1.
5. Charlotte's defection is barely mentioned in Clark's biography, and her revealing autobiography, *Truth Will Out* (1949), is mostly ignored.

1. EARLY DAYS
1. Haldane (1932a), p. 242.
2. According to his mother, Louisa, JBS fagged for Julian's younger brother, Noel Trevelyan ("Trev") Huxley (1889–1914), who hanged himself in 1914. But his sister, Naomi, states several times that it was Julian. See Louisa Kathleen Haldane (1961), p. 223; Mitchison (1968), p. 300; and Mitchison (1973), p. 77.
3. Haldane (1939b). See Appendix 1.
4. Haldane (1939b). The phrase "Pop bitch" appears frequently in Eton memoirs, e.g., Cyril Connolly (1938), p. 242.
5. Quoted in Charles Williams (2010). Note, though, that Williams suspects that malice on the part of JBS may have been involved in the accusation.
6. Julian Huxley (1965).
7. Haldane (1939b).
8. *Eton Chronicle*, November 12, 1908, p. 370.
9. *Eton Chronicle*, November 7, 1907, p. 162.
10. *Eton Chronicle*, July 22, 1909, p. 526; January 5, 1909, p. 406.
11. *Eton Chronicle*, December 3, 1910, p. 810.

12. *Eton Chronicle*, October 20, 1910, p. 782.
13. Haldane (1939b).
14. *Eton Chronicle*, July 27, 1908, p. 316.
15. Haldane (1939b).
16. Haldane (1939b). See also https://grangehill1922.wordpress.com/2014/10/27/alec-rowan-herron/.
17. "My heart doth take in hand
    some godly song to sing;
    The praise that I shall show therein
    pertaineth to the King . . .".
18. Haldane (1939b).
19. "The Disturbance at Oxford," *The Times* (London, England), issue 40213, Friday, May 16, 1913, p. 10.
20. Hollis (1976), p. 24. Hollis wrongly states that Haldane was a Fellow of New College at this stage.
21. M. Newman (1993), p. 26.
22. J. B. S. Haldane, letter to Kingsley Martin, May 3, 1952, Haldane Papers, University College London. HALDANE/5/1/4/172.
23. G. R. Searle (1976), p. 13.
24. The Haldane family home was at the end of Linton Road. It was later demolished so that Woolfson College could be built there.
25. Aldous Huxley to Leonard Huxley, November 11, 1913, in Grover Smith (1969), pp. 55–56. Brackets inserted by Grover Smith.
26. Haldane (1915).
27. Alexander Dalzell Sprunt was born at Hampstead, London, in 1891. The son of John D. Sprunt and Jane Naismith Sprunt, of Montgomerie, Berkhamsted, Hertfordshire, he took a B.A. at Oxford and served in the 4th Bedfordshire as a 2nd lieutenant. He is buried in Lillers Communal Cemetery in France. Sprunt's brothers Gerald Harper and Edward Lawrence were both killed in later actions.
28. J. B. S. Haldane to William Bateson, March 18, 1915, Darlington Papers, Special Collections, Bodleian Library, Oxford University, CSAC106.3.85/J.86. Quoted in Clark (1968), p. 39.
29. Haldane (1939b).
30. Aldous Huxley to Leonard Huxley, April 26, 1915, in Grover Smith (1969), p. 69.
31. Naomi Mitchison (1975a). In this letter Naomi incorrectly claims that Haldane missed the battle of Aubers Ridge.
32. Haldane (1939b).
33. *The Times* (London), issue 40854, Friday, May 14, 1915, p. 5.
34. Haldane (1939b).
35. Blampied (1918), p. 54.
36. Haldane (1939b).

37. Haldane (1939b).
38. Haldane (1939b).
39. Haldane (1932a), p. 237.
40. Haldane (1939b).
41. Julian Huxley (1970), p. 137.
42. Julian Huxley (1965), p. 60.
43. Julian Huxley (1970), p. 114.
44. Aldous Huxley (1923), p. 68.
45. Aldous Huxley (1922), p. 50.
46. Haldane (1939b).
47. Mitchison (1979), p. 62.
48. Montagu (1970), p. 233.
49. May 10, 1926, MI5 Personal File, Ivor Montagu, National Archives KV 2-598/OYA and 1A.
50. For Willi Münzenberg's web of companies, see McMeekin (2003).
51. November 8, 1932, MI5 Personal File, Ivor Montagu, National Archives, KV 2-598/7B.
52. *Daily Express*, December 29, 1931.
53. Leon Trotsky, letter to Reginald Percy Groves (1908–1988), July 13, 1932, MI5 Personal File, Ivor Montagu, National Archives, KV 2-598/0B.
54. MI5 Personal File, Ivor Montagu, National Archives, KV 2-598/35B. For the complicated espionage career of Otto Katz, see Miles (2010).
55. MI5 Personal File, Ivor Montagu, National Archives, KV 2-598/31A.
56. Montagu (1937).
57. Russell (1975), p. 182.
58. Charlotte Haldane (1949), p. 30.
59. Bowker (1993), pp. 101–102, citing a conversation he had with Dr. Ralph Case, a brother of Haldane's former student and friend Martin Case.
60. Adamson (1998), pp. 62, 205, citing her own interview of Betty Burghes, February 4, 1992.
61. Charlotte Haldane (1965), quoted in Adamson (1998), p. 68.
62. Charlotte Haldane (1949), p. 37.
63. Howarth (1978), p. 216.
64. Woolf is misidentified by several sources as either William Jackson or Jack "Bugsy" Wolfe; e.g., in Costello (1988), pp. 216, 648. Haldane called him "Doggy." Apparently the nickname referred to a song Woolf performed.
65. Montagu (1970), p. 196.
66. Charlotte Haldane (1949), p. 54.
67. Haldane (1938c), p. 80.
68. Luck (1999), pp. 23–24.
69. Pirie (1966).
70. Winchester (2009), pp. 200–216.

71. Harland (2001), p. 131. It is possible, though, that the anecdote may refer to the 1950s, when Haldane was married to Helen Spurway, known for her loud and shrill voice.

2. WITH VAVILOV IN THE SOVIET UNION

1. Vavilov's brother Sergei (1891–1951) became a leading physicist and was involved with the Soviet atomic bomb project.
2. Charlotte Haldane (1949), p. 42.
3. Krementsov (1997), p. 22.
4. Charlotte Haldane (1949), p. 46.
5. Charlotte Haldane (1949), p. 49. For the O.G.P.U. (Secret Police), see the Glossary.
6. Keynes (1925), p. 258.
7. Charlotte refers to Levit as a "colleague" of Haldane's, but she appears to have meant only that he was a fellow geneticist. His date of birth is sometimes given as 1894.
8. Charlotte Haldane (1949), p. 51.
9. J. B. S. Haldane, address to the Fabian Society, Thursday, October 25, 1928. Reprinted in Haldane (1932a), pp. 126, 134–135.
10. Douglas Smith (2012).
11. Charlotte Haldane (1949), p. 47.
12. Volkogonov (1994), p. 361.
13. Birstein (2001), p. 42.
14. Haldane (1932a), p. 135.
15. Taylor (1990), p. 152.
16. Mitchison (1974), p. 17.

3. THE THIRTIES

1. Haldane described himself as unmusical. See Haldane (1939b) and Appendix 1.
2. Haldane (1931).
3. Rajani Palme Dutt was born in Cambridge and, like Haldane, read Greats at Oxford. His father was an Indian physician who practiced in the town, but his mother was Swedish, a relative of the future Swedish Prime Minister Olof Palme. His elder brother, Clemens Palme Dutt, was also active in the Communist Party.
4. MI5 Personal File, J. B. S. Haldane, National Archives, KV 2-1832/2A. See also the Andrew Rothstein file, KV 2-1579, National Archives at Kew, quoted in McIlroy (2006).
5. MI5 Personal File, J. B. S. Haldane, National Archives, KV 2-1832/3A. Reported in the *Star*, March 2, 1932.
6. "The Gold-Makers," *The Strand Magazine* 82 (491) (November 1931). The story is reprinted in Haldane (1932a), pp. 271–295.

7. Wiener (1956), pp. 160–162.
8. Howarth (1978), pp. 54, 105, 177–178.
9. Morison (1997), p. 312.
10. For a comprehensive scientific bibliography, see http://jbshaldane.org.
11. Haldane (1924).
12. Haldane (1939b). See Appendix 1.
13. Julian Huxley (1942).
14. Charlotte Haldane (1949), pp. 54–56.
15. Russell (1975), p. 225.
16. Darlington (1978), p. 17. See also Darlington (1962) and Darlington (1969b).
17. Darlington (1968), p. 934.
18. Haldane (1939b). See Appendix 1 for Drs. Grüneberg and Philip.
19. Ernst Boris Chain (1906–1979).
20. Howard Walter Florey (1898–1968).
21. Haldane (1964a).
22. McMeekin (2003). See also Miles (2010).
23. Otto Katz to Harry Pollitt, July 4, 1933, in MI5 Personal File, J. B. S. Haldane, National Archives, KV 2-1832/13A.
24. Charlotte Haldane (1969).
25. *Daily Herald*, November 3, 1934. Reprinted in Haldane (1946a), pp. 18–25.
26. William Ralph Inge (1860–1954), an Anglican priest who appeared regularly in the media.
27. Haldane (1946a), pp. 20–24.
28. Haldane (1946a), p. 19.
29. Haldane (1934).
30. Mitchison (1979), p. 212.
31. Haldane (1964a). See Appendix 3.
32. Mitchison (1979), pp. 191, 211–212.
33. MI5 Personal File, J. B. S. Haldane, National Archives KV 2-1832/19A.
34. Clark (1968), p. 137.
35. Charlotte Haldane (1969).
36. *Daily Worker*, December 24, 1936. Cutting in MI5 Personal File, J. B. S. Haldane, National Archives, KV 2-1832/21A.
37. Letter from unknown person to C. P. Dutt in Paris, January 4, 1937, MI5 Personal File, J. B. S. Haldane, National Archives, KV 2-1832/22A.
38. Cowles (1941), pp. 21–25.
39. Vera Elkan interview, September 11, 1996, Imperial War Museum, Oral History Sound Archive, item 16900, reel 3, 20'53" onward.
40. Haldane (1939a), p. 189.
41. Stewart and Stewart (2011), pp. 173, 415. However, Lethbridge (2013), pp. 115–116, suggests that the Austrian Hermann Hartung "somehow avoided execution."

42. Clark (1968), p. 135, places this incident later, on Haldane's second visit, but does not cite sources. It is also possible, but not very likely, that Clark is referring to some other incident where Haldane was arrested. Clark generally adds considerable confusion to the order of the Spanish incidents.
43. Vera Elkan interview, September 11, 1996, Imperial War Museum, Oral History Sound Archive, item 16900, reel 3, 20'53" onward.
44. MI5 Personal File, J. B. S. Haldane, National Archives, KV 2-1832/31A.
45. Carlson (1981), p. 237.
46. H. J. Muller, letters to Julian Huxley, March 9 and 11, 1937, Lilly Archive of H. J. Muller Papers; cited in Adams (1990), p. 197.
47. *Daily Worker*, April 5, 1937.
48. Foote (1949), p. 11.
49. Fred Copeman interview, 1978, Imperial War Museum, Oral History Sound Archive, item 794, reel 3, 2'54" onward.
50. Haldane (1939b).
51. Fred Copeman interview, 1978, Imperial War Museum, Oral History Sound Archive, item 794, reel 6, 14'44" onward.
52. Walter Krivitsky is now definitely known to have been assassinated by the NKVD in 1941 in his Washington, D.C., hotel room. See Kern (2004).
53. Howson (1998).
54. Vaill (2014), p. 157.
55. Cowles (1941), pp. 33–34.
56. It is not clear within which organization this rank operated. Some MI5 sources cast doubt on it. MI5 Personal File, Hans Kahle, KV 2-1562/57b and 58a.
57. These early details appear in Werner (1991), pp. 312–313.
58. Oddly, MI5 in several places lists 1923 as Gertrude's date of birth. She was surely older than ten when she married Hans.
59. Hemingway (1940).
60. It is reproduced in Radosh, Habeck, and Sevostianov (2001), p. 204.
61. See Regler (1959), p. 297. Regler was obviously a true believer.
62. MI5 Personal File, Hans Kahle, National Archives, KV 2-1562/5a and b.
63. See Krivitsky (1939), pp. 72, 87, 92, 95–96, 198, 298. For full details on Orlov, now confirmed by multiple sources, see Payne (2004), pp. 134–135.
64. Richardson (1982), pp. 174–175.
65. Walter Greenhalgh interview, Imperial War Museum, reel 5, 10'15" onward.
66. MI5 Personal File, J. B. S. Haldane, National Archives, KV 2-1832/42.
67. H. J. Muller, letter to J. Kemeny, October 17, 1963. Cited in Carlson (1981), p. 240.
68. His real name was Arnold Reisky, a Jewish journalist. Richardson (1982), p. 190. See also http://www.alba-valb.org/volunteers/arnold-reid.
69. Charlotte Haldane (1949), pp. 99–116, 126–127.
70. MI5 Personal File, J. B. S. Haldane, National Archives, KV 2-1832/38A.

71. Wiener (1956), pp. 206–207.
72. Krivitsky (1939), p. 87.
73. Payne (2004), p. 228.
74. Haldane (1938b), p. 180.
75. Haldane (1938b), pp. 180–181.
76. Haldane (1938b), pp. 164–166.
77. Haldane (1938b), p. 49.
78. Haldane (1938b), p. 26.
79. Haldane (1938b), pp. 199–200.
80. MI5 Personal File, J. B. S. Haldane, National Archives, KV 2-1832/71A.
81. Haldane (1938b), p. 11.
82. Spender in Crossman (1949), pp. 258–259.
83. Haldane (1939b).
84. Charlotte Haldane (1949), p. 124, introduced some confusion about the dates of this trip, referring only to the invitation as towards the "end of 1937." Baxell (2014), p. 303, wrongly places it in October to November 1937, when Charlotte was definitely in Britain. She arrived sometime in late January 1938 and left "three weeks" later; see Adamson (1998), p. 119, and the *Western Daily Press* (February 14, 1938), p. 11.
85. Mitchison (1979), p. 185.
86. July 21, 1938, MI5 Personal File, J. B. S. Haldane, National Archives, KV 2-1832/87B.
87. November 13, 1938, MI5 Personal File, J. B. S. Haldane, National Archives, KV 2-1832/103A.
88. Fred Copeman interview, 1978, Imperial War Museum, Oral History Sound Archive, item 794, reel 10, 23'04" onward.
89. This was published the following year in Haldane (1940e).
90. Haldane (1938f), p. 24.
91. Haldane (1939b).
92. Haldane (1938d).
93. Maynard Smith (2004).
94. Lerner (1938).
95. Quoted in van Heijenoort (1985), p. 151.
96. Haldane (1940e), pp. xi–xii.
97. For a gentle introduction to this area, see Peter Smith (2013).
98. Feferman (2000), pp. 371–372.
99. J. B. S. Haldane, letter to Dona Torr, November 3, 1938, MI5 Personal File, J. B. S. Haldane, National Archives, KV 2-1832/101A.

4. STALINOPHILIA

1. February 14, 1938, MI5 Personal File, J. B. S. Haldane, National Archives, KV 2-1832/68A.
2. Haldane (1946a), p. 116.

3. Haldane (1946a), p. 97.
4. Haldane (1946a), p. 135.
5. Haldane (1946a), p. 144.
6. Haldane (1946a), p. 189.
7. Haldane (1946a), p. 178.
8. Haldane (1947a), p. 230.
9. Haldane (1947b), p. 43.
10. Haldane (1947b), p. 48.
11. Haldane (1947b), p. 206.
12. Haldane (1947b), p. 209.
13. Haldane (1940b), p. 108.
14. Haldane (1947b), p. 226.
15. Haldane (1947b), p. 52.
16. Haldane (1947a), pp. 120–121.
17. Haldane (1947b), p. 57.
18. Haldane (1947b), pp. 114–115.
19. Haldane (1947a), pp. 84–86.
20. Haldane (1947a), pp. 128–129.
21. Haldane (1947b), p. 15.
22. Haldane (1946a) p. 211.
23. Haldane (1946a), p. 218.
24. Haldane (1946a), pp. 219–220.
25. Haldane (1938e).
26. Haldane (1946a), pp. 229–231.
27. Haldane (1946a), pp. 224–225.
28. Haldane (1940b), p. 241.
29. Haldane (1946a), p. 38.
30. See Adams (1990).
31. See the translations of Koltsov's eugenics articles in Babkov (2013).
32. There is some uncertainty about the exact date that Stalin banned IQ testing or, less formally, transmitted the idea that performing tests would not be welcome.
33. Babkov (2013), p. 476.
34. Fisher (1930).
35. See Birstein (2001), p. 204, and Babkov (2013), p. 647.
36. Circa December 22, 1936.
37. June 23, 1937. The letter is quoted in Soyfer (2003), p. 8, from a copy of the letter supplied to Soyfer by J. Lederberg, who was a close friend of Haldane's in later life.
38. Haldane (1937b).
39. After serving in Spain, Muller briefly went back to Moscow in September 1937 to collect his things and leave for the West; Carlson (1981), pp. 242–244.

40. Charlotte Haldane (1949), p. 318.
41. Popovsky (1984), p. 75, citing a private conversation with Timofeev-Ressovsky. As it happened, Timofeev-Ressovsky remained in Berlin through the Second World War, and after the Soviet occupation of Germany was "liberated" and shipped back to the Gulag anyway.
42. Soyfer (1994), pp. 108–109.
43. Soyfer (2001).
44. Soyfer (1994), pp. 130–131.
45. Haldane to Vavilov, May 4, 1939. Haldane Papers, University College London, HALDANE 5/5/1/14.
46. See, for example, Joravsky (1970), pp. 1–17.
47. Medvedev and Medvedev (2003), pp. 199–205.

5. WAR ON ONE FRONT

1. Haldane (1938b), p. 249.
2. From the script for the BBC talk "A Banned Broadcast" (1934). The talk was not transmitted but was reprinted in Haldane (1946a), p. 19, where Haldane states that it was also printed in the *Daily Herald* on November 3, 1934.
3. Haldane (1938b), pp. 223–224.
4. Graham (1993), pp. 52–53.
5. Haldane (1938b), pp. 206–207.
6. Hyde (1950), p. 92. See below for more on Alan Nunn May.
7. Moorhouse (2014).
8. For a contrary opinion, see Medvedev and Medvedev (2003).
9. Charlotte Haldane (1949), pp. 178–180.
10. Hyde (1950), p. 69.
11. J. B. S. Haldane, "Is There a Russian Enigma?" Quotations here and below are reproduced from Haldane's handwritten manuscript; see Appendix 2 and Haldane (1939d).
12. *Daily Worker*, November 30, 1939.
13. Gollancz (1941).
14. There are notes in Haldane's MI5 file that he received various payments during 1939, including 110 pounds from a "Miss Howard," who frequently sent money to the communists, and some unspecified amount from the American heiress Harriet Shaw Weaver, who had paid 340 pounds in total to communist recipients. MI5 Personal File, J. B. S. Haldane, National Archives, KV 2-1832/22 (summary).
15. Hyde (1950), p. 71.
16. MI5 Personal File, J. B. S. Haldane, National Archives, KV 2-1832/17-18 (summary).
17. See MI5 Personal File, J. B. S. Haldane, National Archives, KV 2-1832/summary for more details on all these meetings, and more.

18. Haldane (1941b).
19. Hyde (1950), pp. 92–94.
20. See, for example, the collected material related to the Convention in the Haldane Papers, University College London, HALDANE 4/19/10.
21. Or so she told Ronald Clark. See Clark (1968), p. 183.

6. IVOR MONTAGU AND THE X GROUP

1. *Sic.* The correct spelling is Ewen.
2. Wright (1987), p. 186.
3. Costello (1988), p. 526, briefly referred to Wright's information about Haldane.
4. West (1999).
5. Haynes and Klehr (1999).
6. Brown (2005), p. 184.
7. For example, Cochran and Harpending (2010), supplementary online materials on Haldane, which elevated Haldane to the head of the X Group. See http://the10000yearexplosion.com/jbs-haldanes/.
8. Semion Dawidowicz Kremer (1900–1991), born in Gomel in Russia, had already been identified as an agent in 1940 by the Soviet defector General Walter Krivitsky. See https://vault.fbi.gov/rosenberg-case/klaus-fuchs/klaus-fuchs-part-62-of. Kremer arrived in England in 1937, but MI5 could not determine his date of departure. Klaus Fuchs knew him as "Alexander." He seems to have returned home in 1942, and to have served from 1943 onward in the Soviet Army, rising to general and Hero of the Soviet Union. He retired in 1956, and died at Odessa in 1991. See also http://www.yadvashem.org/research/research-projects/soldiers/simon-kremer.
9. That is, this part could not be decoded.
10. MI5 noted that Montagu was living in Watford at the time.
11. Not too much weight should be assigned to this. See the remarks below about their remarkable failure to make sense of Montagu.
12. MI5 Personal File, J. B. S. Haldane, National Archives, KV 2-1832/204. The file contains a clipping of Haldane (1946b).
13. The UK.
14. The British.
15. Charlotte Haldane (1949), pp. 178–179.
16. Clark (1968), pp. 164–165.
17. Siebe Gorman was bombed out in May 1941, after which it was relocated to Chessington in Surrey.
18. There are many mentions of Kahle set in this period in Lynton (1995), written by a fellow internee and admirer (Max-Otto Ludwig Loewenstein, a.k.a. "Mark Lynton"). As Lynton's work includes glaring and gross inaccuracies, it is difficult to judge the many novel claims it contains, e.g., that Kahle fought

as a mercenary for Sun Yat Sen and was decorated with the Pour le Mérite in 1918 (he is not in the list of awardees at https://en.wikipedia.org/wiki/List_of_the_Pour_le_M%C3%A9rite_(military_class)_recipients#K).
19. Hemingway to Kahle, 1940, quoted in McLellan (2004), p. 35.
20. Baxell (2014), pp. 422–423.
21. Warren and Benson (1954), p. 302.
22. Clark (1968), pp. 152–161.
23. National Archives ADM 178/313.
24. National Archives AIR 20/4168.
25. See Haldane to Kahle, November 14, 1940. Haldane Papers, University College London, HALDANE 1/5/3/55.
26. Charlotte reported that she could not turn Kahle out when she returned because his wife was expecting a child. If the child survived it would have been Kahle's second. Charlotte Haldane (1949), p. 254.
27. Kahle (1943, 1944a, 1944b, 1945). Several million Soviet soldiers were easily encircled and captured as a result of Stalin's military bungling.
28. See, for example, Kahle to Haldane, March 11, 1943, Haldane Papers, University College London, HALDANE 5/1/1/33.
29. Werner (1991), pp. 242–243, 246, 262, 312–313. "Ruth Werner" was her married name in later years after she fled back to East Germany. Her sources of information included her well-placed economist father, Robert René Kuczynski (1876–1947), himself a communist of long standing.
30. Haynes, Klehr, and Vassiliev (2009), p. 154.
31. Robert Chadwell Williams (1987).
32. Hansard, House of Commons Debates, April 26, 1948, vol. 450, cols. 18–19, http://hansard.millbanksystems.com/commons/1948/apr/26/scientific-committees-professor-haldane#column_19.
33. MI5 Personal File, J. B. S. Haldane, National Archives, KV 2-1832/274.
34. Quoted in Clark (1968), p. 212.
35. MI5 Personal File, J. B. S. Haldane, National Archives, KV 2-1832/318ab.
36. MI5 Personal File, J. B. S. Haldane, National Archives, KV 2-1832/318ab.
37. December 8, 1951, MI5 Personal File, J. B. S. Haldane, National Archives, KV 2-1832/319b.
38. See Andrew and Gordievsky (1990), pp. 317–318, for more on Pontecorvo's espionage.
39. Charlotte Haldane (1949), p. 185.
40. April 4, 1950, MI5 Personal File, J. B. S. Haldane, National Archives, KV 2-1832/276a.
41. April 7, 1950, MI5 Personal File, J. B. S. Haldane, National Archives KV 2-1832/309a.
42. MI5 case note, quoted in Andrew (2009), p. 381.
43. See Andrew and Mitrokhin (2005).

7. THE FATE OF VAVILOV

1. Charlotte incorrectly recalled sailing on the *Windsor Castle*; Charlotte Haldane (1949), p. 192.
2. This letter is cited by Clark (1968), p. 197, without naming the recipient, who is vaguely stated to have been "in Scotland." It may have been Hermann Muller, who was based in Edinburgh at the time.
3. Charlotte Haldane (1949), pp. 195–196.
4. Charlotte Haldane (1949), pp. 198–240.
5. Charlotte Haldane (1949), pp. 254–267.
6. Proposal for Foreign Membership, Royal Society Archives, EC/1942/24. The proposers were H. H. Dale, A. V. Hill, Thomas R. Merton, O. T. Jones, E. J. Russell, J. B. S. Haldane, W. H. Pearsall, H. Munro Fox, J. Herbert Parsons, C. M. Wenyon, C. R. Harington, S. P. Bedson, E. F. Armstrong, I. M. Heilbron, P. M. S. Blackett, A. M. Tyndall, J. E. Lennard-Jones, L. J. Mordell, H. T. Tizard, and A. C. Egerton.
7. Haldane (1942).
8. Haldane (1947b), p. 212.
9. Haldane (1947b), pp. 224.
10. Haldane (1947b), pp. 225–226.
11. Darlington and Harland (1945).
12. Haldane (1947a), p. 151.
13. Haldane (1948), p. 875.
14. As told by Fillipovsky to Mark Popovsky in 1968. See Popovsky (1984), p. 150.
15. Birstein (2001), pp. 236–237.
16. Mitchison (1974), p. 17.
17. Quoted by Clark (1968), p. 199. I have not been able to trace the manuscript of this letter, which Clark does not date.
18. See Paget (1920).
19. See the correspondence from the secretary of the Royal Society in the Haldane Papers, University College London, HALDANE/4/22/2/52,65,85.
20. Birstein (2001), pp. 225, 401.
21. Joravsky (1970), pp. 317–328.
22. Birstein (2001), p. 298.
23. Charlotte Haldane (1949), pp. 42–45.
24. Haldane to Lina Stern, February 9, 1942, Haldane Papers, University College London, HALDANE 5/2/4/164. A letter attached to this one has been misclassified and is actually to a different "Professor Stern."
25. Ro'i (2010).
26. Rapaport (1991), pp. 234–253.
27. Abakumov was under Beria's direction. He was executed in 1954, supposedly for treason.

28. Gordon (1951).
29. Gordon to Haldane, September 24, 1951; Haldane to Gordon, September 27, 1951, Haldane Papers, University College London, HALDANE 5/2/2/71. Jerzy Konorski (1903–1973), a neurophysiologist and critic of Pavlov, had been under sustained attack for several years in the state press.

8. EXPERIMENTS IN THE REVIVAL OF ORGANISMS

1. Haldane Papers, University College London, HALDANE 4/4/20.
2. *Experiments in the Revival of Organisms* (1940), directed by D.I. Yashin (Moscow: Techfilm Studio). The film is freely available at https://archive.org/details/Experime1940.
3. "Notes and News," *The Lancet* 240 (6213) (September 26, 1942), pp. 382–384.
4. *Time* magazine, Monday, November 22, 1943.
5. See Soyfer (1994), pp. 161–162. Lysenko was responsible for the disaster.
6. The patent application was dated November 29, 1928. On December 15, 1934, the patent was issued as USSR patent no. 35976.
7. See Konstantinov and Alexi-Meskishvili (2000).
8. See Bryukhonenko et al. (1929) and Fernan-Perez (1929).
9. Probert and Melrose (1960).
10. See "Experiments in Resuscitation" (1937).
11. Gerya and Yankovsky (1976–1977); Lanovenko, Yankovsky, and Lyavinetz (1976); Lanovenko, and Yankovsky (1972); and Lanovenko and Yankovsky (1973).
12. Negovsky (1995).
13. Krementsov (2009).
14. "Shaw Feels 'Tempted' to Have Head Cut Off," *New York Times*, March 17, 1929, p. 5. Quoted in Krementsov (2009).
15. See Krementsov (2009) for a rapid survey.
16. King (1997) has extensively documented this phenomenon.
17. Even so, the population could hardly keep pace, and would preemptively scratch out photographs of the liquidated in their own copies of the *Great Soviet Encyclopedia*.

9. IT IS YOUR PARTY DUTY, COMRADE!

1. Hogben to Haldane, November 17, 1943, Haldane Papers, University College London, HALDANE 5/1/1/25.
2. Haldane to Hogben, November 22, 1943, Haldane Papers, University College London, HALDANE 5/1/1/26.
3. Hogben to Haldane, November 29, 1943, Haldane Papers, University College London, HALDANE 5/1/1/27.
4. Haldane to Hogben, undated, circa December 1943, Haldane Papers, University College London, HALDANE 5/1/1/28.

5. This and following Woolf quotations are taken from Woolf to Haldane, December 15, 1943, Haldane Papers, University College London, HALDANE 5/1/1/75.
6. Fisher (1925).
7. Haldane to Woolf, December 21, 1943, and January 12, 1944, Haldane Papers, University College London, HALDANE 5/1/1/76 to 5/1/1/78.
8. See Woolf and Waterhouse (1945) and Woolf (1947).
9. Haldane (1945).
10. Emile Burns, February 7, 1945, MI5 Personal File, J. B. S. Haldane, National Archives, KV 2-1832/181a.
11. September 8, 1946, MI5 Personal File, J. B. S. Haldane, National Archives, KV 2-1832/204b.
12. October 28, 1946, MI5 Personal File, J. B. S. Haldane, National Archives, KV 2-1832/208b.
13. This sort of argument was quietly dropped after the USSR acquired nuclear weapons in 1949.
14. March 4, 1947, MI5 Personal File, J. B. S. Haldane, National Archives, KV 2-1832/215a.
15. Haldane (1947b), p. 241.
16. Haldane (1947b), p. 238.
17. Haldane (1946a), pp. 224–225.
18. MI5 Personal File, J. B. S. Haldane, National Archives, KV 2-1832/226a.
19. Our Correspondent, "Intellectuals and Propaganda," *The Times* (London), August 28, 1948, p. 3.
20. September 11, 1948, MI5 Personal File, J. B. S. Haldane, National Archives, KV 2-1832/249a.
21. Solzhenitsyn (1968), p. 388.
22. Mary (Jones?) to Kitty Cornforth, October 13, 1947, MI5 Personal File, J. B. S. Haldane, National Archives, KV 2-1832/226a. Kitty Cornforth was James Klugmann's sister and had married Maurice Cornforth. All of them were Communist Party stalwarts. "Mary (Jones?)" could not be traced.
23. MI5 Personal File, J. B. S. Haldane, National Archives, KV 2-1832/241b.
24. October 24, 1952, MI5 Personal File, Otto Katz, National Archives, KV 2-1384/458B.
25. Not identified. Possibly Bob Jones, the trade union leader and Soviet agent.
26. An abbreviation for "phonetically."
27. Phonetic transcription. Possibly Klement Gottwald (1896–1953); see below.
28. Sir Walter Layton (1884–1966), the newspaper proprietor.
29. Possibly Klement Gottwald, chairman of the Czech Communist Party.
30. *Sic.*
31. Paul Eisler (1919–1966) had lost his job owing to the Slansky show trials and had to work as a lathe-operator. Later he was dismissed there, too. He was

Slansky's economics adviser and may also have been under suspicion because he was Jewish, as was Slansky. He was spared a show trial by the death of Stalin. See Catherine Epstein (2003).
32. The Communist Party newspaper.
33. Pavel Kavan (?–1960), who was merely imprisoned. He had been in London during the war.
34. Medawar (1996) pp. 86–93.
35. Montagu (1956).
36. Montagu (1974).
37. Macintyre (2010) p. 319.
38. Clark (1968), 281.

10. LYSENKO AND LAMARXISM

1. This article was republished in Haldane (1939a), pp. 134–138. The exact date that it appeared in the *Daily Worker* is not known to me.
2. W. W. Garner (1875–1956) and H. C. Allard (1880–1963) discovered photoperiodism. V. I. Razumov was a physiologist and Lysenkoite. V. N. Liubimenko was a botanist.
3. D. A. Dolgushin was a collaborator, close friend, and biographer of Lysenko.
4. In fact, the Soviet delegation had been forbidden to attend by their minders.
5. Contrary to Graham (2016), p. 98, who claims, on p. 82, to have read absolutely everything Lysenko ever wrote and denies that he, *or any Soviet leader*, ever subscribed to the notion of the Soviets making new men!
6. Haldane (1940a), pp. 81–85.
7. Haldane (1947b), p. 76.
8. Haldane (1947b), p. 212.
9. Theodosius Dobzhansky (1900–1975) came to the United States in 1927.
10. Nikolay Timofeev-Ressovsky (1900–1981) remained in Germany during the Second World War and was sent to the Gulag after the war.
11. Sergei Chetverikov (1880–1959) was one of the first Russian geneticists.
12. Haldane (1947b), pp. 223–225.
13. In Haldane (1940a), p. 82, the claim is that Harland lost his job for having a "coloured wife," "among other things."
14. Harland (2001), pp. 95–107.
15. H. J. Muller to H. P. Riley, 1953. Quoted in Carlson (1981), p. 243.
16. See Howell (2008).
17. V. P. Efroimson, "On the Rate of Mutational Processes in Humans" (unpublished manuscript, 1932). Cited by Babkov (2013), p. 543. Babkov does not mention Muller's allegation.
18. Haldane (1935). Consider also V. P. Efroimson, "On Some Problems of Accumulation and Action of Lethals," *Biological Journal* 1 (1932), pp. 87–101. (I have not been able to consult this last reference.)

19. Babkov (2013), p. 543.
20. Haldane (1947b), pp. 223–225.
21. Haldane (1947b), pp. 223–225.
22. Haldane (1941a), pp. 13–14.
23. Haldane (1940d).
24. See *The Oxford Companion to the History of Modern Science*, edited by John L. Heilbron (Oxford and New York: Oxford University Press, 2003), p. 121.
25. Haldane (1941a), p. 110.
26. Haldane (1944).
27. *Sic.*
28. Haldane (1947a), pp. 151–152. Here, S. C. Harland and his move to South America make another appearance, as a domestic example of the sort of thing meted out to geneticists in the USSR.
29. F. G. Gregory to J. B. S. Haldane, August 21, 1947, Haldane Papers, University College London, HALDANE 5/1/2/8/6.
30. L. C. Dunn, "Science in the USSR: Soviet Biology," *Science* 99 (1944), pp. 65–67. Reprinted in Zirkle (1949).
31. Karl Sax, "Soviet Biology," *Science* 99 (1944), pp. 298–299. Reprinted in Zirkle (1949).
32. Anton R. Zhebrak, "Soviet Biology," *Science* 102 (1945), pp. 357–358.
33. *Pravda*, March 6, 1947. Cited in Soyfer (1994), pp. 165–166.
34. *Pravda*, September 2, 1947. Reprinted in Zirkle (1949).
35. K. Mather, "Genetics and the Russian Controversy," *Nature* 151 (1942), pp. 68–74. Reprinted in Darlington and Mather (1950).
36. Lysenko to Haldane, correspondence via the Soviet Embassy, December 27, 1944. Haldane Papers, University College London, HALDANE 4/9/1/1.
37. Haldane to Lysenko, cover letter dated January 3, 1945, Haldane Papers, University College London, HALDANE 4/9/1/1.
38. Letter from Kurt Stern et al. to J. B. S. Haldane, April 17, 1946, Stern Papers. Cited by Krementsov (1997), pp. 122–123 and p. 331 n. 129.
39. Letter from J. B. S. Haldane to Herman Muller, May 15, 1946, Demerec Papers. Cited by Krementsov (1997), pp. 122–123 and p. 331 n. 130.
40. Cyril Darlington, "The Retreat from Science in Soviet Russia," *Nineteenth Century* 142 (1947), pp. 157–168. Reprinted in Zirkle (1949).
41. Angus John Bateman to Haldane, November 6, 1947, Haldane Papers, University College London, HALDANE 5/1/1/2/8/32. The letter is incorrectly dated 1948 there.
42. Genetical Society. Bateman uses "tabled" in the British sense of proposing a resolution for discussion.
43. Cyril Dean Darlington.
44. Soyfer (1994), pp. 183–184.
45. Medvedev and Medvedev (2003), p. 195.

46. Berg (1983). Raissa L'vovna Berg (1913–2006) was a Soviet geneticist who had studied under Hermann Muller in Leningrad. She was dismissed from Moscow University as a result of Lysenkoist pressure and later migrated to the United States in the mid-1970s.
47. Quoted in Soyfer (1994), 190. There is a translated copy in the Haldane Papers at University College London, HALDANE 5/1/2/8/14.
48. *Time* 52 (13) (September 27, 1948), p. 70.
49. Haldane (1945?), pp. 12–13. Due to a mislaid research note the precise source cannot be identified, but image snippets of the original have been retained.
50. Third programme, 18:50. The broadcast was repeated on Friday, January 7, 1949, at 23:10. See Haldane (1948).
51. Letter from W.A. Thompson, February 9, 1952, Haldane Papers, University College London, HALDANE 5/1/4/257.
52. MI5 Personal File, J.B.S. Haldane, National Archives, KV 2-1832/254C and following pages.
53. Angus Bateman, "In Support of Lysenko." Haldane Papers, University College London, HALDANE 4/9/1/7. See Appendix 5.
54. Penrose is known from MI5 reports to have attended a personal meeting with Ivan Glushchenko organized by Haldane. These "insider" details were attributed to an anonymous source in Almond (1954), pp. 314–318, but it follows from there that the same source was also present at the public Glushchenko meeting (of which see more below). Penrose, a colleague of Haldane's at UCL, is the only viable candidate.
55. Haldane (1949).
56. February 11, 1949, MI5 Personal File, J.B.S. Haldane, National Archives, KV 2-1832/258b.
57. John Mahon (1901–1975).
58. March 25, 1949, MI5 Personal File, J.B.S. Haldane, National Archives, KV 2-1832/261a.
59. Bernal (1949).
60. Saturday, October 15, 1949. The meeting was held at Beaver Hall and chaired by J.G. Crowther. *The Anglo-Soviet Journal* (Autumn 1950), pp. 4–17.
61. Not traced. A draft survives in the Haldane Papers, University College London, HALDANE 5/1/2/8/48.
62. Langdon-Davies (1949).
63. Letter from Bernal to Haldane, October 6, 1949, including the draft review. Haldane Papers, University College London, HALDANE/5/1/2/8/48.
64. MI5 Personal File, J.B.S. Haldane, National Archives, KV 2-1832/268Y.
65. MI5 Personal File, J.B.S. Haldane, National Archives, KV 2-1832/268Y.
66. October 27, 1949, MI5 Personal File, J.B.S. Haldane, National Archives, KV 2-1832/269b.
67. Julian Huxley (1949).

68. Penrose seems to have been within the Party orbit in 1949. As stated above, he is almost certainly the source for the anonymous descriptions of these events given in Almond (1954). However, Penrose confused the order of events.
69. December 12, 1949, MI5 Personal File, J. B. S. Haldane, National Archives, KV 2-1832/272a.
70. Maynard Smith incorrectly remembered the date of the meeting as "very soon after I graduated, in about '51, '52." Interview with Richard Dawkins, "Web of Stories," http://www.webofstories.com/people/john.maynard.smith/33?o=SH.
71. Almond (1954), p. 317. We have deduced that Almond's anonymous source here is Penrose, as per above.
72. Soyfer (1994), pp. 91–92.
73. Morton (1951).
74. Fyfe (1950).
75. Bernal (1953).
76. See the letter from Haldane to Bernal, January 4, 1954, Haldane Papers, University College London, HALDANE 5/1/4/17.
77. February 2, 1950, MI5 Personal File, J. B. S. Haldane, National Archives, KV 2-1832/273z.
78. April 4, 1950, MI5 Personal File, J. B. S. Haldane, National Archives, KV 2-1832/276a.
79. Probably the Fabian Arthur Creech Jones (1891–1964).
80. Expletive omitted by the bashful MI5 transcriber.
81. April 11, 1950, MI5 Personal File, J. B. S. Haldane, National Archives, KV 2-1832/280l.
82. July 26, 1950, MI5 Personal File, J. B. S. Haldane, National Archives, KV 2-1832/286a.
83. Bernal may have been boasting here, as his own private life included numerous conquests without any repercussions whatsoever, apart from several stray children.
84. Macleod (1997), pp. 26–27.
85. December 12, 1950, MI5 Personal File, J. B. S. Haldane, National Archives, KV 2-1832/299b.
86. Presumably Emile Burns (1889–1972) and his wife Margaret Elinor Burns (1888–1978).
87. December 21, 1952, MI5 Personal File, J. B. S. Haldane, National Archives, KV 2-1832/335A.
88. Macleod (1997), pp. 26–27.
89. Macleod (1997), p. 55.
90. Paul Kammerer (1880–1926), who is generally believed to have faked experiments showing inheritance of acquired characters.

91. *Sic.*
92. Letter to the Society for Cultural Relations, June 1, 1951, Haldane Papers, University College London.
93. Haldane (1953), p. xi.
94. Schaechter (2009), p. 44.
95. Haldane may have meant Øjvind Winge (1886–1964), the Danish yeast geneticist, and Sol Spiegelman (1914–1983), the American molecular biologist.
96. Haldane (1954b).
97. Haldane (1954c), pp. 112–114.
98. Haldane (1959), p. 111.
99. Haldane (1964a). See Appendix 3 for the complete text.

11. SOCIAL BIOLOGY

1. Haldane (1923).
2. Muller (1925); Horatio Newman (1929).
3. The reader should note the careful use of *differences* in this discussion. This was well understood by Haldane himself, though he does not dwell on it. Logically, no outcome at all is possible without the presence of both nature and nurture; the question is which *differences* matter in producing outcome *differences*.
4. Burks (1927, 1928).
5. Spearman (1927).
6. Haldane (1932a), p. 21.
7. Haldane (1932a), pp. 137–138.
8. Muller (1935).
9. Loren Graham's attempted explanation for this rejection—that "Stalin had by this time discarded visions of a radical transformation of Soviet Society and had moved toward greater attention to religion, established social norms, and patriotism in the face of the looming threat of Nazi Germany"—is unfathomable. This was 1936. See Graham (2016), p. 59.
10. Haldane (1938a), p. 128. JBS often repeated this joke. In 1963 at a CIBA foundation symposium, Julian Huxley immediately interjected "*Ceteris paribus!*"
11. Crew et al. (1939).
12. Haldane (1947a), p. 110.
13. Haldane (1947a), p. 116.
14. Haldane (1946c).
15. Haldane (1954b).
16. See Herrnstein (1973).
17. Haldane (1962).
18. Haldane (1963).
19. Haldane (1964b).

12. ANIMAL BEHAVIOR FROM LONDON TO INDIA

1. Maynard Smith to Haldane, October 1, 1947, Haldane Papers, University College London, HALDANE/5/2/4/144.
2. Maynard Smith (1985). But see Maynard Smith's recollections during a prior era, in his preface to Haldane (1968), p. vii, where he recalled first reading *The Inequality of Man*, not *Possible Worlds*, at Eton.
3. Maynard Smith (1985).
4. Maynard Smith (2004).
5. Haldane Papers, University College London, HALDANE 3/5/1.
6. Morris (2006), p. 112.
7. Kruuk (2003), p. 197. Note that Kruuk gives no reason for the antipathy.
8. Mayr told this story many years later to S. Sarkar (2005).
9. Medawar (1996), p. 129.
10. Lorenz (1952); Spurway (1952).
11. Morris (2006), pp. 108–123.
12. Medawar (1996), pp. 86–93.
13. Marler (1985), p. 322.
14. "Lecturer Stamped on Dog's Tail," *The Times* (London), November 14, 1956, p. 7.
15. "Mrs. Haldane Freed: Fine Paid," *The Times* (London) November 19, 1956, p. 6. Waller is identified by Clark (1968), pp. 235–237, who gives a very inaccurate account of this incident.
16. Bonner (2002), p. 119.
17. MI5 Personal File, J. B. S. Haldane, National Archives, KV 2-1832/276B.
18. Haldane (1963), p. 337.
19. Volkogonov (1994), p. 361.
20. Lederberg interview with Steven J. Dick, November 12, 1992, National Institutes of Health, National Library of Medicine.
21. Lewontin (1985), p. 682.
22. Haldane to Lederberg, March 7, 1959, The Joshua Lederberg Papers, NIH National Library of Medicine, http://profiles.nlm.nih.gov/ps/retrieve/Collection/CID/BB.
23. Haldane to Lederberg, October 6, 1962, NIH National Library of Medicine.
24. Bonner (2002), pp. 119–120.
25. *Singapore Free Press*, January 19, 1961, p. 1.
26. Kosygin was First Deputy Premier to Khrushchev at this time.
27. Letter from Haldane to Lederberg, May 16, 1960. NIH National Library of Medicine, https://profiles.nlm.nih.gov/ps/retrieve/ResourceMetadata/BBAEAL.
28. Clark (1968), pp. 259–260.

29. Haldane (1964c).
30. Haldane to Kosygin, circa October 24, 1964. Reproduced in Clark (1968), p. 299.
31. See, for example, Dronamraju (1968, 1985, 1995).

13. A CERTAIN AMOUNT OF MURDER

1. See Hollander (2006).
2. A fresh perspective on this is given in McMeekin (2009).
3. Conquest (1990).
4. Maynard Smith (2004).
5. *The Times* (London), issue 45825, May 18, 1931, p. 13; issue 45826, May 19, 1931, p. 17; issue 45827, May 20, 1931, p. 15.
6. *The Times* (London), issue 45765, March 7, 1931, p. 14.
7. Wiener (1956), pp. 206–207.
8. Dronamraju (1995), p. 89.
9. Dronamraju (1995), p. 90.
10. Dronamraju (1995), p. 139.
11. Julian Huxley (1965), p. 60.
12. Pirie (1966).
13. *The Times* (London), issue 56184, Wednesday, December 2, 1964, p. 13.
14. Preface to Clark (1968).
15. Kevles (1985), p. 215.
16. Preface to Majumder (1993), p. vi.
17. Neel (1993), p. 331.
18. Roberts (1997), p. 207.

APPENDIX 1. WHY I AM [A] COOPERATOR

1. Haldane Papers, University College London, HALDANE 1/2/63.
2. [Ed.] Sir Clarence Henry Kennett Marten (1872–1948).
3. [Ed.] H. C. Hollway-Calthrop.
4. [Ed.] Haldane's stray quote marks in the manuscript have been corrected, for clarity.
5. [Ed.] Sir Arthur Grenfell Wauchope (1874–1947), who had first risen to prominence in the Boer War.
6. [Ed.] Chandra Shamsher Jang Bahadur Rana (1863–1929).
7. [Ed.] Kairouan.
8. [Ed.] Leslie Hore-Belisha (1893–1957).
9. [Ed.] Thomas Hunt Morgan (1866–1945), evolutionary biologist and geneticist, and winner of the 1933 Nobel Prize in Physiology or Medicine.
10. [Ed.] Francis William Aston (1877–1945), winner of the 1922 Nobel Prize in Chemistry.

11. [Ed.] Sir James Hopwood Jeans (1877–1946), the cosmologist.
12. [Ed.] Harold Whitridge "Pete" Davies (1894–1946), professor of physiology at the University of Sydney, 1930–1946. After a spell at New College, Oxford, assisting John Scott Haldane, he taught briefly at the University of Adelaide. In 1920 he went on to various positions in Britain and the United States before accepting a chair at Sydney in 1930.
13. [Ed.] Donald Dexter Van Slyke (1883–1971).
14. [Ed.] Ernest Laurence Kennaway (1881–1958).
15. [Ed.] The gods decided otherwise.
16. Henri Barbusse, *L'Enfer*, 1.101.
17. Jeremiah 17:9.
18. [Ed.] Here Haldane leaves the rest of the page blank and abandons the topic of Aldous Huxley.
19. [Ed.] Walter Russell Brain, first Baron Brain (1895–1966), and Eric Benjamin Strauss (1894–1961).
20. [Ed.] Dame Janet Maria Vaughan (1899–1993).
21. Please Miss Antivivisectionist, this is not cruelty to a dog. The dog's gastric juice is three times as acid as a man's. You can therefore give him stronger acid to drink without hurting him. He also secretes more gastric juice than a man in proportion to his size, so you can give him a relatively larger volume to drink.
22. [Ed.] Carl Erich Correns (1864–1933), a pioneering German geneticist, one of several, including the more well-known Hugo de Vries (1848–1935), to independently rediscover Mendel's laws at the turn of the century.
23. [Ed.] Alfred Henry Sturtevant (1891–1970), an American geneticist who was the first to map the chromosome. Calvin Blackman Bridges (1889–1938). Sturtevant's colleague Hermann Joseph Muller is mentioned often in the text and notes above.
24. [Ed.] Alexandra A. Prokofyeva-Belgovskaya (1903–1984), a Soviet geneticist.
25. [Ed.] J. B. Bradbury (1841–1930), Downing Professor of Medicine at Cambridge. His assistant, Walter Ernest Dixon (1871–1931), was his junior by thirty years and the first pharmacologist at Cambridge.
26. [Ed.] Alfred James Lotka (1880–1949), an American mathematician chiefly remembered for his contribution, with Vito Voltera (1860–1940), to the so-called Lotka-Voltera differential equations, which describe the dynamic behavior of predator-prey relationships.
27. [Ed.] Jan Tinbergen (1903–1994), a mathematical economist educated at Leyden. The ethologist Niko Tinbergen (1907–1988) was a brother. Tinbergen held simultaneous posts at the University of Amsterdam and the Netherlands School of Economics. Both brothers won the Nobel Prize in their respective disciplines.
28. [Ed.] John Maynard Keynes (1883–1946), the influential economist.

29. [Ed.] Sir Horace Edmund Avory (1851–1935).
30. [Ed.] Stuart James Bevan (1872–1935), a King's counsel at the bar, later a member of Parliament.
31. [Ed.] Friedrich Engels, "Anti-Dühring" (1878). This is a contraction of the seldom-used complete title, *Herr Eugen Dühring's Revolution in Science*. Engels also wrote *Ludwig Feuerbach and the End of Classical German Philosophy* (1878), to which he appended Karl Marx's *Theses on Feuerbach* (written in 1845 but never published by Marx himself).
32. [Ed.] Dorothea de Winton (1890–1982) and Rose Scott-Moncrieff (1903–1991) were British geneticists. The cytologist Cyril Dean Darlington is mentioned often in the text and notes above.
33. [Ed.] Otto Heinrich Warburg (1883–1970) was the author of *The Chemical Constitution of Respiration Ferment* (1928) and the Nobel Prize winner for Medicine in 1931. Robert Percival Cook (1906–1989) and Leslie William Mapson FRS (1907–1970) were alumni of the Dunn Biochemistry Laboratory run by F. G. Hopkins at Cambridge.
34. [Ed.] Hans Grüneberg FRS (1907–1982) and Anna-Ursula Philip (1908–1995?) were both German-born geneticists of Jewish origin.
35. [Ed.] Cecil Gordon (1906–1960) and P. A. R. Street (no further details traced) collaborated with Helen Spurway on studies of fly genetics in the late 1930s. Gordon, who was one of Haldane's students at UCL, became a pioneer in the field of operations research during the Second World War.
36. [Ed.] Edwin Stephen Goodrich FRS (1868–1946), a former artist who became the Linacre Professor of Zoology at Oxford, trained at the Slade School in London.
37. [Ed.] Henry Edward Armstrong FRS (1848–1937) was a distinguished chemist. Haldane may have been unaware of his death at the time he was writing.
38. [Ed.] The manuscript ends abruptly here.

APPENDIX 2. HALDANE ON THE NAZI SOVIET PACT

1. This letter was published in the *New Statesman and Nation*; see Haldane (1939d).
2. This sentence appears in an asterisked footnote in Haldane's manuscript but is shown inline here.
3. From Lewis Carroll's *Alice's Adventures in Wonderland* (1865):

> *'Tis the voice of the Lobster: I heard him declare*
> *"You have baked me too brown, I must sugar my hair."*
> *As a duck with its eyelids, so he with his nose*
> *Trims his belt and his buttons, and turns out his toes.*
> *When the sands are all dry, he is gay as a lark,*
> *And will talk in contemptuous tones of the Shark;*

> *But, when the tide rises and sharks are around,*
> *His voice has a timid and tremulous sound.*
> *I passed by his garden, and marked, with one eye,*
> *How the Owl and the Panther were sharing a pie:*
> *The Panther took pie-crust, and gravy, and meat,*
> *While the Owl had the dish as its share of the treat.*
> *When the pie was all finished, the Owl, as a boon,*
> *Was kindly permitted to pocket the spoon;*
> *While the Panther received knife and fork with a growl,*
> *And concluded the banquet by —*

APPENDIX 3. SELF OBITUARY

1. Haldane (1964a). The transcript was also published as "I've Always Been Something of a Dabbler," *Daily Worker*, December 14, 1964. I have not seen the latter version.

APPENDIX 4. VENONA INTERCEPTS

1. Pencil annotation: issued 23-1-1967.
2. Pencil annotation: issued 23-1-1967.
3. Pencil annotation: Issued 23-1-1967.
4. Pencil annotation: 16.8.40 PFDT/71.
5. Pencil annotation: Issued 23-1-1967.
6. Pencil annotation: business.
7. Pencil annotation: ?
8. Pencil annotation: Issued 23-1-1967.

APPENDIX 5. IN SUPPORT OF LYSENKO

1. Haldane Papers, University College London, HALDANE 4/9/1/7.
2. [JBS] ? 1. Biology limited pessimistic sterile. Harland puts up cotton production Peru. Vitamins. Penicillin.
3. [JBS] No.
4. [JBS] ? 2. E.g. [rabbits] phototropism. Egg laying by parasites. [Amt. of coat colour]
5. [JBS] No. 3 You can't select genes for high milk yield on a poor diet.
6. [JBS] No. 4 mutations always known to be occurring. Not rare.
7. [JBS] Evidence. 5. Evidence! What racial characters inherited are. Mendel's laws. At end.
8. [JBS] Nonsense. 6. Evidence! Clearly e.g. maternal instincts → birth injury. Do you say 3 generations [as slaves] causes bad heredity? At end.
9. [JBS] Where. 7. This is plain nonsense. Read [Dreen] (1945). Varley (1947) J An. Ecol. Portères [R. A. Soc] 1948. Darwinian sexual selection. ([Ed.] Hal-

dane often referred to G. C. Varley, "The Natural Control of Population Balance in the Knapweed Gall-Fly [*Urophora jaceana*]," *Journal of Animal Ecology* 16 [2] [November 1947], pp. 139–187. It is not clear which paper by Roland Portères is meant, or what the indecipherable annotation for 1945 refers to.)
10. [JBS] No. 8. Utterly incomplete. Does not tell you how this property inherited. E.g. in some cases through mother only, in others father only, in others dominant, recessive etc. How get better bananas.
11. [JBS] No.
12. [JBS] 9. Who says it is destructive? Even X-Rays can reverse mutations.
13. [JBS] Evidence. 10. Evidence for yield per acre or per cow > England or Denmark.
14. [JBS] Not always. 11. You can compete for other things than food. E.g. resistance to frost, disease.
15. [JBS] ? 12. Have you read e.g. Knight on cotton breeding. ([Ed.] R. L. Knight, "Theory and Application of the Backcross Technique in Cotton Breeding," *Journal of Genetics* 47 (1945), pp. 76–86.)
16. [JBS] ? 13. Int. gen. cong. Abt. ¼ papers on how to change genes. ([Ed.] International Genetics Congress.)
17. [JBS] 14. As a Leninist I don't believe real development can be imposed on animals or plants from outside. Can do so by introducing internal contradictions or making them more important. Ignorant, sectarian. Brown. ?orthodox manner.

EARLY GULAG MEMOIRS AND DESCRIPTIONS

A cut-off date of 1961 has been used in this list, which is certainly not complete. Most of the items are firsthand accounts, a few are secondary. As noted in the text, Haldane was alive when every single one of these was published.

1918 Bariatinsky, Marie. *Diary of a Russian Princess in a Bolshevik Prison.* Berlin: Buchdrukerei Press.

1919 Naudeau, Ludovic. Five Months in Moscow Prisons. *Current History Magazine of the New York Times* (October 1919), pp. 127–136, and (November 1919), pp. 318–321.

1920 Kalpashnikoff, A. *A Prisoner of Trotsky's.* New York: Doubleday.

1920 Naudeau, Ludovic. *En prison sous la terreur russe.* Paris: Librairie Hachette.

1922 McCullagh, F. *A Prisoner of the Reds: The Story of a British Officer Captured in Siberia.* New York: E.P. Dutton.

1925 Harding, Mrs. Stan [Sedine Milana]. *The Underworld of State.* London: G. Allen and Unwin.

1926 Doubassoff, Irene. *Ten Months in Bolshevik Prisons.* Edinburgh: Blackwood.

1926 Malsagoff, S.A. *An Island Hell.* London: Philpot.

1926 Melgunov, Serge P. *Red Terror in Russia.* London: J.M. Dent.

1927 Duguet, Raymond. *Un Bagne en Russie Rouge.* Paris: J. Tallandier.

1928 Bezsonov, J.D. *My Twenty-Six Prisons and My Escape from Solovetski.* London: Jonathan Cape.

1928 Bezsonov, Yuri. *Mes vingt-six prisons et mon évasion de Solovki.* Paris: Payot.

1928 Klinger, Anton. *Solovetskaia katorga: Zapiski bezhavshego.* Berlin: Arkhiv russkoi revoliutsii.

1928 Shirvindt, Evsei Gustavovich. *Russian Prisons.* London: International Class War Prisoners' Aid.

1929 Buxhoeveden, Baroness Sophie. *Left Behind: Fourteen Months in Siberia during the Revolution, December 1917–February 1919.* London: Longmans, Green.

1929 Cederholm, Boris. *In the Clutches of the Cheka.* New York: Houghton Mifflin.

EARLY GULAG MEMOIRS AND DESCRIPTIONS

1931   Atholl, Katharine Marjory Stewart-Murray, Duchess of. *Conscription of a People*. London: P. Allen.

1931   Brunovsky, Vladimir. *Methods of the OGPU*. London: Harper and Brothers.

1931   Eccard, Frédéric. "Le Travail forcé en Russie soviétique." *Revue Hebdomadaire*, April 25.

1931   Grady, Eve G. *Seeing Red: Behind the Scenes in Russia Today*. New York: Brewer, Warren and Putnam.

1931   Pim, Alan, and Edward Bateson. *Report on the Russian Timber Camps*. London: E. Benn.

1931   *Times* [London]. Articles on forced labor in Soviet Russia. May 18, 19, 20.

1932   Atholl, Katharine Marjory Stewart-Murray, Duchess of. *The Truth about Forced Labour in Russia*. London: P. Allen.

1933   Anonymous. *Out of the Deep: Letters from Soviet Timber Camps*. London: Geoffrey Bless.

1933   Martsinkovski, Vladimir Filimonovich. *With Christ in Soviet Russia*. Prague: Knihtiskárna V. Horák.

1933   Nussimbaum, Lev [Essad Bey]. *OGPU: The Plot against the World*. New York: Viking.

1934   Anonymous. "Les Camps de concentration de l'URSS." *Etudes*, March 20.

1934   Tchernavin, Tatiana. *Escape from the Soviets*. New York: Dutton.

1935   Danzas, Julia [Anonymous]. *Red Gaols*. London: Burns, Oates and Washbourne.

1935   Kitchin, George. *Prisoner of the OGPU*. London: Longman, Green and Co.

1935   Solonevich, Ivan. *Escape from Russian Chains*. London: Williams and Norgate.

1935   Solonevich, Ivan. *Russia in Chains*. London: Williams and Norgate.

1935   Tchernavin, Vladimir. *I Speak for the Silent Prisoners of the Soviets*. Boston: Ralph T. Hale.

1938   de Beausobre, Julia [Julia Namier]. *The Woman Who Could Not Die*. London: Chatto and Windus.

1938   Littlepage, John D. *In Search of Soviet Gold*. New York: Harcourt, Brace and Co.

1940   Ciliga, Anton. *The Russian Enigma*. London: George Routledge and Sons.

1940   Maksimov, Grigorii Petrovich. *The Guillotine at Work: Twenty Years of Terror in Russia: Data and Documents*. Chicago: Alexander Berkman Fund.

1940   Utley, Freda. *The Dream We Lost*. New York: John Day.

1941   Mower, Lilian T. *Arrest and Exile*. New York: Morrow.

1942   Scott, John. *Behind the Urals*. Boston: Houghton Mifflin.

1945 Barmine, Alexander. *One Who Survived: The Life Story of a Russian under the Soviets.* New York: G. P. Putnam's Sons.
1945 Czapski, Joseph. *Souvenirs de Starobielsk* (Collection Témoignages cahier 1). Pamphlet. Paris: N.p.
1945 Halpern, Ada. *Conducted Tour.* New York: Sheed and Ward. Also published as *Liberation—Soviet Style.*
1945 Mora, Sylvester, and Peter Zwierniak. *La Justice soviétique.* Rome: Magi-Spinetti.
1945 White, William L. *Report on the Russians.* London: Eyre and Spottiswoode.
1946 Kravchenko, Victor. *I Chose Freedom.* New York: Charles Scribner's Sons.
1947 Dallin, David J., and Boris I. Nicolaevsky. *Forced Labor in Soviet Russia.* New Haven, Conn.: Yale University Press.
1947 Dangerfield, Elma. *Beyond the Urals.* London: British League for European Freedom.
1947 Zajdlerowa, Zoe [Anonymous]. *The Dark Side of the Moon.* New York: Scribners.
1948 Beausobre, Julia De. *The Woman Who Could Not Die.* London: Victor Gollancz.
1948 Gliksman, Jerzy. *Tell the West, by Jerzy Gliksman; An Account of His Experiences as a Slave Laborer in the Union of Soviet Socialist Republics.* New York: Gresham.
1948 Koriakov, Mikhail. *I'll Never Go Back: A Red Army Officer Talks.* E. P. Dutton.
1949 Buber-Neumann, Margarete. *Under Two Dictators.* London: Edward Fitzgerald.
1951 Czapski, Joseph. *The Inhuman Land.* London: Chatto and Windus.
1951 Fehling, Helmut. *One Great Prison: The Story behind Russia's Unreleased POWs.* Boston: Beacon Press.
1951 Herling, Gustav. *A World Apart.* New York: Roy.
1951 Lipper, Elinor. *Eleven Years in Soviet Prison Camps.* Chicago: World Affairs Book Club.
1951 Orr, Charles. *Stalin's Slave Camps: An Indictment of Modern Slavery.* Brussels, Belgium: International Confederation of Free Trade Unions.
1951 Petrov, Vladimir. *It Happens in Russia: Seven Years' Forced Labour in the Siberian Goldfields.* London: Eyre and Spottiswoode.
1951 Stypulkowski, Zbigniew. *Invitation to Moscow.* London: Thames and Hudson.
1951 Vogeler, Robert I. *I Was Stalin's Prisoner.* New York: Harcourt.
1951 Weissberg, Alexander. *The Accused.* New York: Simon and Schuster.
1952 Gonzalez, Valentin [El Campesino]. *Listen Comrades: Life and Death in the Soviet Union by El Campesino.* London: Heinemann.

1952 Prychodko, Nicholas. *One of the Fifteen Million*. London: Dent.
1952 Weissberg, Alexander. *Conspiracy of Silence*. London: Hamish Hamilton.
1954 Ekart, Antoni. *Vanished without Trace: Seven Years in Soviet Russia*. London: M. Parrish.
1954 Scholmer, Joseph. *Vorkuta*. London: Weidenfeld and Nicolson.
1954 Smith, C. A. *Escape from Paradise*. London: Hollis and Carter.
1956 Rawicz, Slavomir. *The Long Walk: The True Story of a Trek to Freedom*. London: Constable.
1957 Bauer, Josef M. *As Far as My Feet Will Carry Me*. Translated by Lawrence Wilson. London: Andre Deutsch.
1957 Begin, Menachem. *White Nights: The Story of a Prisoner in Russia*. Translated by Katie Kaplan. New York: Harper and Row.
1958 Fittkau, Gerhardt. *My Thirty-Third Year: A Priest's Experience in a Russian Work Camp*. New York: Farrar, Straus and Cudahy.
1958 Roeder, Bernhardt. *Katorga: An Aspect of Modern Slavery*. London: Heinemann.
1960 Noble, John. *I Was a Slave in Soviet Russia*. New York: Devin-Adair.
1960 Parvilahti, Unto. *Beria's Gardens: A Slave Laborer's Experiences in the Soviet Utopia*. New York: Dutton.
1961 Armonas, Barbara. *Leave Your Tears in Moscow*. Philadelphia: Lippincott.

BIBLIOGRAPHY

ARCHIVAL SOURCES
Haldane Papers, University College London.
National Archives, London.
   MI5 Personal File, J. B. S. Haldane, KV 2-1832.
   MI5 Personal File, Hans Kahle, KV 2-1562, KV 2-1563, KV 2-1564, KV 2-1565, KV 2-1566.
   MI5 Personal File, Otto Katz, KV 2-1382, KV 2-1383, KV 2-1384.
   MI5 Personal File, Ivor Montagu and the Film Society, KV 2-598, KV 2-599, KV 2-600, KV 2-601.
   Signals Intelligence, VENONA Intercepts. HW 15/43.

PUBLISHED SOURCES
Adams, Mark B. (editor). 1990. *The Wellborn Science: Eugenics in Germany, France, Brazil and Russia.* London: Oxford University Press.
Adamson, Judith. 1998. *Charlotte Haldane: Woman Writer in a Man's World.* London: Macmillan.
Almond, Gabriel. 1954. *Appeals of Communism.* Princeton, N.J.: Princeton University Press.
Andrew, Christopher. 2009. *Defend the Realm: The Authorized History of MI5.* New York: Alfred Knopf.
Andrew, Christopher, and Oleg Gordievsky. 1990. *KGB: The Inside Story.* New York: Harper Collins.
Andrew, Christopher, and Vasili Mitrokhin. 2005. *The World Was Going Our Way: The KGB and the Battle for the Third World.* New York: Basic Books.
Applebaum, Anne. 2003. *Gulag: A History.* New York: Doubleday.

Babkov, V. V. 2013. *The Dawn of Human Genetics*. Cold Spring Harbor, N.Y.: Cold Spring Harbor Laboratory Press.

Baxell, Richard. 2014. *Unlikely Warriors: The British in the Spanish Civil War*. London: Aurum Press.

Bell, P. B. (editor). 1959. *Darwin's Biological Work: Some Aspects Reconsidered*. London: Cambridge University Press.

Benton, Jill. 1990. *Naomi Mitchison: A Biography*. London: Pandora.

Berg, Raissa. 1983. *On the History of Genetics in the Soviet Union: Science and Politics; The Insight of a Witness*. St. Louis, Missouri: Washington University, 1983.

Bernal, J. D. 1949. "The Biological Controversy in the Soviet Union and Its Implications." *The Modern Quarterly* 4 (3) (Summer).

———. 1953. "Stalin as Scientist." *Modern Quarterly* 8 (3) (Summer).

Birstein, Vadim J. 2001. *The Perversion of Knowledge: The True Story of Soviet Science*. Cambridge, Mass.: Westview.

Blampied, H. J. 1918. *With a Highland Regiment in Mesopotamia: 1916–1917*. Bombay: Times Press.

Bonner, John Tyler. 2002. *Lives of a Biologist*. Cambridge, Mass.: Harvard University Press.

Bowker, Gordon. 1993. *Pursued by Furies: A Life of Malcolm Cowley*. New York: St. Martin's Press.

Brown, Andrew. 2005. *J. D. Bernal: The Sage of Science*. London: Oxford University Press.

Bryukhonenko, S., et al. 1929. "Expériences avec la tête isolée du chien. I. Technique et conditions des experiences." *Journal de Physiologie et de Pathologie Générale* 27, pp. 31–45.

Burks, Barbara Stoddard. 1927. "Foster Parent-Foster Child Comparisons as Evidence upon the Nature-Nurture Problem." *Proceedings of the National Academy of Sciences* 13 (12) (December 15), pp. 846–848.

———. 1928. "The Relative Effects of Nature and Nurture upon Mental Development: A Comparative Study on Foster Parent-Foster Child Resemblance and True Parent-True Child Resemblance." *Pedagogical Seminary* 32, pp. 389–413. Also published in *27th Yearbook of the National Society for the Study of Education,* part 1 (1928), pp. 219–316.

Carlson, Elof Axel. 1981. *Genes, Radiation and Society: The Life and Work of H. J. Muller*. Ithaca, N.Y.: Cornell University Press.

Caute, David. 1988. *The Fellow Travellers: Intellectual Friends of Communism*. New Haven, Conn.: Yale University Press.

Clark, Ronald W. 1968. *JBS: The Life and Work of J. B. S. Haldane*. New York: Coward-McCann.

Cochran, Greg, and Henry Harpending. 2010. *The 10,000 Year Explosion: How Civilization Accelerated Human Evolution*. New York: Basic Books.

Conquest, Robert. 1990. *The Great Terror: A Reassessment*. London: Oxford University Press.

Connolly, Cyril. 1938. *Enemies of Promise*. London: Routledge. Reprinted 1960.

Conway, Flo, and Jim Siegelman. 2005. *Dark Hero of the Information Age: In Search of Norbert Wiener, the Father of Cybernetics*. New York: Basic Books.

Costello, John. 1988. *Mask of Treachery*. London: Collins.

Cowles, Virginia. 1941. *Looking for Trouble*. New York: Harper.

Crew, F. A. E, et al. 1939. "Social Biology and Population Improvement." *Nature* 144 (September 16), pp. 521–522.

Crossman, Richard (editor). 1949. *The God That Failed*. New York: Harper and Row. Republished by Regnery Gateway, 1983.

Darlington, Cyril Dean. 1953. *The Facts of Life*. London: Allen and Unwin.

———. 1962. *Genetics and Man*. London: Allen and Unwin. Revised edition of *The Facts of Life*.

———. 1968. "Determined but Lonely." Review of Clark (1968). *Nature* 220 (November 30), pp. 933–934.

———. 1969a. "Haldane's Influence" (Review of Dronamraju [1968]). *Nature* 222 (April 5), pp. 56–57.

———. 1969b. *The Evolution of Man and Society*. New York: Simon and Schuster.

———. 1978. *The Little Universe of Man*. London: Allen and Unwin.

Darlington, Cyril Dean, and S. C. Harland. 1945. "Prof. N. I. Vavilov." *Nature* 156 (3969) (November 24), pp. 621–622.

Darlington, Cyril Dean, and K. Mather. 1950. *Genes, Plants and People: Essays on Genetics*. London: Allen and Unwin.

Dewsbury, Donald A. (editor). 1985. *Studying Animal Behavior: Autobiographies of the Founders*. Chicago: University of Chicago Press.

Dronamraju, K. R. (editor). 1968. *Haldane and Modern Biology*. Baltimore, Md.: Johns Hopkins University Press.

———. 1985. *The Life and Work of J. B. S. Haldane with Special Reference to India*. Aberdeen: Aberdeen University Press.

——— (editor). 1995. *Haldane's Daedalus Revisited*. London: Oxford University Press.

Epstein, Catherine. 2003. *The Last Revolutionaries: German Communists and Their Century*. Cambridge, Mass.: Harvard University Press.

"Experiments in Resuscitation." 1937. *Nature* 139 (3515) (March 13), p. 441.

Feferman, Anita Burdman. 2000. *From Trotsky to Gödel: The Life of Jean van Heijenoort*. Natick, Mass.: A. K. Peters.

Fernan-Perez, J. 1929. "La survivance du système nerveux central dans des conditions artificielles." *La Presse Médicale* 37 (February 6), pp. 183–184.

Fisher, R. A. 1925. *Statistical Methods for Research Workers*. Edinburgh: Oliver and Boyd.

———. 1930. *The Genetical Theory of Natural Selection*. Oxford: Clarendon Press.

Foote, Alexander Allan. 1949. *Handbook for Spies*. London: Doubleday.

Fyfe, James. 1950. *Lysenko Is Right*. London: Laurence and Wishart.

Gerya, Yu. F., and V. D. Yankovsky. 1976–1977. "Use of Artificial Circulation in Resuscitation of Drowned Dogs." *Resuscitation* 5 (3), pp. 145–152.

Gollancz, Victor (editor). 1941. *Betrayal of the Left*. London: Left Book Club.

Goodman, Martin. 2007. *Suffer and Survive: Gas Attacks, Miners' Canaries, Spacesuits and the Bends: The Extreme Life of Dr. J. S. Haldane*. New York: Simon and Schuster.

Gordon, W. W. 1951. "The Pavlov Conference." *Soviet Studies* 3 (1), pp. 34–59.

Graham, Loren. 1993. *The Ghost of the Executed Engineer: Technology and the Fall of the Soviet Union*. Cambridge, Mass.: Harvard University Press.

———. 2016. *Lysenko's Ghost*. Cambridge, Mass.: Harvard University Press.

Haldane, Charlotte. 1949. *Truth Will Out*. London: Weidenfeld and Nicolson.

———. 1965. "My Husband the Professor." BBC Radio Interview, Third Programme, September 11. Held in the British Library Sound Archive, M516W BD 1.

———. 1969. "J.B.S." *The Listener*, January 30.

Haldane, J. B. S. 1915. "Reduplication in Mice." With A. D. Sprunt and N. M. Haldane. *Journal of Genetics* 5 (2) (December), pp. 133–135.

———. 1923. *Daedalus, or Science and the Future*. London: Kegan Paul.

———. 1924. "A Mathematical Theory of Natural and Artificial Selection I." *Transactions of the Cambridge Philosophical Society* 23, pp. 19–41.

———. 1925. *Callinicus: A Defence of Chemical Warfare*. New York: Dutton.

———. 1927. *Possible Worlds*. London: Chatto and Windus.

———. 1930. *Enzymes*. London: Longmans. Republished by the MIT Press, 1965.

———. 1931. "What I Think About." *The Nation* (May 13), p. 525.

———. 1932a. *The Inequality of Man*. London: Chatto and Windus.

———. 1932b. *The Causes of Evolution*. London: Longmans.
———. 1934. Letter to the Editor. *Nature* 133 (3350) (January 13), p. 65.
———. 1935. "The Rate of Spontaneous Mutation of a Human Gene." *Journal of Genetics* 31, pp. 317–326.
———. 1937a. "A Dialectical Account of Evolution." *Science and Society* 1 (4) (Summer).
———. 1937b. "The Position of Genetics." *Nature* 140 (3540) (September 4).
———. 1938a. *Heredity and Politics*. London: George Allen and Unwin.
———. 1938b. *A.R.P.* London: Victor Gollancz.
———. 1938c. *My Friend Mr. Leakey*. New York: Harper.
———. 1938d. Reply to "Is Professor Haldane's Account of Evolution Dialectical?" by A.P. Lerner. *Science and Society* 2 (2) (Spring).
———. 1938e. Letter to the Editor. *Nature* 142 (3602) (November 12), p. 851.
———. 1938f. *The Marxist Philosophy and the Sciences*. London: Allen and Unwin.
———. 1939a. *Science and Everyday Life*. London: Lawrence and Wishart.
———. 1939b. *Why I Am [a] Cooperator*. Unpublished autobiography, Haldane Papers, University College London, HALDANE 1/2/63. The date is uncertain.
———. 1939c. Review of *Principles of Genetics* (1939) by E.W. Sinnott and L.C. Dunn. *Biometrika* 31(1) (July), p. 220.
———. 1939d. "Is There a Russian Enigma?" *New Statesman and Nation*, September 30.
———. 1940a. *Science in Peace and War*. London: Lawrence and Wishart.
———. 1940b. *Keeping Cool and Other Essays*. London: Chatto and Windus.
———. 1940c. *Adventures of a Biologist*. New York: Harper. American edition of *Keeping Cool*.
———. 1940d. "Lysenko and Genetics." *Science and Society* 4 (4) (Fall), pp. 433–437.
———. 1940e. Preface to F. Engels, *The Dialectics of Nature*. London: Lawrence and Wishart.
———. 1941a. *New Paths in Genetics*. London: George Allen and Unwin.
———. 1941b. "A.R.P. Today." *The Labour Monthly* (December), pp. 498–499.
———. 1942. "Soviet Biology." *The Labour Monthly* (November), pp. 347–348.
———. 1944. Letter to the Editor. *Nature* 154 (3909) (September 30), p. 429.
———. 1945. *Why Professional Workers Should Be Communists*. London: Communist Party. Pamphlet.

———. 1946a. *A Banned Broadcast and Other Essays*. London: Chatto and Windus.

———. 1946b. "The Case of Dr. Allan Nunn May." *Daily Worker*, May 18.

———. 1946c. "The Interaction of Nature and Nurture." *Annals of Human Genetics* 13 (1), pp. 197–205.

———. 1947a. *What Is Life?* New York: Boni and Gaer.

———. 1947b. *Science Advances*. London: Macmillan.

———. 1948. BBC Panel Discussion on Lysenko. Transcribed in *The Listener* 30 (December 9), pp. 873–875.

———. 1949. "In Defence of Genetics." *Modern Quarterly* 4 (3) (Summer), pp. 194–202.

———. 1951. *Everything Has a History*. London: George Allen and Unwin.

———. 1953. Foreword to *Symposia of the Society for Experimental Biology Number VII: Evolution*. Cambridge: Cambridge University Press.

———. 1954a. *The Biochemistry of Genetics*. London: George Allen and Unwin.

———. 1954b. Review of Cyril Darlington, *The Facts of Life*. *The British Journal of Sociology* 5 (1) (March), pp. 88–89. See Darlington (1953).

———. 1954c. "Statics of Evolution." In *Evolution as a Process*, edited by Julian Huxley, A. C. Hardy, and E. B. Ford. London: George Allen and Unwin.

———. 1959. "Natural Selection." In Bell (1959), pp. 101–149.

———. 1962. Review of Francis Galton, *Hereditary Genius* (1869; 1962 reissue). *Journal of Genetics* 58 (2) (March), pp. 271–272.

———. 1963. "Biological Possibilities for the Human Species in the Next Ten Thousand Years." In *Man and His Future*, edited by G. E. W. Wolstenholme. London: J. and A. Churchill.

———. 1964a. "J. B. S. Haldane's Self-Obituary." *The Listener* (December 10), pp. 934–935.

———. 1964b. "The Proper Social Application of the Knowledge of Human Genetics." In *The Science of Science*, edited by M. Goldsmith and A. Mackay, pp. 150ff. London: Souvenir Press and Penguin Books.

———. 1964c. "Cancer Is a Funny Thing." *New Statesman* (February 21).

———. 1968. *Science and Life: Essays of a Rationalist*. Edited by John Maynard Smith. London: Pemberton.

———. 1976. *The Man with Two Memories*. London: Merlin Press.

———. 1985. *On Being the Right Size and Other Essays*. Edited by John Maynard Smith. London: Oxford University Press.

———. 2009. *What I Require from Life: Writings from Science and Life.* Edited by K. R. Dronamraju. London: Oxford University Press.
Haldane, J. B. S., and John Baker. 1933. *Biology in Everyday Life.* London: George Allen and Unwin.
Haldane, J. B. S., and Julian Huxley. 1927. *Animal Biology.* London: Oxford University Press.
Haldane, John Scott. 1935. *The Philosophy of a Biologist.* London: Oxford University Press.
Haldane, Louisa Kathleen. 1961. *Friends and Kindred: Memoirs of Louisa Kathleen Haldane.* London: Faber and Faber.
Harland, Sydney Cross. 2001. *Nine Lives: The Autobiography of a Yorkshire Scientist.* Edited by Max Millard. New York: Boson Books.
Haynes, John, and Harvey Klehr. 1999. *Venona: Decoding Soviet Espionage in America.* New Haven, Conn.: Yale University Press.
Haynes, John, Harvey Klehr, and Alexander Vassiliev. 2009. *Spies: The Rise and Fall of the KGB in America.* New Haven, Conn.: Yale University Press.
Hemingway, Ernest. 1940. *For Whom the Bell Tolls.* New York: Charles Scribner's.
Herrnstein, Richard J. 1973. *IQ in the Meritocracy.* New York: Little, Brown.
Hollander, Paul. 1981. *Political Pilgrims: Travels of Western Intellectuals to the Soviet Union, China and Cuba.* New York: Harper. 1983 printing.
———. 2006. *The End of Commitment.* Chicago: Ivan R. Dee.
Hollis, Christopher. 1976. *Oxford in the Twenties: Recollections of Five Friends.* London: Heinemann.
Howarth, T. E. B. 1978. *Cambridge between Two Wars.* London: Collins.
Howell, Yvonne. 2008. "Efroimson, Vladimir Pavlovich." *The Supplement to the Modern Encyclopedia of Russian, Soviet and Eurasian History.* Edited by Bruce F. Adams. Vol. 9. Gulf Breeze, Fla.: Academic International Press. http://scholarship.richmond.edu/cgi/viewcontent.cgi?article=1046&context=mlc-faculty-publications.
Howson, Gerald. 1998. *Arms for Spain: The Untold Story of the Spanish Civil War.* New York: St Martin's Press.
Huxley, Aldous. 1922. *Crome Yellow.* London: Chatto and Windus.
———. 1923. *Antic Hay.* London: Chatto and Windus.
Huxley, Julian. 1942. *Evolution: The Modern Synthesis.* London: Allen and Unwin.
———. 1949. "Lysenko Once More." *Spectator* (November 11), p. 7.
———. 1965. Obituary of J. B. S. Haldane. *Encounter* (October), p. 60.

———. 1970. *Memories*. New York: Harper and Row.
Hyde, Douglas. 1950. *I Believed*. London: William Heinemann.
Joravsky, David. 1970. *The Lysenko Affair*. Cambridge, Mass.: Harvard University Press.
Kahle, Hans. 1943. *Under Stalin's Command*. London: Caledonian Press.
———. 1944a. *One Triumphant Year: A Unique Survey of Red Army Successes*. London: Russia Today Society.
———. 1944b. *They Plotted against Hitler: The Story behind the Attempt on Hitler's Life*. London: I.N.G. Publications.
———. 1945. *Stalin, the Soldier*. London: Russia Today Society.
Kahle, Hans, and John Heartfield. 1943. *Know Your Enemy! Aspects of the German Army's Strategy and Morale*. London: Inside Nazi Germany Publications.
Kern, Gary. 2004. *A Death in Washington: Walter G. Krivitsky and the Stalin Terror*. New York: Enigma Books.
Kevles, Daniel. 1985. *In the Name of Eugenics*. Berkeley and Los Angeles: University of California Press.
Keynes, John Maynard. 1925. "A Short View of Russia." Reprinted in *Essays in Persuasion*. London: Macmillan, 1931. Reissued 2010.
King, David. 1997. *The Commissar Vanishes: The Falsification of Photographs and Art in Stalin's Russia*. New York: Metropolitan Books.
Kolchinsky, Eduard I. 2014. "Nikolai Vavilov in the Years of Stalin's 'Revolution from Above' (1929–1932)." *Centaurus* 56 (4) (November), pp. 330–358. doi:10.1111/1600-0498.12059.
Konstantinov, Igor E., and Vladimir V. Alexi-Meskishvili. 2000. "Sergei S. Brukhonenko: The Development of the First Heart-Lung Machine for Total Body Perfusion." *The Annals of Thoracic Surgery* 69 (3), pp. 962–966. http://dx.doi.org/10.1016/S0003-4975(00)01091-2.
Krementsov, Nikolai. 1997. *Stalinist Science*. Princeton, N.J.: Princeton University Press.
———. 2009. "Off with Your Heads: Isolated Organs in Early Soviet Science and Fiction." *Studies in History and Philosophy of Biological and Biomedical Sciences* 40, pp. 87–100.
Krivitsky, Walter. 1939. *In Stalin's Secret Service*. New York: Harper. Reprinted by Enigma Books, 2000.
Kruuk, Hans. 2003. *Niko's Nature: The Life of Niko Tinbergen and His Science of Animal Behaviour*. London: Oxford University Press.
Langdon-Davies, John. 1949. *Russia Puts the Clock Back: A Study of Soviet Science and Some British Scientists*. London: Victor Gollancz.

Lanovenko, I. I., and V. D. Yankovsky. 1972. "Haemodynamic Changes During Reanimation after Prolonged Clinical Death Caused by Blood Loss." *Resuscitation* 1 (4) (December), pp. 311–325.

Lanovenko, I. I., and V. D. Yankovsky. 1973. "Influence of Previous Adaptation to High Altitude on the Cardiovascular System of Dogs during Resuscitation after Prolonged Clinical 'Death' Evoked by Blood Loss." *Resuscitation* 2 (3) (September), pp. 207–220.

Lanovenko, I. I., V. D. Yankovsky, and A. S. Lyavinetz. 1976. "Artificial Circulation for Resuscitation after Prolonged Clinical Death Caused by Hypoxia." *Resuscitation* 5 (2), pp. 75–84.

Lerner, A. P. 1938. "Is Professor Haldane's Account of Evolution Dialectical?" *Science and Society* 2 (2) (Spring).

Lethbridge, David. 2013. *Norman Bethune in Spain: Commitment, Crisis, and Conspiracy*. Portland, Oregon: Sussex Academic Press.

Lewontin, Richard C. 1985. Review of *On Being the Right Size and Other Essays*, by J. B. S. Haldane, edited by John Maynard Smith. *Nature* 314 (6013) (25 April 1985): 682–683.

Lorenz, Konrad. 1952. *King Solomon's Ring*. Translated from the German edition of 1950. New York: Thomas Crowell.

Luck, James Murray. 1999. *Reminiscences*. Palo Alto, Calif.: Annual Reviews. http://dynamics.org/Altenberg/PAPERS/BCLSFV/REFS/jmluckremin.pdf.

Lynton, Mark. 1995. *Accidental Journey: A Cambridge Intern's Memory of World War II*. Woodstock, N.Y.: Peter Mayer.

Lysenko, Trofim D. 1951. *Heredity and Its Variability*. Moscow: Foreign Languages Publishing House.

Macleod, Alison. 1997. *The Death of Uncle Joe*. London: Merlin Press.

Macintyre, Ben. 2010. *Operation Mincemeat*. New York: Harmony.

Majumder, P. P. (editor). 1993. *Human Population Genetics: A Centennial Tribute to J. B. S. Haldane*. New York: Springer.

Marler, Peter. 1985. "Hark Ye to the Birds: Autobiographical Marginalia." In Dewsbury (1985), pp. 315–346.

Martin, Kingsley (editor). 1958. *New Statesman Profiles: Drawings by Vicky*. London: Readers Union.

Maynard Smith, John. 1985. "In Haldane's Footsteps." In Dewsbury (1985), pp. 347–356.

———. 2004. Interview with Richard Dawkins. *Web of Stories*, http://www.webofstories.com/people/john.maynard.smith/33?o=SH.

McIlroy, John. 2006. "The Establishment of Intellectual Orthodoxy and the

Stalinization of British Communism, 1928–1933." *Past and Present* 192 (August), p. 218.

McLellan, Josie. 2004. *Antifascism and Memory in East Germany*. New York: Oxford University Press.

McMeekin, Sean. 2003. *The Red Millionaire. A Political Biography of Willi Münzenberg, Moscow's Secret Propaganda Tsar in the West*. New Haven, Conn.: Yale University Press.

———. 2009. *History's Greatest Heist*. New Haven, Conn.: Yale University Press.

Medawar, Peter. 1986. *Memoir of a Thinking Radish: An Autobiography*. London: Oxford University Press.

———. 1996. *The Strange Case of the Spotted Mice: And Other Classic Essays on Science*. London: Oxford University Press.

Medvedev, Roy, and Zhores Medvedev. 2003. *The Unknown Stalin*. Woodstock, N.Y.: Overlook Press.

Medvedev, Zhores. 1969. *The Rise and Fall of T. D. Lysenko*. New York: Columbia University Press.

Miles, Jonathan. 2010. *The Dangerous Otto Katz: The Many Lives of a Soviet Spy*. New York: Bloomsbury.

Mitchison, Naomi. 1968. "Beginnings" in Dronamraju (1968), pp. 299–305.

———. 1973. *Small Talk: Memories of an Edwardian Childhood*. London: Bodley Head.

———. 1974. "The Haldanes: Personal Notes and Historical Lessons." *Proceedings of the Royal Institution* 47, pp. 1–21.

———. 1975a. "Gas Masks." *New Scientist* (May 1), p. 279.

———. 1975b. *All Change Here: Childhood and Marriage*. London: Bodley Head.

———. 1979. *You May Well Ask: A Memoir, 1920–1940*. New York: Harper.

Montagu, Ivor. 1937. Letter to the Editor. *New Statesman and Nation* (February 1).

———. 1956. *Land of Blue Sky: A Portrait of Modern Mongolia*. London: Denis Dobson.

———. 1969. Letter. *Labour Monthly* (March), p. 140.

———. 1970. *The Youngest Son: Autobiographical Chapters*. London: Lawrence and Wishart.

———. 1974. "Solzhenitsyn, Gulag and the Solzhenitsyn Industry." *Marxism Today* (December), pp. 367–376.

Moorhouse, Roger. 2014. *The Devils' Alliance: Hitler's Pact with Stalin*. New York: Basic Books.

Morison, Patricia. 1997. *J. T. Wilson and the Fraternity of Duckmaloi*. Amsterdam: Rodopi.

Morris, Desmond. 2006. *Watching: Encounters with Humans and Other Animals*. London: Little Books.

Morton, Alan. 1951. *Soviet Genetics*. London: Laurence and Wishart.

Muller, Hermann. 1925. "Mental Traits and Heredity as Studied in a Case of Identical Twins Reared Apart." *Journal of Heredity* 16, 433–448.

———. 1935. *Out of the Night*. New York: Vanguard Press. Republished by Victor Gollancz, London, 1936.

Neel, James V. 1993. "How Would Haldane Have Viewed the Societal Implications of Today's Genetic Knowledge?" In Majumder (1993).

Negovsky, V. 1995. "The Neurological Stage in Reanimatology." *Resuscitation* 29 (2) (April), pp. 169–176.

Newman, Horatio. 1929. "Mental and Physical Characteristics of Twins Raised Apart." *Journal of Heredity* 20, pp. 49–64, 97–104, 153–166.

Newman, M. 1993. *Harold Laski*. New York: Springer.

Paget, Stephen. 1920. *Sir Victor Horsley: A Study of His Life and Work*. New York: Harcourt.

Payne, Stanley. 2004. *The Spanish Civil War, The Soviet Union, and Communism*. New Haven, Conn.: Yale University Press.

Pirie, N. W. 1966. "John Burdon Sanderson Haldane, 1892–1964." *Biographical Memoirs of Fellows of the Royal Society* 12 (November), pp. 218–249.

Popovsky, Mark. 1984. *The Vavilov Affair*. Hamden, Conn.: Archon.

Pringle, Peter. 2008. *The Murder of Nikolai Vavilov*. New York: Simon and Schuster.

Probert, W. R. and Melrose, D. G. 1960. "An Early Russian Heart-Lung Machine." *British Medical Journal* (April 2), pp. 1047–1048.

Provine, William B. 1971. *The Origins of Theoretical Population Genetics*. Chicago: University of Chicago Press.

Radosh, Ron, Mary R. Habeck, and Grigory Sevostianov. 2001. *Spain Betrayed: The Soviet Union in the Spanish Civil War*. New Haven, Conn.: Yale University Press.

Rapaport, Yakov. 1991. *The Doctor's Plot of 1953: A Survivor's Memoir of Stalin's Last Act of Terror, Against Jews and Science*. Cambridge, Mass.: Harvard University Press.

Regler, Gustav. 1959. *The Owl of Minerva*. Trans. Norman Denny. New York: Farrar, Straus and Cudahy.

Richardson, R. Dan. 1982. *Comintern Army: The International Brigade and the Spanish Civil War*. Lexington: Kentucky University Press.

Roberts, Edwin. 1997. *The Anglo Marxists: A Study in Ideology and Culture.* New York: Rowman and Littlefield.

Ro'i, Yaacov. 2010. "Shtern, Lina Solomonovna." YIVO Encyclopedia of Jews in Eastern Europe. http://www.yivoencyclopedia.org/article.aspx/Shtern_Lina_Solomonovna.

Russell, Dora. 1975. *The Tamarisk Tree: My Quest for Liberty and Love.* New York: Putnam's.

Sarkar, S. 2005. "In Memoriam: Ernst Mayr (1904–2005)." *Journal of Biosciences* 30 (4), pp. 415–418.

Schaechter, M. 2009. *Encyclopedia of Microbiology.* London, Oxford: Academic Press.

Searle, G.R. 1976. *Eugenics and Politics in Britain, 1900–1914.* London: Noordhof.

Smith, Douglas. 2012. *Former People.* New York: Farrar, Straus and Giroux.

Smith, Grover (editor). 1969. *Letters of Aldous Huxley.* New York: Harper.

Smith, Peter. 2013. *An Introduction to Gödel's Theorems.* Cambridge: Cambridge University Press.

Solzhenitsyn, Alexander. 1968. *The First Circle.* New York: Harper and Row.

———. 1974. *The Gulag Archipelago.* New York: Harper and Row.

Soyfer, Valery N. 1994. *Lysenko and the Tragedy of Soviet Science.* New Brunswick, N.J.: Rutgers University Press.

———. 2001. "The Consequences of Political Dictatorship for Russian Science." *Nature Reviews Genetics* 2 (September), p. 723.

———. 2003. "Tragic History of the VII International Congress of Genetics." *Genetics* 165 (September), pp. 1–9.

Spearman, Charles. 1927. *Abilities of Man.* London: Macmillan.

Spurway, Helen. 1952. "Behold, My Child, the Nordic Dog." *The British Journal for the Philosophy of Science* 3 (11) (November), pp. 265–272. Review of Lorenz (1952).

Stewart, Roderick, and Sharon Stewart. 2011. *Phoenix: The Life of Norman Bethune.* Toronto: McGill Queen's Press.

Stigler, Stephen. 1986. *The History of Statistics.* Cambridge: Harvard University Press.

Taylor, S.J. 1990. *Stalin's Apologist: Walter Duranty, The* New York Times*'s Man in Moscow.* London: Oxford University Press.

Vaill, Amanda. 2014. *Hotel Florida: Truth, Love, and Death in the Spanish Civil War.* New York: Farrar, Straus and Giroux.

van Heijenoort, Jean. 1985. "Friedrich Engels and Mathematics." In *Selected Essays*, pp. 123–151. Naples: Bibliopolis.

Volkogonov, Dmitri. 1994. *Lenin: A New Biography*. New York: Free Press.
Warren, C. E. T., and James Benson. 1954. *The Midget Raiders*. New York: William Sloane.
Werner, Ruth. 1991. *Sonya's Report: The Fascinating Autobiography of One of Russia's Most Remarkable Secret Agents*. London: Chatto and Windus.
Werskey, Gary. 1978. *The Visible College: The Collective Biography of British Scientific Socialists of the 1930s*. New York: Holt, Rinehart and Winston.
West, Nigel. 1999. *Venona: The Greatest Secret of the Cold War*. London: HarperCollins.
Wiener, Norbert. 1956. *I Am a Mathematician: The Later Life of a Prodigy*. Cambridge, Mass.: MIT Press.
Williams, Charles. 2010. *Harold Macmillan*. London: Phoenix.
Williams, Robert Chadwell. 1987. *Klaus Fuchs, Atom Spy*. Cambridge, Mass.: Harvard University Press.
Winchester, Simon. 2009. *The Man Who Loved China*. New York: Harper.
Woolf, Barnet. 1947. "Studies on Infant Mortality, Part II. Social Aetiology of Stillbirths and Infant Deaths in County Boroughs of England and Wales." *British Journal of Social Medicine* 1 (2) (April), pp. 73–125.
Woolf, Barnet, and John Waterhouse. 1945. "Studies on Infant Mortality, Part I. Influence of Social Conditions in County Boroughs of England and Wales." *The Journal of Hygiene* 44 (2) (April), pp. 67–98.
Wright, Peter. 1987. *Spycatcher: The Candid Autobiography of a Senior Intelligence Officer*. New York: Viking.
Zirkle, Conway (editor). 1949. *Death of a Science in Russia: The Fate of Genetics as Described in* Pravda *and Elsewhere*. Philadelphia: University of Pennsylvania Press.

INDEX

Abakumov, Viktor, 132, 350n27
Academy of Sciences, 86, 131
Agol, Israel, 50, 79–80, 84–85, 164–165
agriculture: grafting and, 167, 169, 198; the Soviet Union and, 80–82, 160–163, 170, 188, 198–199; vernalization and, 81–82, 123, 161–164, 168, 170, 185–186, 201
air-raid shelters. *See* ARP (air-raid precautions)
Alexander, Bill, 112
Allason, Rupert, 104
All Russian Physiological Congress, 134
animal revival, 133–142
Antic Hay (Huxley), 22
Anti-Dühring (Engels), 61
anti-Soviet Tactical Center, 77
Arndt, H.W., 112
ARP (air-raid precautions), 89–91, 119
A.R.P. (Haldane), 89
Attlee, Clement, 111, 115
autojektor, 133, 135–116, 136, 137–139. *See also* Bryukhonenko, Sergey

Bach, Aleksei Nikolaevich, 35–36, 86, 122, 129–130
Bach, Lydia, 122
Baltic States, 92
Barbarians at the Gate (Woolf), 95
BARChmer, 105
BARON, 105, 109–110, 156. *See also* VENONA; X Group
Bateman, Angus, 173–174, 181, 183, 189
Bateson, William, 17, 34
Battle of Aubers Ridge, 18
Battle of Festubert, 18
Battle of Teruel, 58
BBC, 4–5, 28, 44–45, 180, 183, 198, 232, 347n2
Beliaev, Aleksandr, 141
Bell, Clive, 22
Berg, Lev, 86
Berg, Raissa L'vovna, 176, 355n46
Beria, Lavrenti, 127–128, 131, 350n27
Bernal, John Desmond, 29, 105, 185, 187, 189, 193, 356n83
Bethune, Norman, 48–50, 56
biological heredity, 1
"Biological Possibilities for the

384   INDEX

Human Species in the Next Ten Thousand Years" (Haldane), 214
Biometrika, 143–144, 146
Birstein, Vadim, 127
Black Watch, 18–19, 249–253, 257–258
Blunt, Anthony, 27, 104
Bolsheviks, 34, 37–38, 88, 122, 234
Bondarenko, Aleksandr, 86
Bonner, John Tyler, 224–225
Brecht, Bertolt, 151
Brezhnev, Leonid, 88, 199
Britain, 89–90, 150, 153, 236. *See also* United Kingdom
British Medical Journal, The, 139
British Roumanian Friendship Association, 153
British Soviet Society, 153
Brown, Andrew, 105
Bryukhonenko, Sergey, 133–142
Burdon, Richard (great grandfather), 10
Burdon-Sanderson, John Scott (great uncle), 10
Burdon-Sanderson, Mary Elizabeth (grandmother), 9
Burgess, Guy, 104
Burghes, Betty, 28
Burghes, Charlotte (first wife). *See* Haldane, Charlotte (first wife)
Burghes, John McLeod, 26–27
Burghes, Ronnie (stepson), 27, 47
Burks, Barbara, 205
Burns, Emile, 69, 148, 180

Caballero, Francisco Largo, 53, 58
Cairncross, John, 104
calculus, 65–67, 69–70
Callinicus (Haldane), 47
Cambridge, 29, 40–41, 43. *See also* Trinity College
Cambridge Five, 104

Campbell, Johnny, 180, 187, 190–191
capitalism, 71–74, 94
Carlson, Elof Axel, 239
Carlyle, A. J., 13
Case, Martin, 27, 43, 113
Case, Ralph, 43
Cauchy, Augustin-Louis, 65, 67
Causes of Evolution (Haldane), 42, 145
Central Bombing and Stokes Howitzer School, 20
Chain, Ernst Boris, 44, 308
Chambers, Whittaker, 233
Cheka. *See* Soviet Security Police
chemical weapons, 18, 47–48
Chesterton, G. K., 204
Chetverikov, Servei, 79
child mortality, 143
chromosomes, 17
Churchill, Winston, 236
CIA, 103
Clark, Ronald, 5, 344n42
Clarke, William Carey, 223–24
class, 36–37, 70, 75, 77, 79, 175, 234, 244. *See also* eugenics
Cockburn, Claud, 50
code cracking. *See* cryptanalysis
Cold War, 148, 151
Cole, G. D. H., 13–14
communism: Haldane and, 5–6, 8, 29, 60, 74, 206, 227, 238–239, 244–245; Muller and, 80, 207; VENONA and, 104. *See also* Communist Party; Soviet Union
Communist Party: ARP and, 91–92; Charlotte and, 57, 118, 122–123; disillusionment with, 233–235, 237–238; genetics and, 77; the Gulag and, 237; Haldane and, 6, 26, 99, 153, 159, 238–240; Haldane on, 75, 151; Levit and, 35, 79; Lysenkoism and, 176; Marxism and, 88; Montagu and, 24;

INDEX 385

Nazi-Soviet Pact and, 93; Nin and, 58; propaganda and, 237; Second World War and, 97; Smith and, 235; Stalin and, 70; Stern and, 130; Woolf and, 29; X Group and, 120
Communist Party of Great Britain. See CPGB (Communist Party of Great Britain)
Congress of Genetics, 83–84
Copeman, Fred, 51–52, 60–61
Cornforth, Kitty, 153
Cornforth, Maurice, 183–184
CORPORATION. See CPGB (Communist Party of Great Britain)
Council of Scientific and Industrial Research (CSIR), 229–230
Cowles, Virginia, 48, 53
CPGB (Communist Party of Great Britain), ix; Darlington and, 173; Dutt and, 40; Haldane and, 61, 148, 190–195; headquarters of, 1; Kahle and, 111, 114; Lysenko and, 181–185, 188–189, 331–337; Nazi-Soviet Pact and, 93; purpose of, 8; Reid and, 57; the Spanish Civil War and, 46; X Group and, 101, 107. See also Communist Party; Daily Worker
Crome Yellow (Huxley), 22
cryptanalysis, 102–104, 109

Daedalus (Haldane), 22, 203
"Daedalus, or Science and the Future" (Haldane), 26
Daily Express, 26
Daily Herald, 45
Daily Sketch, 121
Daily Worker, 187; China and, 60; Haldane and, 5, 51, 69–70, 95, 97, 108, 148, 190, 195, 211; Kahle and, 114; People's Convention and, 98; Spanish Civil War and, 48, 50, 56–57; Stalin and, 192. See also CPGB (Communist Party of Great Britain); Pollitt, Harry
Darbishire, A. D., 17
Darlington, Cyril, 43–44, 124, 169, 173–714, 177–179, 197, 212–213
Darwin, Charles, 5, 42
Davies, John Langdon, 186
Dawkins, Richard, 5, 219
D-Day, 113
Dependents Aid Committee Fund, 60
"Dialectical Account of Evolution, A" (Haldane), 48, 61
dialectical materialism, 62–65
Dialectics of Nature, The (Engels), 62, 66–68
Dobb, Maurice, 29, 40
Dobzhansky, Theodosius, 164, 171
Driberg, Tom, 111
Dronamraju, K. R., 239
Dubinin, N. P., 179, 186
Duebener, Maria Caroline, 54
Duff, Patrick, 112
Dunn, L. C., 170
Duranty, Walter, 38
Dutt, Clemens Palme, 48, 342n3
Dutt, Rajani Palme, 40, 93, 342n3

Efroimson, Vladimir Pavlovich, 165–166
Eisenstein, Sergei, 25
Eisler, Paul, 352n31
Elkan, Vera Ines Morley, 49–50
embassy communications, 102–104, 109. See also one-time pad; VENONA
Engels, Friedrich, 61–62, 65–67
Engels Society, 180, 184
Eton, 11–12, 246–249
eugenics: Darlington and, 212–213;

Galton and, 213–214; Haldane and, 203–206, 209, 211–215; Muller and, 207–209, 211; Soviet Union and, 77–79, 207, 346n32. *See also* genetics
evolutionary biology, 62–64
Evolution of Man and Society (Darlington), 44
Experiments in Bringing the Dead to Life, 133
Experiments in the Revival of Organisms, 133–142, 134, 136–137

Facts of Life, The (Darlington), 197, 212
Farnham Left Book Club, 60
fascism, 96
Fillipovsky, Grigori, 126
Finland, 96, 102–103
First Circle, The (Solzhenitsyn), 152
First World War, 17–20
Fisher, R. A., 42–43, 63, 80, 145, 177–179, 208–209, 213
Florey, Howard Walter, 44, 308
Foote, Alexander Allan, 51
Forster, E. M., 238
Frisch, Karl von, 221
Fuchs, Klaus, 103, 112, 114, 348n8
Fyfe, James, 189

Galton, Francis, 143, 203, 213–214
GCHQ (Government Communications Headquarters), ix, 104, 106
Gellhorn, Martha, 53
general intelligence, 205. *See also* eugenics
genes, 17, 203. *See also* eugenics; genetics
Genetical Theory of Natural Selection (Fisher), 80
genetics: Communist Party and, 77; CPGB and, 181–183; Haldane and, 17, 42–43; Marxism and, 87–88;

newts and, 196; plant, 80–83, 124, 160–163, 167–170, 178; Soviet Union and, 50, 159–160, 166, 171, 197–198; study of, 200. *See also* eugenics; vernalization
Genetics of Genius, The (Efroimson), 166
Germany, 92–93, 108–110, 114–115, 126, 131, 150
Glushchenko, Ivan Yevdokimovich, 187–188, 355n54
Gödel, Kurt, 67
"GoldMakers, The" (Haldane), 41
Gollan, John, 189
Gollancz, Victor, 14, 95–96, 233
Goodrich, Edward Stephen, 12, 17
Gordon, W. W., 132
Gorky, Maxim, 37, 78, 80, 236
Gossett, W. S., 145
Govorov, Leonid Ipatevich, 129
GPU. *See* Soviet Security Police
grafting, 167, 169, 198. *See also* agriculture; Lysenko, Trofim Denisovich
Graham, Loren, 353n5, 357n9
Great Depression, 31
"Great Soviet Biologist, A" (Haldane), 159–160
Great Soviet Encyclopedia, 79–80, 209, 238, 351n17
Greenhalgh, Walter, 56
Gregory, F. G., 169–170
GRU (Soviet Military Intelligence), 26, 55, 102, 105, ix
the Gulag, 38, 127, 152, 165, 173, 219, 233–234, 236–237
Gulag Archipelago, 236

Haldane, Charlotte (first wife), 6, 85, 204; autobiography of, 60, 118; China and, 60; Communist Party and, 57; death of, 45; first visit to the Soviet Union, 33–37, 345n84;

Kahle and, 111; London and, 43; MI5 and, 57, 120; Nazi-Soviet pact and, 93; political views of, 28–29; relationship with Haldane, 26–28, 60–61, 290–292; second visit to the Soviet Union, 121–124, 130; Spanish Civil War and, 46–48, 56–57, 59. *See also* Spurway, Helen (second wife)

Haldane, Elizabeth Sanderson (aunt), 9

Haldane, John Burdon Sanderson, 230; animal behavior and, 219, 221, 227; ARP and, 89–91, 97; autobiography of, 6, 8, 246–298; on *Barbarians at the Gate*, 95; Berkeley and, 43; the Black Watch and, 18–19, 249–253, 257–258; Bonner on, 224–225; Cambridge and, 26–27, 29, 41, 43, 143; cancer diagnosis, 230–232; Carlson on, 239; chemical weapons and, 18, 47; childhood of, 10, 16; communism and, 5–6, 8, 29, 60, 74, 206, 227, 238–239, 244–245; Copeman on, 51–52; Cowles on, 53–54; CPGB and, 1, 6, 47, 99, 148–149, 152–153, 180, 190–195, 217; CSIR and, 229–230; the *Daily Worker* and, 148, 190, 195; Darlington on, 44; death of, 198, 232; Dronamraju on, 239; on economic systems, 39–40, 94; emigration to India, 225–227, 228; on Engels, 66–68; Eton and, 11–12, 246–249, 339n2; eugenics and, 203–206, 209, 211–215; experiments and, 16–17, 30, 273–281; family of, 9; father's death and, 46; financial troubles and, 190, 195, 347n14; on Finland, 96; genetics and, 17, 42–43; Greenhalgh on, 56; Harland on, 31; Huxley on, 239; impotency and, 28, 220; India and, 263–271; Indian Statistical Institute and, 229; John Innes Institute and, 293–294; Kahle and, 111–112, 114; Kevles on, 240; London and, 43; Luck on, 30; Lysenko and, 4, 159–160, 163, 166–169, 171–172, 176–180, 182–184, 186, 188–189, 195–198; Majumder on, 240; Marler on, 223; Marxism and, 6, 61–62, 70, 72–73, 146; on May, 117; Mayr on, 239; Medawar on, 222–223, 240; on medicine, 283–284; MI5 and, 2–3, 6–8, 33, 40, 45, 50, 57, 60, 69, 108, 184, 193–194, 195, 347n14; military service and, 17–20, 251–272; *The Modern Quarterly* and, 48; Montagu on, 5; mother's passing and, 230; move to Hungary, 151; Muller on, 51; on the Nazi-Soviet Pact, 299–305; Neel on, 241; New College and, 249–251; obituary of, 30, 44, 230, 240, 306–311; Oxford and, 12–15, 20, 229; the People's Convention and, 98; Philby and, 116; Pirie on, 239; political agitation, 13–14; political propaganda and, 49–51, 70–76, 84–85, 94–99, 130, 222; politics and, 5, 12–13, 29, 44–46; on Pontecorvo, 117; population growth and, 288–289; relationship with Charlotte, 26–28, 60–61, 290–292; relationship with Spurway, 61, 99, 193, 218; on research, 296–298; resignation from the Admiralty, 115, 117; Roberts on, 241; the Royal Society and, 41; trip to Russia, 33–38; scientific credentials of,

5–6; scientific papers and, 42, 216; scientific propaganda and, 133–142, 134, 164, 199; Smith and, 217–218; Soviet Academy of Science and, 2–4; on the Soviet Union, 39–40; Spanish Civil War and, 46–53, 56, 58, 60; Stalin and, 156–157, 238; Stern and, 130–132; submarine research and, 108, 112–113; Trinity College and, 23, 285–286; twenty-first birthday of, 15; UCL and, 43, 225, 294–295; United States and, 229; on universities, 286–287; Vavilov and, 123–125, 128; Vavilov on, 87; VENONA and, 101; on war, 45, 271–272; Wiener on, 41, 238; Woolf and, 143–148; X Group and, 105–108, 110, 118–119, 195, 238, 348n7. *See also* Haldane Papers; *specific works*

Haldane, John Scott (father), 9, 15–16, 46, 112, 246, 248–249

Haldane, Louisa Kathleen (mother), 9–10, 47, 219, 339n2

Haldane, Naomi (sister), 9, 17, 23, 38, 127, 219, 339n2

Haldane, Richard Burdon (uncle), 9

Haldane, Robert (grandfather), 9

Haldane, Sir William (uncle), 9

Haldane Papers, 8, 76

Hamnett, Nina, 43

Hardy, G. H., 12, 72

Harland, Sydney Cross, 31, 124, 128, 164–165, 177–179, 354n28

Haynes, John, 104

Head of Professor Dowell, The (Beliaev), 141

Heartfield, John, 111

Heijenoort, Jean van, 67

Hemingway, Ernest, 53–55, 112, 114

Herbert, Godfrey, 113

Hereditary Genius (Galton), 213–14

Heredity and Its Variability (Lysenko), 171–73

Heredity and Politics (Haldane), 209, 211

Herrnstein's syllogism, 213

Herron, Alec Rowan, 13–14

Hill, A. V., 42

Hinduism, 226–227

Hinshelwood, Cyril, 196–197

Hiss, Alger, 103

History of the Communist Party (Stalin), 70

Hitchcock, Alfred, 25

Hitler, Adolph, 44, 91–92, 244

HMS *Thetis*, 112

Hogben, Lancelot, 143–414

Hollis, Christopher, 14

homosexuality, 11

Hopkins, Frederick Gowland, 29, 42

Horsley, Victor, 128

Hungary, 233–234, 237

Hutt, Allen, 29, 193

Huxley, Aldous, 14–15, 22–23, 27. *See also specific works*

Huxley, Julian, 11, 21, 43, 187, 220, 239, 339n2

Huxley, Noel Trevelyan, 339n2

Hyde, Douglas, 91, 93, 96–97

Ibárruri, Dolores, 60

"In Defence of Genetics" (Haldane), 183

India, 225–226, 263–271

Indian Statistical Institute, 226

Inequality of Man, The (Haldane), 205, 218

Inge, Dean, 45

Institute for Biochemistry in Moscow, 35, 86

Institute for Experimental Biology, 77

Institute of Applied Botany, 35

INDEX   389

Institute of Experimental Physiology, 140
Institutes of Plant Breeding and Genetics, 34
INTELLIGENSTIA. *See* Montagu, Ivor
International Aid to Democratic Greece, 153
International Medical Congress, 151
interrogations, 125–126
Island Hell, An (Malsagoff), 236
Ives, George, 112

JBS. *See* Haldane, John Burdon Sanderson
Jefferis, Millis Rowland, 113
Jermyn, Elizabeth, 113
Jerzfeld, Helmut Franz Josef, 111
Jewish Anti-Fascist Committee, 131–132
John, Augustus, 43
John Innes Horticultural Institute, 17, 34–35, 43, 293–294
Johnson, Hewlett, 151
Joliot-Curie, Irène, 151
Jones, Mary, 153
Joravsky, David, 129
Joseph, Harriet Augusta (great-grandmother), 10
Joseph, Samuel (great-great-grandfather), 10
Journal of Genetics, 218, 226
Journal of Heredity, 125
Journal of Hygiene, 144, 148

Kahle, Gertrude Ernestina, 54
Kahle, Hans, 7, 54–55, 93, 111–115, 348n18, 348n26
Kammerer, Paul, 356n90
Karpechenko, Georgii Dmitrievich, 35, 127–128
Katz, Otto, 25, 45, 110–111, 153–156

Keatinge, Harriet Augusta (grandmother), 10
Keatinge, Richard (great grandfather), 10
Kevles, Daniel, 240
Keynes, John Maynard, 22, 35
KGB. *See* Soviet Security Police
Khrushchev, Nikita, 88, 199, 231, 238
Khvat, Aleksandr, 125–126, 129
King Solomon's Ring (Lorenz), 221
Kinsey Report, 215
Klehr, Harvey, 104
Klugmann, James, 180
Koestler, Arthur, 237
Koltsov, Nikolai Konstantinovich, 77–80, 85–86. *See also* eugenics
Korean War, 30
Kosygin, Alexei, 229, 231
Krementsov, Nikolai, 141
Kremer, Simon Davidovich (Semion Dawidowicz), 105–109, 348n8
Krivitsky, Walter, 53, 344n52, 348n8
Kuczynski, Jürgen, 111
Kuczynski, Ursula, 114, 349n29

Labour Monthly, 5, 114, 123
Lamarck, Jean-Baptiste, 64, 88, 166, 334
Lamarckism, 79, 211
Lamarxism, 159, 219. *See also* Lysenko, Trofim Denisovich
Lange, Johannes, 204–205
Laptev, Ivan, 171, 173
Laski, Harold, 13–14, 203
Lawrence, D. H., 22
Lederberg, Joshua, 227–229
Left Book Club, 89, 95, 207
Lenin, Vladimir, 35, 37, 69, 149–150, 226, 234, 236
Leninism, 72
Lenin Memorial Meeting, 149
Lerner, A. P., 65

Levit, Solomon Grigorievich, 35, 50, 79–80, 86, 164–165
Levy, Hyman, 45
Lewontin, Richard, 228
Life, 114
limpet mines, 113
linkage, 17
Little-Bittner anomaly, 168
Llanstephan Castle (ship), 121
Lorenz, Konrad, 219–221
Lowry, Malcolm, 27, 43
Luck, James Murray, 30
Lysenko, Trofim Denisovich: Bernal and, 185–186; concept of variety and, 83; CPGB and, 181–185, 188–189, 331–337; criticism of, 131, 165, 170–171, 175–179, 197, 200–201; death of, 200; Haldane's support for, 1, 4, 159–160, 163, 166–169, 171–172, 178–180, 182–184, 186, 188–189, 195–198; Stalin's support of, 76, 80; VASKhNIL and, 86; Vavilov and, 84–85, 87; vernalization and, 81–82, 161–163, 165, 168, 170, 185, 201. *See also* agriculture; genetics; Lysenkoism
Lysenkoism, 129, 170, 188. *See also* Lysenko, Trofim Denisovich
Lysenko Is Right (Fyfe), 189

Maclean, Donald, 104
Macleod, Alison, 193, 195
Macmillan, Harold, 11
Mahalanobis, Prasanta Chandra, 226, 229
Majumder, P. P., 240
Manhattan Project, 103, 114, 117
Mannerheim, Carl, 96
Man Who Knew Too Much, The (film), 25

Margolin, D. S., 86
Marler, Peter, 223
Martin, Kingsley, 14
Marty, André, 55
Marx, Karl, 65–67
Marxism: genetics and, 87–88; Haldane and, 6, 61–62, 69–70, 72–73, 146; Kolstov and, 79; Lysenko and, 175; van Heijenoort and, 67; Vavilov and, 88
Marxism and Science (Haldane), 87
Marxist Philosophy and the Sciences, The (Haldane), 61–63
mathematics, 65–67, 72
Mather, Kenneth, 171
Mathews, Herbert, 53
May, Alan Nunn, 92, 108, 115, 117
Mayr, Ernst, 220, 239
Mechnikov Institute of Infectious Diseases in Moscow, 36
Medawar, Peter, 156, 218, 220, 222–223, 228, 240
Medical Biological Institute, 80
medicine, 283–84
Meister, G. K., 86
Mendel, Gregor, 5, 17, 178–79
Mendelism, 42–43, 181, 182, 196, 198, 333
Merseyside Aid to Spain Committee, 60
Mesopotamia, 19, 128
MGB. *See* Soviet Security Police
MI5 (UK domestic military intelligence), ix; Bernal and, 187; Charlotte and, 57; Communist Party of Great Britain and, 1, 8, 189; Haldane and, 2–3, 6–8, 33, 40, 45, 50, 56–57, 60, 69, 108, 184, 192–194, 194, 195, 347n14; Kahle and, 54–55, 111, 114; Montagu and, 7, 24–26; Philby and, 104; Pon-

tecorvo and, 117; *Spycatcher* and, 101; VENONA and, 8, 101–110; X Group and, 118, 120
MI6, 7
Michurin, Ivan, 82, 196–198
Mikhoels, Solomon, 131
Mitchison, Naomi. *See* Haldane, Naomi (sister)
MMTV virus, 169
Modern Quarterly, The, 30, 48, 88, 183, 185, 189
"Modern Synthesis" (Huxley), 43
Montagu, Ivor: career of, 25; on Haldane, 5; Haldane and, 23–24; MI5 and, 7, 24–26; propaganda and, 60, 130, 133, 141, 153–156, 237; X Group and, 26, 101–102, 105–108, 110, 120
Montagu, Owen, 101
Morrell, Lady Ottoline, 21–22
Morris, Desmond, 220
Morton, A. L., 29, 189
Moscow Society of Physiology, 131
Moscow University, 77
Moscow University Clinic, 35
Moscow Zoo Biological Institute, 79
Muller, Hermann Joseph: Agol and, 80; on Haldane, 51; Haldane and, 85, 173; Levit and, 80, 165; Soviet Academy and, 176; eugenics and, 205, 207–209, 211; Spanish Civil War and, 50, 56
Münzenberg, Willi, 24, 45, 154–155
Muralov, A. I., 86
Mussolini, Benito, 244
My Friend Mr. Leakey (Haldane), 29, 195

Nature, 84, 124, 140, 171
Nazis, 44–45
Nazi-Soviet Pact, 92–94, 96, 122, 130, 233, 299–305

Needham, Joseph, 30
Neel, James V., 241
Negrín, Juan, 52–53, 58, 113
New Century, 26
New College, 12–15, 18, 249–251, 273–281. *See also* Oxford
Newman, Horatio, 205
New Masses, 57
New Statesman, 25, 94–95, 230
newts, 196
New York Times, 38, 84
Nin, Andreu, 58
Nineteenth Century, 173
NKVD. *See* Soviet Security Police
NOBILITY, 105, 108–109, 120, 156

Odessa Institute of Selection and Genetics, 82
OGPU. *See* Soviet Security Police
one-time pad, 102–103. *See also* embassy communications; VENONA
Oparin, Aleksandr, 35
Orbeli, Leon, 132
Orlov, Alexander, 55, 58
Orwell, George, 233
Out of the Night (Muller), 207, 209, 211
Oxford, 203, 229. *See also* New College

Palchinsky, Peter, 38, 90
Pavlov, Ivan, 131
Pearson, Egon, 143
Pearson, Karl, 143, 145
Penrose, Lionel Sharples, 183, 187–188, 355n54, 356n68
People's Convention of January 1941, 97–98
Philby, Kim, 104, 116
Picasso, Pablo, 151

Pirie, N. W., 30, 239
Poland, 92, 94
Pollitt, Harry: Charlotte and, 60, 119; Haldane and, 1–4, 45, 47, 50, 57, 69, 184–185, 189–192; Nazi-Soviet Pact and, 93; Spanish Civil War and, 58. See also *Daily Worker*
Pollitt, Marjorie, 69
Pontecorvo, Bruno, 115, 117–118
Pontecorvo, Gillo, 117
Pontecorvo, Guido, 117
population genetics, 42–43, 206, 213, 216
Possible Worlds (Haldane), 204–205, 218
POUM (the Workers' Party of Marxist Unification), 53, 58, 233
Pravda, 171, 176
Prezent, Isak, 79, 82–84, 86
Prince of Wales (Edward VIII), 19
prison camps. See the Gulag
propaganda: capitalist, 235, 244; CPGB and, 93, 111; Haldane and, 49–51, 70–77, 84–85, 94–99, 130, 133–142, 164, 191, 199, 222; Montagu and, 60, 130, 133, 141, 153–156, 237; Soviet Union and, 24, 30, 33, 76, 133–142, 152; YMCA and, 41
Pryanishnikov, Dmitry, 128–129
Punnett, R. C., 42

race, 210, 214
Redgrave, Michael, 98
reduplication, 17
Reichstag Fire trial, 44–45
Reid, Arnold, 57
Relief Committee for the Victims of German Fascism, 44–45
Renton, Donald, 112
RESERVIST, 105. See also VENONA

RESIDENT, 107. See also VENONA
"Retreat from Science in Soviet Russia, The" (Darlington), 173
Roberts, Edwin, 241
Robeson, Ellie, 59
Robeson, Paul, 59
Romania, 92
Rosenberg, Ethel, 103
Rosenberg, Julius, 103
Rothman, Kajsa, 50
Royal Society, 41, 123, 127–128
Russell, Bertrand, 22, 43, 128, 237
Russell, Dora, 27, 43
Russia: eugenics and, 77–79; Haldane's visit to, 33–35, 38; Hungary and, 233–234, 237; interrogations and, 125–126; punishment of scientists in, 36–37, 76, 79, 83–86, 123, 125, 127, 129, 131, 171, 175–176, 234; religious freedom and, 76, 123; science in, 34; Second World War and, 91–94, 96; show trials and, 25, 37–38, 77–78, 110, 132, 233–235, 352n31; timber exports, 236; torture and, 55. See also Communist Party; the Gulag; Soviet Union
Russian Eugenics Society, 78. See also eugenics
Russia Puts the Clock Back (Davies), 186
Rust, Bill, 57, 59, 61, 114, 123, 180

Sabotage (film), 25
Saratov, 127
SAUSAGE-DEALERS, 106, 110. See also VENONA
Sax, Karl, 170–171
Science (journal), 171
Science and Society (journal), 48, 167
Science of Science, The, 215

INDEX 393

Science Service of Washington, D.C., 210
Second World War, 89–94, 96–99
Secret Agent (film), 25
Shakhty Trial, 37–38
Shaw, George Bernard, 141
Sherrington, Charles, 42
show trials, 25, 37–38, 77, 79, 110, 132, 233–235, 352n31. *See also* the Gulag; Russia; Soviet Security Police; Soviet Union
Simone, Andre. *See* Katz, Otto
skew distributions, 144
Slansky, Rudolf, 153–154, 352n31
SMERSH, 7. *See also* VENONA
Smith, John Maynard, 64, 188, 217–218, 235, 237
Smith, William Robertson, 10
socialism, 69, 73
Society for Cultural Relations with the USSR, 153
Society for the Study of Racial Pathology, 78. *See also* eugenics
Society of Materialist Biologists (OBM), 79
Society of Materialist Physicians, 79
Solovetsky Islands, 236. *See also* the Gulag
Solovki (film), 237
Solzhenitsyn, Alexander, 152
Soviet Academy of Science: Haldane and, 2–3, 232
Soviet Genetics (Morton), 189
Soviet Science Snipits (film), 133
Soviet Security Police: arrests of scientists by, 79, 84, 86; Charlotte's first visit to the Soviet Union and, 35; Charlotte's second visit to the Soviet Union and, 122–123; espionage and, 102; Kahle and, 55, 111; Katz and, 45; Muller and, 50; names for, ix; Negrín and, 58; passports and, 57; Stern and, 131; Vavilov and, 88, 125, 127. *See also* the Gulag; Russia; Soviet Union
Soviet Studies (journal), 132
Soviet Union: agriculture and, 80–82, 160–163, 170, 188, 198–199; Civil War of, 233–234; coded communications and, 102–103; disillusionment of Charlotte with, 122–123; disillusionment with, 233–235, 237–238; espionage and, 101–102, 105, 114; eugenics and, 207; executions and, 55, 58, 85–86, 126–127, 129, 131–132, 165, 209, 234; genetics and, 50, 159, 166, 171, 197–198; Germany and, 92–93, 108; Haldane on, 39–40, 69–73; propaganda and, 24, 30, 33, 76, 133–142, 152. *See also* Communist Party; the Gulag; Russia; Soviet Security Police
Soviet War News Film Agency, 133
Soyfer, Valery, 188
Spanish Civil War, 46–60, 93, 154
Spanish Earth (film), 60
Spectator, 187
Spender, Stephen, 59
Spiegelman, Sol, 357n95
Spratt, Philip, 29
Sprunt, Alexander Dalzell, 17, 340n27
Spurway, Helen (second wife): affairs and, 220; animal behavior research and, 219, 221–222, 227; emigration to India, 226; Indian Statistical Institute and, 229; newts and, 196; police dog incident, 223–224; relationship with Haldane, 61, 99, 193, 218;

submarine testing and, 113; trip to Russia, 121. *See also* Haldane, Charlotte (first wife)
Spycatcher (Wright), 101
Stalin, Joseph, 58, 69, 74, 76, 88, 92, 131; death of, 132, 189; Doctors' Plot and, 132; eugenics and, 79–80, 346n32; Haldane and, 156–157, 238; Lysenko and, 83, 175; Second World War and, 92–93; Spanish Civil War and, 53; VENONA and, 104
"Stalin as a Scientist" (Bernal), 189
Stalin Peace Prize, 189
Stasi, 115
Statistical Methods for Research Workers (Fisher), 145
Stern, Lina Solomonovna, 36, 129–132
stick-bombing, 113
Strachey, Lytton, 22
Strand Magazine, 41
submarines, 108, 112–113
Sudoplantov, Pavel, 127
Suez Crisis of 1956, 225
Swedish intelligence, 104
Sylvester, James Joseph, 10

Thälmann, Ernst, 54
39 Steps, The (film), 25
Thompson, E. P., 237
timber exports, 236
Time magazine, 139
Times, The (London), 236
Timofeev-Ressovsky, Nikolay, 79, 85, 164, 347n41, 353n10
Tinbergen, Niko, 219–220
Tirpitz (ship), 113
Tolstoy, Aleksandra, 78
Torr, Dona, 68
Towndrow, M. B., 116

Trinity College, 23, 29, 285–286. *See also* Cambridge
Trotsky, Leon, 25, 234
Trotter, Coutts (grandfather), 10
Truman, Harry, 103–104
Truth Will Out (C. Haldane), 60
twin studies, 78–80. *See also* eugenics

UCL (University College London), 1, 6, 43, 110, 218, 225, 230, 294–295
Ultra project, 109–110
United Kingdom, 102, 104. *See also* Britain
United States, 74, 102–104, 153
University of California, Berkeley, 43
University of Moscow, 139
University of Texas, 80
U.S. Department of Agriculture, 188

Varley, C. H., 113
VASKhNIL (Lenin All-Union Academy of Agricultural Sciences), 83, 85–88, 131, 163, 175–176, ix
Vavilov, Nikolai, 33–34, 51, 76, 82–88, 121–125, 126, 173, 209
VENONA, ix, 8, 101–110, 314–329. *See also* embassy communications
Venona (Allason), 104
vernalization, 81–82, 123, 161–164, 168, 170, 185–186, 201
Vernalization (journal), 82
VOKS (All-Union Society for Cultural Relations with Foreign Countries), 33, 35

Weaver, Harriet Shaw, 347n14
Weisband, William, 104
Wells, H. G., 177
Werner, Ruth. *See* K2ynski, Ursula
West, Nigel, 104–105

"Why I am [a] Cooperator" (Haldane), 244–98
Wiener, Norbert, 41, 57, 238
Wilkinson, "Red Ellen," 25, 111–112
Wilson, James Thomas, 41
Winge, Øjvind, 357n95
Women in Love (Lawrence), 22
Woolf, Barnet, 29, 45, 143–148
Woolf, Leonard, 95
Woolf, Virginia, 22
Workers Gazette, The, 141
World Congress of Intellectuals for Peace, 151, 176
World War I, 17–20
World War II, 89–94, 96–99

Wright, Peter, 101, 104
Wright, Sewall, 43, 213
Wyatt, Woodrow, 11

X-3. *See* submarines
X-craft. *See* submarines
X Group, 8, 26, 102, 104–107, 109–110, 118–120, 195, 238, 348n7

YMCA, 40–41
Younger Son (Montagu), 156

Zhdanov, Andrei, 175
Zhdanov, Yury, 175–176
Zhebrak, Anton, 171, 174